Flavius Josephus
The Jewish Wars
Or The History of the Destruction of Jerusalem

A Complete & Unabridged Paraphrase

Bob Beasley

Living Stone Books

FLAVIUS JOSEPHUS: THE JEWISH WARS
or A History of the Destruction of Jerusalem

A Paraphrase by Bob Beasley
© 2015 Living Stone Books

All rights reserved. No part of this book may be reproduced, stored on a retrieval system, or transmitted in any form or by any means—electronic, mechanical, photocopy, recording, or otherwise—except for brief quotations for the purpose of review or comment, without the prior permission of the publisher.

Printed in the United States of America

ISBN-13: 978-0-9890922-2-7

Beasley, Bob 1938-

Ancient History/Church History

Hartville, Ohio

"Until recently, no ancient historian was more widely read than the first-century soldier-statesman Flavius Josephus, the greatest source of our knowledge of the Holy Land 'between the Testaments.' In many a Victorian home and American Sunday-school library, a copy of William Whiston's translation of Josephus' works, first published in 1737 and reprinted more than 200 times since, held pride of place next to the Scriptures."

Benjamin Balint - The Wall Street Journal, January 18, 2013

"And when ye shall see Jerusalem compassed with armies, then know that the destruction thereof is nigh. Then let them which are in Judea flee to the mountains; and let them which are in the midst of it depart out; and let not them that are in the countries enter thereunto. For these will be the days of vengeance, that all things that are written may be fulfilled.

But woe to them that are with child, and to them that give suck, in those days! For there shall be great distress in the land, and wrath upon this people. And they shall fall by the edge of the sword, and shall be led away captive into all nations: and Jerusalem shall be trodden down of the Gentiles, until the times of the Gentiles be fulfilled."

Jesus the Christ - The Gospel of Luke 21:20-24 (KJV)

FLAVIUS JOSEPHUS

Titus Flavius Josephus (37 – c. 100 AD), was born Joseph ben Matityahu. He was a first century Romano-Jewish scholar, historian and hagiographer, who was born in Jerusalem—then part of Roman Judea—to a father of priestly descent and a mother who claimed royal ancestry.

He initially fought against the Romans during the First Jewish–Roman War as head of Jewish forces in Galilee, until surrendering in 67 AD to Roman forces led by Vespasian after the six-week siege of Jotapata. Josephus claimed the Jewish Messianic prophecies refferred to Vespasian becoming Emperor of Rome. In response, Vespasian decided to keep Josephus as a slave and interpreter. After Vespasian became Emperor in 69, he granted Josephus his freedom, at which time Josephus assumed the emperor's family name of Flavius.

Flavius Josephus fully defected to the Roman side and was granted Roman citizenship. He became an advisor and friend of Vespasian's son Titus, serving as his translator when Titus led the siege of Jerusalem, which resulted—when the Jewish revolt did not surrender—in the city's destruction and the looting and destruction of Herod's Temple. Josephus recorded Jewish history, with special emphasis on the first century AD and the First Jewish–Roman War, including the siege of Masada.

His most important works were *The Jewish War* (c. 75) and *Antiquities of the Jews*. *Antiquities of the Jews* recounts the history of the world from a Jewish perspective. These works provide valuable insight into first century Judaism and the background of Early Christianity. Additional works include *Against Apion* and *The Life of Flavius Josephus*. The short, latter book has been paraphrased by Bob Beasley and is available as an eBook.

BOB BEASLEY

Bob is a graduate of Westminster Theological Seminary in California and lives in Hartville, Ohio with his wife Amy. He is the author of six other published books including *The Wisdom of Proverbs* and *101 Portraits of Jesus in the Hebrew Scriptures*. Bob has had a long-time interest in church history and ancient history. His blog, www.todaysproverb.com, appears each weekday.

Table of Contents

FOREWORD: vi

JOSEPHUS'S PREFACE: x

BOOK I - (33 Chapters) 1
Containing the Timeframe of 167 Years from the Capture of Jerusalem by Antiochus Epiphanes (168 BC) to the Death of Herod the Great (9 AD).

BOOK II: (22 Chapters) 105
From Herod's Death until Nero Sent Vespasian to Suppress the Jewish Uprising. Containing the Interval of 69 years.

BOOK III: (10 Chapters) 183
Containing the Interval of about One Year from Vespasian's Coming to Conquer the Jews to the Taking of Gamala.

BOOK IV: (11 Chapters) 231
Containing the Timeframe of about One Year from the Siege of Gamala to the Coming of Titus to Besiege Jerusalem.

BOOK V: (13 Chapters) 289
Containing the Interval of about Six Months from Titus's Arrival to Lay Siege to Jerusalem to the Great Extremity to which the Jews were Reduced.

BOOK VI: (10 Chapters) 347
Containing the Interval of about One Month from the Great Extremity to which the Jews Were Reduced, to the Taking of Jerusalem by Titus.

BOOK VII: (11 Chapters) 389
Containing the Interval of about Three Years from the Taking of Jerusalem by Titus, to the Sedition of the Jews at Cyrene.

MAPS & CHARTS: 427
Containing Maps of Israel in New Testament Times; A Map of Old Jerusalem; Herod's Temple illustration; The Hasmonaean or Maccabean Family Tree; and Three Charts of the Antipater-Herodian Descendants.

DEDICATION

This book is dedicated to the many historians down through the ages who, like Josephus, have provided succeeding generations with a legacy of world history in the written word.

1 Corinthians 10:31

Foreword

The Jewish Wars is an amazing book. It is the story of the destruction of Jerusalem and its holy Temple by the Romans under Titus in A.D. 70. Flavius Josephus was an eyewitness of the tragic events as a Jewish commander and governor, and as a Roman prisoner of war. Captured following the siege of Jotapata in 67 AD, he became a slave, then a friend of Emperor Vespasian and his son, Titus. He was then required to travel with the Roman army and was able to record these events. His unique record has been preserved over thousands of years, hand-copied by monks much like the pages of another ancient manuscript—the Bible. Originally written in Hebrew, *The Jewish Wars* manuscript was translated into the Greek during Josephus's lifetime.

I first became acquainted with *The Jewish Wars* more than 30 years ago when I first attempted to read William Whiston's 1737 Greek-to-English translation. It was tough going for me. Not only was it a very wooden and antique translation of the Greek text, but the sentences were very long and convoluted, and the pronouns were copious. It was difficult to decipher their referents. The language Whiston used was of his own time and slow going for the modern reader. I began to think about paraphrasing the public domain book in 1997, and did some work on Book I at that time. But the project was soon abandoned in favor of other, more pressing undertakings. In 2013, I ran across a copy of my earlier work and began again, using the free public domain Kindle version. What you have in your hands is the culmination of that effort.

My paraphrase simply seeks to tell Josephus's story of the terrible events of that day in modern language. You won't find any footnotes, but each sub-paragraph is numbered, so footnotes in the public domain edition will be easy to find. I'm also not going to argue about the veracity of Josephus's account. I'm simply a writer who wants others to read what I believe to be one of the most fascinating stories ever told. For further research and study, simply go on the web and search for "Josephus." I refferred to several of the translations you'll find there in my attempt to tell the story correctly—to be faithful to the text—but my paraphrase is solely from the Kindle book that is available free of charge at Amazon.com.

I've been told that if a Christian family had only two books at the turn of the 20th century, one was the Bible and the other a copy of *The Complete Works of Josephus*, of which *The Jewish Wars* is but a major section. Josephus's *Jewish Antiquities* is his magnum opus, taking the reader from the creation of the world through Herod the Great and shortly thereafter. Other shorter works are also included in that volume, the hard bound copy of which is available from Hendrickson Publishers, Inc.

While it is true that I have an interest in these ancient times from a Christian perspective, *The Jewish Wars* is for anyone interested in ancient history, and in particular the history of the ancient near east.

Of course, Christians believe that Jesus accurately predicted the events that led up to Jerusalem's destruction. In Matthew's gospel, chapter 24, beginning at verse 1, we read in the New King James Version:

> *"Then Jesus went out and departed from the temple, and His disciples came up to show Him the buildings of the temple. And Jesus said to them, "Do you not see all these things? Assuredly, I say to you, not one stone shall be left here upon another, that shall not be thrown down."*

Jesus then went on in Matthew 24 to speak of the events surrounding the Temple's destruction. Parallel accounts may be found in Mark chapter 13 and Luke 21. Josephus also describes the geographical features of the Holy Land at that time. The land seems to be quite different from the arid, desert region we know today. His lengthy description of the Essenes, an ancient Hebrew sect from whom came The Dead Sea Scrolls, will be of particular interest to Christian and Jewish readers. The book also provides excellent background for the events in Judea and Samaria surrounding the early Church and the travels of Paul and the other apostles.

A brief note about some of the conventions I have used: You will find a few clarifying notes in the text. If they are mine, they will be surrounded by parentheses. If they are from Whiston, they will be encompassed by brackets. The numbering of sub-chapters is Whiston's convention and I have continued their use for those who might want to double-check certain issues with other translations. I have not adopted the more precise numbering system that occurs in other more scholarly editions. Josephus's unusual book and chapter captions have been retained from Whiston's translation.

I want to thank my wife, Amy, for putting up with the many hours I've spent in these labors, as well as for her support and help. Nat Belz offered inspiration and encouragement, as did my good friends Bob Drake, Merle Messer, Phil Reinheimer, Jerry Archer, and Conley Brown, along with many others. Elizabeth Vanderpool and Londa and George Miladin greatly encouraged me in the project's early days. Beth De Bona was an excellent editorial assistant. Her help in preparing the finished manuscript was invaluable. I also appreciate the folks at *www.bible-history.com* for allowing me to use their maps and Herodium and Hasmonaean family trees. You'll find them at the back. My thanks to you all.

The Jewish Wars is a brutal book. If it were a movie, I suppose it would carry a PG-13 rating. It is not for young children. Many images are nightmarish. It's a book about ancient warfare, and like any warfare, it is violent and often very cruel. But it is truly one of the great and unforgettable stories of all time. All I have attempted to do is to make it accessible to today's reader. I hope you will enjoy *The Jewish Wars* as much as I have in preparing it for you.

Bob Beasley
Hartville, Ohio
January, 2015

Josephus's Preface

1. The war pitting the Jews against the Romans was the greatest of all wars. I don't mean just in my own time, but of all past wars whether between cities or nations. Of this great war, armchair historians have collected worthless hearsay and contradictory falsehoods and have edited them in a pseudo-scholarly fashion. Furthermore, some of those who had no part in the Jewish wars have misinterpreted what went on, either out of an attempt to flatter the Romans or from their hatred of the Jews. Their writings are simply not historically accurate, sometimes accusing and sometimes applauding. I am Joseph, son of Matthias, a Hebrew by birth and also a priest. I fought against the Romans. I was then captured and forced to witness what went on. I am the author of this work. I wrote the original volume in Hebrew and sent it to the Upper Barbarians. This translation into Greek is for the benefit of all those living under Roman rule.

2. At the time of the great upheaval, Rome was having internal troubles, and the Jews who wanted to throw off the Roman yoke took advantage of the situation. Some Jews were motivated by lust for gain while others feared that the troublesome times might force the loss of their possessions. The Jews in Judea were hoping that many expatriate Jews from Syria and beyond the Euphrates would join them in the insurrection. The Gauls, who lived near Rome, were also deploying for battle, as were the Celts. In fact, the whole world was in an uproar following Nero's death (in 68 AD). The times afforded an opportunity for some to grab power, and the military underwent change as many saw armed conflict as a way to line their own pockets. I thought it absurd that the truth was so misrepresented by these so-called historians in such important events. I thought

it best to let the Greeks and Romans who were not engaged in the Jewish war to remain ignorant of these facts, leaving them to their flatteries and fictional accounts. I knew that by my writings in Hebrew that the Upper Barbarians, the Parthians (from present-day Iran) and Babylonians — knew with accuracy how the war began, what misery it brought to our race, and how it all ended. Even the most remote Arabians and those Jews beyond the Euphrates and the Adiabene knew about the war.

3. While it is true that these writers call their accounts of the war "history," it seems that they fail miserably and tell nothing that even resembles the truth. For they want to demonstrate the greatness of Rome while at the same time diminishing the actions of the Jews. They fail to understand that greatness cannot be attributed to a people who have conquered only a tiny land. These writers also overlook the length of the war and the vast numbers of Roman troops who suffered in it. Ironically, the brilliance of Rome's military leaders at Jerusalem will be thought of as inept if what they attained is considered inconsequential.

4. However, I will not go to the other extreme while opposing those who magnify the Romans. I will not exalt the actions of my own people too highly. I will merely seek to report the deeds of both sides accurately. My feelings regarding the affairs I describe will undoubtedly be revealed between the lines. I must be allowed to mourn over the misery undergone by my countrymen. For it was our own rebellious temperament that destroyed Judea. The Jewish tyrants brought the force of Roman arms upon us. Rome unwillingly attacked us and burned our holy temple. Titus Caesar, whose forces destroyed the temple, pitied the people who were under the thumb of the rebels there. He delayed his siege of the city time after time, hoping that more time would give the Jewish leaders opportunity to repent. I also speak passionately about the Jewish tyrants and thieves. And I deeply lament the misfortunes of Judea. Please indulge my expressions of emotion, even though they break the rules of writing history. For it has come to pass that Jerusalem—the city that had achieved the greatest prosperity of any other city in the Roman Empire—has, in the end, fallen victim to the worst of catastrophes. Accordingly, it seems that all of the misfortunes that have fallen on men since the creation of the world are not worthy to be compared with those of the Jews, who fell at the hands of foreigners. Because of this it is impossible for me top contain my grief.

If anyone wants to be inflexible in his criticism of me, let him then attribute the facts to history and the mourning to the writer alone.

5. However, I must justly accuse the Greek scholars who, when such great circumstances arise in their own times, eclipsing the old wars by comparison, they sit as judges, bitterly condemning the labors of the best ancient historians. These modern scholars may well surpass the old in their eloquent use of language, yet they are inferior in the execution of their intentions. For instance, these modern writers compose new histories of the Assyrians and Medes as if the ancient writers had not described their histories accurately enough. They do this even though their abilities are inferior to the older historians. Their thoughts are far removed from the ancient viewpoint. For in older times, everyone wrote in the present tense. Their immediate concern for current events makes their histories invaluable. If they had told lies, such concoctions would have been immediately exposed by their readers. Preserving the memory of previously unrecorded events and setting forth the affairs of one's own time for posterity is worthy of real praise and commendation. Esteemed is he who has conducted his original research in earnest, not rewriting the histories that other men have written, but who composes an entire body of history on his own.

Accordingly, I have been charged with great responsibility and have taken very great pains regarding the history of the Jewish wars. Though I am a foreigner to you, I dedicate this work to both the Greeks and the Romans as a memorial to its great events. Some of the mouths of our prominent Greek writers are open, and their tongues are wagging for their own greedy ends. But they are muzzled and silent when it comes to writing history, where they must speak the truth and gather facts with a great industriousness. So they leave such historical writing to inferior people, and to those unacquainted with the exploits of their rulers. Instead, though neglected by the Greeks, we prefer the real truth of historical facts.

6. Now is not a fitting time to speak of the ancient history of the Jews— who they were originally; how they revolted against the Egyptians; what country they travelled over; what countries they later overran; and how they suffered exile from the land. Nor will I write about other events of long ago. Many Jews before me have composed exacting histories of our ancestors. Some of the Greeks have done it as well and have also translat-

ed our history into their own language. They have not wandered far from the truth in those histories. But I will begin my history where these other historians and our own prophets leave off. I will go over in detail those events of the war that occurred in my own time, and with all the diligence I can muster. But I shall deal only briefly with the events that preceded my time.

7. I will begin by telling how Antiochus Epiphanes took Jerusalem by force and held it for three years and three months until he was ejected by the Hasmonaeans. I shall then tell how their descendants bickered about governmental organization and consequently brought in the Romans and Pompey. I shall discuss how Herod (the Great), the son of Antipater, ended the (Maccabean) dynasty with the help of Sossius, and also how our people rebelled upon Herod's death while Augustus was emperor of Rome. I have written about how the war broke out in the 12th year of Nero's reign, about the fate of Cestius, and about what places the Jews raided in the first skirmishes of the war.

8. I shall also describe how the Jews built walls around neighboring cities, and how Nero, upon Cestius's defeat, was in fear of losing the war and appointed Vespasian as commanding general. I shall also portray how this Vespasian with his eldest son, Titus, made an expedition into Galilee and Judea, the size and composure of his army, and the great number of auxiliary troops killed in Galilee. Vespasian took some of its cities by force, though some surrendered. When I have done all of this, I shall describe the discipline of the Roman soldiers, and the training of their legions, together with a description of the size and nature of the two Galilees and the territory of Judea. And besides this, I shall pay particular attention to what is unique about this country, the lakes and springs that are there, and what misery conquered cities suffered.

9. After all this, I shall set forth the events surrounding Nero's death when Jewish sufferings had worsened. As Vespasian was proceeding to attack Jerusalem, he was called back to Rome to assume supreme power there, having been unwillingly made emperor by his soldiers. When he left for Egypt to restore order in the realm, the civil war in Judea became extremely violent. Tyrants arose, and their forces split into feuding factions.

10. Then I shall relate how Titus marched out of Egypt into Judea for a second time. I will describe the state of his army; the sieges Titus laid upon Jerusalem; the ramparts he built; the walls of the city and their measurements; and the structure of the holy temple and house, and their measurements, including the altar. I have also included descriptions of several Jewish festivals; the seven days of purity; the sacred ministrations of the high priests; their garments; and the very nature of the holiest place in the temple. I have not hidden anything, nor added anything to the truth already known about these things.

11. Following all of this, I will tell of the tyrants' barbarous treatment of their own people within Jerusalem's walls. In contrast, the Romans showed mercy in sparing foreigners. Titus frequently invited the rebels to come to terms of surrender out of his desire to preserve the city and its temple. I shall also distinguish between the sufferings of the people and their calamities. The people of Jerusalem were greatly afflicted by the Jewish rebellion and subsequent famine, and were finally overwhelmed. Nor shall I fail to mention the misfortunes of those who chose to desert, nor the punishments inflicted on those who were caught running away.

The temple was ultimately burned to the ground against Caesar's wishes, though snatched out of the fire were many sacred articles stored there. I won't leave out how the entire city was destroyed with the signs and wonders that preceded its fall, nor of the tyrants taken captive, the enslavement of multitudes of inhabitants, nor all the misfortunes into which the people of Judea fell. Finally, I will tell of what the Romans did to the remaining walls of the city; how they demolished the other fortifications in Judea; how Titus rode over the entire country to settle the peace; and how he returned triumphantly to Italy.

12. I have set all of these things down in seven books, and have left no door open for either complaint or accusation of anyone acquainted with this war. I have written it all down for the sake of those who love truth, but not for those who are more pleased with fictitious fables. And I now begin my account of these things with what I call my, "First Book."

The Jewish Wars

Book I

Containing the Timeframe of 167 Years
from the Capture of Jerusalem
by Antiochus Epiphanes (apx. 168 BC)
to the Death of Herod the Great (9 AD).

Chapter 1

How the City of Jerusalem Was Conquered, and the Temple Pillaged (by Antiochus Epiphanes). Additionally, the Exploits of the Maccabeans, Matthias and Judas, and the Latter's Death

1. When Antiochus Epiphanes was negotiating with Ptolemy VI (King of Egypt) about his right to the Syrian throne, some powerful Jews became subversive in Judea and wanted to overthrow the government. These high-ranking men could not stand for the indignity of being subject to their equals. However, one of the high priests, Onias, took over the government and threw the sons of Tobias out of Jerusalem. They fled to Antiochus Epiphanes (in Syria) and appealed to him to invade Judea under their leadership. The king agreed, as he was already predisposed to attacking the Jews. Antiochus then marched out a great army. He took Jerusalem by storm and killed a great number of those who sided with Ptolemy. His soldiers also plundered the city without mercy, and destroyed the holy temple. For three years and six months, he put a stop to the Jews' daily sacrifice for the expiation of sins. Meanwhile, Onias, the high priest, escaped to Ptolemy and received a parcel of land from him in Heliopolis. There, Onias built a city that resembled Jerusalem and a temple that re-

sembled its temple. We shall speak more of this temple later.

2. The sudden seizure of Jerusalem, with a great slaughter and pillage of its citizens, failed to satiate Antiochus. Overcome with a passion for violence and seeking to avenge his army's losses during the siege, he compelled the Jews to do away with the laws of their country. He stopped the practice of infant circumcision and then sacrificed the flesh of pigs upon their sacred altar. All of the Jews opposed these abominations, and Antiochus executed their leaders as a consequence.

One man, Bacchides, who commanded the fortresses of the city, added his own innate cruelty to Antiochus's evil commands. He indulged in all sorts of extreme wickedness. Bacchides tormented the most respectable citizens, man by man, and daily threatened Jerusalem with destruction. Ultimately, though, he provoked those who suffered his tortures to avenge themselves.

3. Matthias, the son of Asemonaeas, (ben Asemonaeas), was one of the priests who lived in a village called Modein. He armed himself and his family of five sons and they assassinated Bacchides by stabbing him with daggers. Then, out of fear of reprisal by the many enemy troops in Jerusalem, Matthias and his sons escaped to the mountains. There, a great many Jews joined them. Encouraged by their numbers, Matthias came down out of the mountains to fight Antiochus's army. He soon defeated them and drove the Syrians out of Judea. So Matthias (Maccabeus) came to lead the Jews due to his success against Antiochus. He became the people's ruler by acclamation. But Matthias died soon after and left the leadership of the government to his eldest son, Judas.

4. Believing that Antiochus Epiphanes would not sit still for long, Judas (Maccabeus) formed an army of Jews and became the first Jewish leader to make a treaty with the Romans. After Epiphanes had led a second campaign into Judea, he drove the Syrian king out of the country, defeating his army. Encouraged by his great success, Judas then assaulted the Syrian garrison in Jerusalem, for it still controlled the upper city. He drove the troops into the lower city—called the Citadel, and regained control of the temple. Judas first cleansed the entire temple area then built a wall around the complex. He then made new vessels for the sacred rites and brought them into the temple, for the former vessels had been desecrated. Judas

also built another altar and resumed the holy sacrifices. No sooner were the city and its temple operational again than Antiochus Epiphanes died. His son, Antiochus, then succeeded him as king, retaining his father's hatred of the Jews.

5. This Antiochus then put together a force of 50,000 foot soldiers and 5,000 cavalries. Then, with 80 elephants, he marched through Judea into its mountainous regions. He captured a small city called Bethsuron, but in a narrow valley at a place called Bethzacharia, Judas met Antiochus and his fighting men. Before the battle began, Judas's brother, Eleazar, spotted the largest of the Syrian elephants bearing a great tower on its back and wearing golden armor. Supposing that Antiochus was on the beast, Eleazar ran across the no-man's-land separating the two armies. Cutting his way through the Syrian troops, Eleazar reached the elephant, but because of the animal's great height he could not reach its rider, whom he thought to be the Syrian king. So Eleazar ran his sword through the elephant's belly and the giant animal fell, crushing Eleazar to death.

As it turned out, the elephant's rider was but a private citizen. Had he, in fact, turned out to be Antiochus, Eleazar still would have proven nothing, except perhaps his boldness. He had chosen to die when he only had the slimmest hope of killing the king. To Judas, this disappointment became an omen as to how the entire battle would end. The Jews fought bravely for a long time, but Antiochus's forces won the victory, with greater numbers and good fortune on their side. Losing a great many of his men, Judas took those who remained and escaped to the Gophna toparchy. Meanwhile, Antiochus went on to Jerusalem but stayed there only a few days. He did leave a garrison behind which he thought large enough to control the city, but took the rest of his forces to their winter quarters in Syria.

6. After Antiochus had left Judea, Judas Maccabeus wasn't idle. Many of the Jews had rallied around him and together with those who had escaped the earlier battle at Bethzacharia, he attacked Antiochus's generals at a village called Acedasa. Judas succeeded in killing a great number of Syrians, but at last Judas himself was slain. A few days later, Judas's brother John was also murdered by Antiochus's men, concluding a plot to take his life.

Chapter 2

Judas's Successors – Jonathan Simeon and John Hyrcanus

1. Judas's brother Jonathan succeeded him. Jonathan was cautious in his dealings with his people as well as in all other matters. He reinforced his authority by preserving friendship with the Romans, and he also formed a treaty with Antiochus's son. But the alliances were not enough to ensure his security. The tyrant Trypho—guardian of the young Antiochus—plotted against Jonathan. While the Jewish leader was journeying to Ptolemais to visit Antiochus, Trypho's men ambushed Jonathan's small party and put Jonathan in chains. Trypho then began a campaign against the Jews, but was driven away by Jonathan's brother Simon. Enraged at his defeat, Trypho executed his prisoner.

2. Simon proved very efficient in his management of the country. He took Gazara, Joppa, and Jamnia—all cities near his home. He also captured the Syrian garrison in Jerusalem and demolished the citadel. Afterward, his army served as auxiliary troops for Antiochus in his internal quarrel with Trypho, whom he attacked in Dora before going on an expedition against the Medes. But Simon could not quell Antiochus's ambition even though he had assisted him in killing Trypho. It wasn't long before Antiochus sent his general Cendebaeas with an army to lay waste to Judea and to overthrow Simon. Simon, though up in years, conducted the war as if he were a much younger man. He sent his sons with a force of his most able-bodied men against Cendebaeas while he took another part of the army and attacked the Syrian king on his flank. He also carried out a great number of ambushes in the mountains and continually held the upper hand in all his assaults upon invading forces. When Simon returned a glorious conqueror, he was made high priest. He had freed the Jews from 170 years of Macedonian domination.

3. But Simon, too, was the victim of a plot upon his life. His assassination occurred at a festival given by his son-in-law, Ptolemy. This Ptolemy also put Simon's wife and two of his sons in prison and sent some men to kill John—also called Hyrcanus. But when Simon received advance word of the plot against him he fled to Jerusalem, having confidence in the protec-

tion of the people there. He knew that they revered the glorious victories of his father, Simon, and at the same time, they detested the wickedness of Ptolemy. Ptolemy also tried to get into the city by way of another gate, but was repelled by the people who had just received Hyrcanus. So Ptolemy retired to one of the fortresses near Jericho called Dagon. When Hyrcanus had received the high priesthood that his father had held before him and had offered sacrifice to God, he hurried off to attack Ptolemy in order to gain the release of his mother and brothers.

4. John Hyrcanus laid siege to the fortress at Dagon and seemed to have the upper hand, but was overcome with sympathy for his mother and brothers. For when Ptolemy feared defeat was imminent he brought out John's mother and brothers and sat them on the fortress wall. He proceeded to beat them with rods for all to see. He also threatened that unless Hyrcanus withdraw immediately, he threatened to throw John's loved ones off of the wall and to their deaths. On seeing the torture and hearing the threats, John's pity and concern grew greater than his anger. His mother, on the other hand, was not afraid, neither of the beating she received nor of the threat of death. Rather, she stretched out her hands and prayed that her son not be moved by her situation to spare the wretch, Ptolemy. She said that it was better for her to die at his hand than to live a long life if John would punish Ptolemy for the injuries he had inflicted upon their family. When John saw his mother's courage and heard her appeal, he remounted the attack. But when he saw her subsequently beaten, torn by the stripes she had received, he grew weak and was overcome by his love for her. As the siege was delayed for some time because of these events, the year of rest for the Jews came. The Jews rest every seventh year as they also rest every seven days. Therefore, the siege against Ptolemy ceased. Ptolemy then murdered John's mother and brothers and made his escape to Zeno—also known as Cotulas—who was tyrant of Philadelphia.

5. Meanwhile, the Syrian Antiochus was so angered by what he had suffered at the hands of Simon he mounted a campaign into Judea, making camp outside Jerusalem and laying siege to Hyrcanus. But John Hyrcanus persuaded Antiochus to stop the siege with the offer of 300 gold talents that John Hyrcanus procured from the 3,000 talents he had removed from the tomb of David [the richest of the Hebrew kings]. Hyrcanus then used what was left of the gold to hire foreign mercenaries, becoming the

first Jewish leader to do so.

6. At another time, Antiochus had gone on a further expedition against the Medes, giving Hyrcanus an opportunity for revenge. So John quickly attacked some Syrian cities, thinking them undefended. Such was the case. He took Medabe and Samaga together with the towns around them. He also captured Shechem and Argatain. Besides that, he subdued the Cuthaean nation that lived near the temple built in imitation of the temple at Jerusalem. Hyrcanus also took a great many of the cities of Idumea, including Adoreon and Marisa.

7. John's campaign proceeded as far as Samaria—where today king Herod's city called Sebaste exists—which he surrounded with a wall. He then appointed his sons Aristobulus and Antigonus as commanders over the siege. They pushed the assault so hard that famine broke out in Sebaste, and the residents of the city were forced to eat things they had never before considered food. The Samaritans then invited Antiochus — also called Aspendius—to come to their rescue. Antiochus prepared his men and answered the city's request but was beaten back by Aristobulus and Antigonus. They pursued him as far as Scythopolis, but Antiochus managed to escape. So the brothers returned to Samaria and again shut its citizens inside the wall. When they had taken the city, they demolished it and made slaves of all its former inhabitants. Given courage by their success at Samaria, the brothers then marched their army back to Scythopolis. They raided the city and laid waste to all the countryside around Mount Carmel.

8. But the successes of John Hyrcanus and his sons aroused envy at home, and the smell of treason was in the air. Many Jews formed together to oppose John and would not rest until war broke out. They soon got their wish but were defeated by the Maccabees. So John lived the rest of his life happily at peace. He administered the government of Judea for 33 years in the most extraordinary way. Then he died (in 106 BC), leaving five sons. He was certainly a most happy man, and afforded no one the opportunity to complain of his rule. John alone possessed three of the most desirable things in the world: the government of his nation, the high priesthood, and the gift of prophecy. For the Deity spoke with him so that John was not ignorant of any future event. He even foresaw and foretold

the fall from power of his two eldest sons. Their story of catastrophe highly deserves telling. But they were men who were far inferior to their father in aptitude, prosperity, and happiness.

Chapter 3

Aristobulus Is the First to Crown Himself King After He Had Executed His Mother and Brother.
He Died After Reigning No More Than a Year.

1. After the death of their father, the eldest son, Aristobulus, changed the government of Judea into a monarchy, setting himself up as its king. He became the first Jewish leader to place a royal crown upon his head in the 471 years and three months since the Jews were set free from slavery in Babylon. Aristobulus apparently loved his brother, Antigonus, who was the second born, and elevated him to a level of seniority in the kingdom, almost equal to his own rank. But Aristobulus arrested his remaining brothers and put them in prison. He also put his mother in chains for disputing his authority. Her late husband, John Hyrcanus, had made her governess of the region. She was to die of starvation while in prison due to the barbarity of her son.

2. The king would later also mistakenly kill his brother, Antigonus, whom he loved and had made his partner in the kingdom. It happened like this. Evil men in the kingdom had plotted to slander Antigonus. Believing their lies, the king took vengeance on his brother. At first, Aristobulus could not believe what he was hearing, partly out of affection for his brother and partly because he thought the stories against him were jealously concocted. But on the occasion of the Feast of Tabernacles (during which it was the Jews' ancient custom to live in tents in remembrance of their ancestors' flight from Egypt) Antigonus came from his army post dressed very ostentatiously. It so happened that the king was sick in bed. Antigonus arrived at the climax of the festival in order to pray and offer sacrifices for his brother at the temple. He was surrounded by his armed guard and dressed very flamboyantly. Antigonus's visit became the occasion for those plotting against him to go to king Aristobulos and complain of the pompous manner in which the king's brother and his personal guard had marched into the city. They claimed that his insolence was too great for a mere private citizen and that he had no doubt come

with a band of armed men to assassinate the king. It was obvious to his accusers that Antigonus could not endure the mere enjoyment of royal honors when it was in his power to seize control of the kingdom itself.

3. So Aristobulos, little by little, gave credence to their accusations. But he was careful not to openly share his suspicions, and instead increased his personal security measures. He placed bodyguards in a certain dark subterranean passage adjacent to where he lay ill. It was in a part of Jerusalem known as the Citadel in those days. Later on it was to be known as Antonia (named for Mark Antony). The king gave orders that if Antigonus came to visit him unarmed they should let his brother pass through safely. But if he came clad in his armor the bodyguards should kill him. Aristobulus also took the precaution of sending messengers to his brother telling him to arrive unarmed. But the queen cunningly manipulated the situation with those plotting against Antigonus. She persuaded the messengers not to deliver the king's communication, but instead to tell Antigonus that his brother had heard of his new ornamental suit of armor purchased in Galilee. Because the king's sickness prevented him from coming to see all of this military finery, he very much wished for Antigonus to wear it when he visited, particularly since his departure from Jerusalem was imminent.

4. As soon as Antigonus heard the message, he sensed that his brother was in a good mood and didn't fear any harm from him. So he set out for the king's bedside in his armor to show the king. But as Antigonus was passing through the dark corridor called Strato's Tower he was ambushed and slain by the king's bodyguards. Antigonus thus became a prominent example of how slander destroys the good will of men and even brotherly affection. Even our most tender feelings for a loved one are not strong enough to resist the onslaught of jealousy.

5. An astonishing thing happened to a man named Judas that same day. He was an Essene, a prophet from a sect of the Jews that lived in solitude in desert regions. His predictions of the future had never before failed to come to pass, nor had they ever been proven erroneous. When Judas saw Antigonus passing by the temple that day on his way to the king's chambers, he shouted to an acquaintance nearby — probably one of his students. "O how strange!" he cried out, "It is best that I die now, for the truth lies dead before me. A prediction I have made has proven false, for

Antigonus is still alive on the day he ought to have died. The place where I foresaw him slain was in Strato's Tower — some 75 miles from Jerusalem. Since four hours of the day are already past it will be impossible for the prophecy's fulfillment." When the old man had said this, he became very depressed. But the news came a short time later that Antigonus had met his end in an underground passageway, also known as Strato's Tower, the same name of the place the prophet had in mind, but a place far off by the (Mediterranean) sea in Caesarea. This ambiguity had caused the prophet's confusion.

6. On hearing of his brother's death, Aristobulus blamed himself for it and his condition grew instantly worse. His fragile health declined steadily. His very soul trembled at the thought of what evil he had done. His stomach was torn to pieces by his intolerable grief, and he coughed up a great quantity of blood. As one of the king's servants carried the blood from the room he slipped, as if by some supernatural providence, and fell down in the very place where Antigonus's blood lay spilled. So it happened that some of the murderer's blood was mingled with the blood of the murdered. A deep mournful cry arose from those who witnessed the servant's fall, as if the servant had done it on purpose. The king heard all the wailing and asked what had happened. Nobody wanted to tell him. But the king threatened them and forced them to speak up. Hearing of the mingled blood, the king burst into tears and said sadly, "I see that my crime has not escaped the all-seeing eye of God. The vengeance of my kinsman is hotly pursuing me. O you most impertinent body! How long will you retain the soul that ought to die in retribution for the murders of a mother and a brother? Let them take it all now, so that their ghosts will not be disappointed by the offering of but a few morsels of my bowels." No sooner had these words come out of the king's mouth than he dropped dead. Aristobulos had reigned over Judea for less than a year when he died in 104 BC.

Chapter 4

Alexander Janneus Becomes King and Reigns for 27 Years

1. Upon Aristobulos's death, his widow released his brothers from prison.

She made Alexander (Janneus) king, who was both the oldest boy, and had the mildest temperament. When he came to power, he had one of his brothers executed, sensing this brother's personal aspirations to the throne. But he held the other brother in high regard. The surviving brother only wanted to live a quiet life and was not prone to meddling in the affairs of state.

2. It came to pass that war broke out between Alexander and Ptolemy—also called Lathyrus—after Ptolemy had taken the town of Asochis. Alexander's troops killed many of the Egyptians, but Ptolemy ultimately won the victory. But when his mother, Cleopatra, called Ptolemy back to Egypt Alexander assaulted and captured Gadara and Amatheus. The latter was the strongest of all the fortresses along the river Jordan. In Amatheus were stored the most precious belongings of Theodorus, son of Zeno. But suddenly, Theodorus appeared on the scene, recapturing his treasures and putting 10,000 Jews to the sword. Alexander soon recovered from this defeat and turned his attention to the coast. He succeeded in capturing Rapphia, Gaza, and Anthedon, which would later be named Agrippias by King Herod.

3. When Alexander had enslaved all the residents of these cities, the Jewish nation rose up against him at one of its festivals. Interestingly, it was on the occasion of these Jewish feast days that most insurrections occurred in Judea. Alexander felt doomed to fall victim to the conspiracy set against him, but his mercenary troops from Pisidia and Cilicia came to his rescue. No Syrians were among these troops because of their hatred of the Jews. When Alexander had killed more than 6,000 of the rebels, he attacked Arabia. He overran that country, as well as Gilead and Moab, and forced them to pay tribute. He then returned to Amatheus, but the fortress had been abandoned and then demolished by Theodorus, who was apparently discouraged when news reached him of Alexander's victories.

4. So Alexander turned back toward Arabia to fight Obodas, its king. Obodas had set up an ambush for the Jewish forces near Golan. Alexander fell into it and lost his entire army. They became trapped in a narrow ravine and were crushed to death under the charge of a great herd of camels. Alexander escaped to Jerusalem, but the magnitude of his defeat at Golan aggravated the hatred that already boiled up among the peo-

ple, and they rebelled against him. But, yet again, Alexander was able to match their fury. Over a six year long series of battles fought between the people and Alexander's troops and mercenaries, at least 50,000 Jews died. These victories gave Alexander little reason to rejoice as he found himself killing off his own subjects. He therefore stopped fighting and instead tried to negotiate a peaceful settlement with the Jews. But this personal inconsistency and change of policy only made them hate Alexander all the more. When he finally asked them what he could do to pacify them, they replied, "Kill yourself! But even your death would not reconcile us to one who did such terrible things to us." While this was going on, the Jews invited the Assyrian king Demetrius—called Eucurus ["The Untimely"]—to come to their aid. Demetrius agreed, hoping that such an alliance would serve to bring him increased power and possessions. So, with his army of Assyrian soldiers, he joined the Jews and their auxiliary troops near Shechem.

5. Alexander met these forces with 1,000 horsemen and 8,000 mercenary foot troops. He also brought about 10,000 Jews with whom he had found favor. Opposing him were 3,000 horsemen and 14,000 troops on foot. Before the battle began, the two kings made speeches attempting to draw their opponent's soldiers into their own camp, or to make them revolt. Demetrius hoped to induce Alexander's mercenaries to leave him, while Alexander hoped to influence the Jews who had joined with Demetrius to desert the Assyrian king. But the Jews with Demetrius could not forget their rage against Alexander. Neither could the Greeks with Alexander choose to be unfaithful. So the battle was on, and soon swords clashed in close quarters. Demetrius's forces won the battle although Alexander's mercenaries distinguished themselves in both strength and bravery. But the real outcome of the battle was different from what anyone could have expected. The very men who had summoned Demetrius to assist them abandoned him. Meanwhile, Alexander was soon joined by 6,000 of the Jews who had just opposed him, after he tried to take refuge in the hills. Apparently the Jews were moved by their compassion for his defeat. This turn of events was more than Demetrius could stand. Figuring that Alexander was again on equal military terms with him and that all Judea would ultimately gather behind the Jewish king, Demetrius left the country and returned to Syria.

6. The remainder of the Jews who had fought against Alexander did not set aside their dispute when the Syrians withdrew. They waged an ongoing war with Alexander until the king had killed the majority of them and had driven the rest into Bemeselis. But soon, Alexander also took that city and brought all those he captured back to Jerusalem as prisoners. Upon his arrival in the city, Alexander was in such a fit of rage that he took his savagery beyond the limits of blasphemy. He ordered the crucifixion of 800 of his prisoners in the middle of the city, and then had their wives and children butchered before their eyes. All the while, Alexander watched the spectacle, drinking wine and laying openly with his concubines. In fear of Alexander's wickedness, 8,000 of those hostile to him fled across the Judean borders on the following night. Their self-imposed exile ceased only upon Alexander's death. Meanwhile, by such atrocities, Alexander secured peace in the kingdom and the fighting ceased.

7. But a new disturbance was to arise, fostered by a Syrian ruler named Antiochus — also called Dionysius. He was last in the line of Seleucid kings and Demetrius's brother. Antiochus was preparing to attack the Arabians. Alarmed, Alexander dug a deep trench as an obstacle between Antipatris, near the mountains, all the way to the coast at Joppa. He also erected a high wall in front of the trench and placed wooden towers at regular intervals along it, hoping to defend against the attack of the Syrian king. But the wall and trench failed to stop Antiochus, who burned the towers and leveled the ditch before marching past it with his army. Figuring he would return to have his vengeance upon the builder of the obstruction, Antiochus pushed on toward the Arabs. As the Arabian king maneuvered his forces into terrain suitable for battle, he suddenly wheeled his 10,000 strong cavalry unit around and attacked the surprised and unprepared Syrian troops. A fierce battle ensued. As long as their king was alive, the Syrian forces held their ground, although mercilessly cut up by the Arabs. But when Antiochus fell, exposing himself to danger on the front lines while trying to rally his troops, the Syrians gave ground and were driven from the field. The majority of Antiochus's army was slaughtered either in the battle or while running away. The rest managed to take shelter in a village called Cana, but all but a few ultimately starved to death.

8. Soon after Antiochus's death, the citizens of Damascus, out of their hatred for Ptolemy, son of Mennaeus, brought in the Arabian king, Aretas,

and made him king of Coele-Syria. Aretas led a military expedition into Judea and defeated Alexander in battle. However, after the battle he withdrew by mutual consent. But Alexander went on to capture Pella and then marched toward Gerasa again, coveting Theodorus's treasures once more.

At Gerasa, he built three walls around the fortress and succeeded in taking it without a battle. Alexander then conquered Golan and Seleucia and took the so-called "Ravine of Antiochus." He also captured the stronghold at Gamala and then relieved its governor, Demetrius, because of the many accusations against him. Alexander then returned to Judea after a military campaign of three full years, and was cordially received by its citizens as a result of the successes he had achieved. Although he was finally at rest from war, disease then attacked Alexander's body. Afflicted by malaria, he tried to alleviate its symptoms by returning to lead his army. He proceeded to plunge into ill-timed military campaigns that hastened his death. He forced his body to endure greater hardship than it was able to bear. Alexander died in 78 BC, having ruled Judea for 27 years.

Chapter 5

Alexandra Becomes Queen and Reigns for Nine years, but the True Rulers Are Pharisees

1. Alexander left the kingdom to his widow, Alexandra. He convinced himself that the Jews would submit to her rule inasmuch as she possessed none of his own innate cruelty. In fact, she had consistently opposed Alexander's crimes against his people. She, therefore, had the goodwill and affection of the nation. Alexander made no mistake in expecting Alexandra to be a good queen. She firmly handled the reins of government thanks to the godly reputation she held among the people. She meticulously studied the ancient traditions and customs of the Jewish people and fired those members of her government who violated the holy laws. Out of the two sons she had by Alexander, she appointed the elder, Hyrcanus, to be high priest, both because of his age and his disposition. He was mild mannered and apathetic, unconcerned about governmental affairs. The younger boy, Aristobulos, was hot-tempered. Alexandra wisely thought it best to confine him to private life.

2. Growing in power in the government alongside Alexandra were the

Pharisees. This sect of Jews had the reputation for excelling all others in their strict observance of Jewish religious decrees and as exacting interpreters of the law. As the queen was intensely religious, she gave too much credence to the Pharisees' advice. Slowly, the Pharisees took advantage of the queen's naiveté and soon became the true administrators of the country. They exiled and recalled whomsoever they pleased and enslaved or freed men at their pleasure. In other words, they enjoyed the power of royalty in Judea while Alexandra had to deal with the nation's finances and other problems. She proved, however, to be equal to the task of administering the affairs of state. By continually recruiting soldiers, she doubled the size of her army and gathered a large force of foreign mercenaries. In strengthening her nation, Judea became a formidable foe to foreign powers. But while Alexandra ruled her people the Pharisees ruled her.

3. The Pharisees had Diogenes executed. He was a very distinguished citizen who had been Alexander's friend. They accused him of having advised the erstwhile king to crucify the 800 rebels. They had also urged Alexandra to do away with the rest of the men involved in inspiring Alexander to punish the rebels with such cruelty. She was too superstitious not to let them have their way. So the Pharisees proceeded to kill whomever they pleased. The most prominent of the citizens under the threat of death took refuge with Aristobulus, who persuaded his mother to spare their lives in deference to their high standing in the community. If she weren't convinced of their innocence, she could expel them from Jerusalem. So these men were allowed to go unpunished and subsequently dispersed around Judea.

At one point, Alexandra sent a force to assist Damascus under the pretext that Ptolemy was continually harassing the city. The troops later returned to Jerusalem without having accomplished much one way or the other. On the other hand, she was successful in causing Tigranes, king of Armenia, to withdraw from his positions around Ptolemais. By entreaties and presents she won him over, and he departed. It is just as likely, however, that Tigranes was recalled by domestic problems in his homeland, that Lucullus had invaded.

4. In the meantime, Alexandra was stricken ill and her younger son Aristobulus seized his opportunity to gain power. With the aid of his numerous followers, all devoted to him because of his fiery disposition, Aris-

tobulos took control of all the fortresses in Judea. With the money he found he hired mercenary troops and proclaimed himself king. His older brother, Hyrcanus, complained bitterly to his mother about his brother's arrogance, so the queen arrested Aristobulos's wife and children and had them locked up in Antonia. Antonia was a fortress adjoining the north side of the temple, which, as I said, was formerly called the Citadel. It took its name Antonia when Mark Antony came to power. In the same way, Augustus and Agrippa had given their names to the cities Sebaste and Agrippias. But before Alexandra could take measures against Aristobulos for disinheriting his brother, she died in 69 BC, having reigned for nine years.

Chapter 6

Aristobulus Becomes King Following Hyrcanus's Abdication, but Is Brought Back to Power through Antipater and Aretas. Pompey Arbitrates the Brothers' Disputes

1. Hyrcanus was the eldest son and true heir to the throne of Judea, and indeed, his mother had committed the crown to him before she died. But Aristobulos surpassed his brother in power, courage, and ability. When the battle for the throne erupted near Jericho, most of Hyrcanus's army deserted him and went over to Aristobulos. So Hyrcanus and those who remained loyal to him quickly made their way to Antonia. They took Aristobulos's wife and children hostages, who were still imprisoned there in order to ensure their own safety. Fortunately, before any real harm occurred, the brothers reached a peaceful settlement. Aristobulus became king, and Hyrcanus abdicated any further pretensions to the throne. The elder would still receive all the honor due to the brother of the king. Cordially embracing one another after they reached final agreement in the temple, they then switched homes. Aristobulus moved into the royal palace while Hyrcanus took up residence in the house formerly occupied by his brother.

2. Aristobulus's enemies were shocked and fearful at his sudden ascension to the throne. Particularly concerned was Antipater, Aristobulus's long time and bitter enemy. (Antipater was to become Herod the Great's father). Antipater was an Idumean by birth and one of the leading men of

that nation by virtue of his ancestry and wealth. He persuaded Hyrcanus to flee to Aretas, king of Arabia, and to announce there his intention of regaining the kingdom. At the same time, Antipater had persuaded Aretas to receive Hyrcanus and to assist him in reestablishing his rightful claim. He pressed the issue with King Aretas by slandering Aristobulos's character, all the while praising Hyrcanus as the firstborn and, therefore, the rightful heir. Antipater further argued that it fitting that Aretas use his great kingdom to protect those unjustly deprived of their rights.

Having persuaded each man to do as he had planned, Antipater spirited Hyrcanus from Jerusalem one night, fleeing with him from the city. Using all haste they soon reached the Arabian capital, Petra. There, Hyrcanus presented King Aretas with many gifts and, after much discussion, convinced the king to give him an army of 50,000 mounted and foot soldiers to reinstate him as the rightful king of Judea. With this large force, Hyrcanus attacked the Aristobulus's army and was victorious, driving his brother back behind the walls of Jerusalem. Were it not for the intervention of the Roman general Scaurus, Hyrcanus's force would have quickly stormed and captured the city. But Pompey the Great, then at war with Tigranes, had sent Scaurus into Syria from Armenia. When he reached Damascus, which had recently been taken by the Roman commanders Metellus and Lollius, Scaurus took command of all Roman forces in that area. Then, hearing of the state of affairs in Judea he hurried to Jerusalem to seize what seemed to be a fabulous opportunity.

3. Sure enough, no sooner had Scaurus arrived on Jewish soil than he received ambassadors from both of the brothers, each asking for Scaurus's assistance against the other. The 300 gold talents sent to him by Aristobulus outweighed Scaurus's sense of justice, so he sent a message to Hyrcanus and the Arabians threatening them with the displeasure of Rome and Pompey if they did not withdraw their siege of Jerusalem. Terror stricken, the Arabian king immediately withdrew his forces to Philadelphia. So Scaurus then returned to Damascus. But Aristobulus saw his opportunity. Having escaped certain capture by the Arabians, he gathered his troops together and pursued the Arabian retreat, overtaking them at a place called Papyron. He killed about 6,000 of Aretas's men together with Phalion, Antipater's brother.

4. Now bereft of allies, Hyrcanus and Antipater redirected their hopes

from the Arabians to their Roman adversaries. Pompey had just passed through Syria and had stopped at Damascus. Hyrcanus and Antipater there sought his assistance. Offering him no bribes they made the same arguments they had used to persuade Aretas. They urged Pompey to detest the violence of Aristobulos and to help restore the Judean throne to the brother to whom it justly belonged, both by reason of his birthright and of his excellent character. But Aristobulos had also made his way to the Roman camp. Relying on Scaurus's susceptibility to a bribe, he dressed himself in his utmost royal finery to seek Rome's favor. But his pride got the better of him, and he soon changed his mind. Aristobulos apparently could not endure acting beneath his royal dignity to woo the Romans. So he left Damascus for Diospolis and then returned to Judea.

5. Pompey was furious at Aristobulus's behavior. His anger and the appeals of Hyrcanus and his friends prompted Pompey to pursue Aristobulus with the Roman army, augmented by a large force of Syrian auxiliary troops. Passing Pella and Scythopolis, Pompey reached Corea, where the road from Damascus ascends into Judea. At Corea, he heard that Aristobulus had altered his route, fleeing instead to Alexandrium, a well-equipped stronghold situated high up on a mountain. Pompey sent a message to him there ordering him to come down. Aristobulus was inclined to fight rather than to obey Pompey's order since Pompey gave it in such an arrogant way. But when he observed the Alexandrium citizens' terror, and as his friends urged him to consider the irresistible power of the Romans, Aristobulos gave in and came down to meet Pompey. When he had made a long-winded defense of his claims to the Judean throne, he returned to the stronghold. At his brother's invitation, Aristobulus descended a second time and spoke again about the justness of his cause. He then returned to the mountain unimpeded by Pompey. And so, torn between hope and fear, Aristobulos descended in hopes of persuading Pompey to let him have the crown. He then ascended, fearing that if he did not they might think he had given up his claim. Soon, however, Pompey ordered Aristobulus to evacuate the fortress and insisted that all of his governors throughout the land do the same. Knowing that the Jewish officials would not act upon an order not written in Aristobulus's handwriting, Pompey forced the king to write out each order. Aristobulus complied with Pompey's demands, but then angrily left for Jerusalem to prepare for war with the Romans.

6. But Pompey was determined not to let Aristobulos make any siege preparations, so he followed close on his heels. He was further pressed on by news of the death of Mithridates that came to him near Jericho. The soil in that region is the most fertile in Judea and produces huge numbers of palm trees as well as balsam trees. The branches of this latter tree are cut with sharp stones, and the sap that trickles forth is harvested from the incisions. Pompey set up camp in this territory for one night and then marched quickly to Jerusalem. When Aristobulos saw Pompey's forces approaching the city, he was so terrified that he went out to meet him. He offered Pompey both gold and the surrender of himself and the city, and succeeded in pacifying Pompey's wrath. But none of Aristobulos's promises was fulfilled. When Gabinius was sent to the city to bring back the gold, Aristobulos's friends would not so much as open the gate for him.

Chapter 7

How Pompey Captured Jerusalem, Taking the Temple [by Force] and Entered the Holy of Holies; and His Other Exploits in Judea.

1. Further angered by this rebuke, Pompey kept Aristobulus under arrest. He then advanced toward Jerusalem analyzing the best route and method of attack. Pompey observed the strength of the city walls and realized the difficulty of overcoming them. Additionally, a steep and tortuous ravine stood between the Romans and the city walls, within which rose the temple mount, itself strongly fortified. It would provide a second line of defense should Pompey be successful in his siege.

2. As Pompey deliberated his strategy, a fight broke out in the city between the two factions. Those supporting Aristobulus wanted to defend the city while those for Hyrcanus wanted to throw open the gates for Pompey and the Romans. The number of people supporting Hyrcanus was steadily increased by the fear inspired by the spectacle and the discipline of the Roman troops. Those backing Aristobulus found themselves defeated, and soon retreated into the temple. They cut the bridge connecting it to the city and prepared to defend it unto their deaths. The remaining populace admitted the Romans into the city and surrendered the palace. Pompey

then sent a garrison of troops to occupy Jerusalem under the command of Piso—one of his lieutenant generals. Piso distributed guards around the city, then turned his attention to the temple. Since those holding the sacred ground would not listen to terms, Piso began to prepare the surrounding land for an assault. Those who had sided with Hyrcanus gladly offered him both their advice and their assistance.

3. Pompey supervised the work on the north side of the temple, filling up the ditches and the entire ravine with materials gathered by his troops. The vast depth of the ravine, combined with the fact that the temple's defenders used their higher position to impede the Romans, made the work extremely difficult. In fact, the Romans might never have been able to complete the task had not Pompey taken note of the Sabbath day. On each seventh day, the Jews abstain from all work because of their religious beliefs. On that day, Pompey filled the ravine but kept his soldiers from fighting, for the Jews would fight in self-defense if the need arose. Once he filled up the ravine, Pompey built high towers on the filled earth and brought forward those battering rams that had been pulled up from Tyre. With the rams, he attempted to bring down the temple walls while the missiles from his catapults beat back the Jewish defenders. But the magnificent temple walls in this sector were extraordinarily strong and resisted the pummeling for a long time.

4. While the Romans were suffering under these many hardships, Pompey developed a great admiration for the Jewish courage and tenacity. He was particularly impressed that they did not cease to observe their religious practices even in the midst of the missiles that rained down on their heads. Just as if the city were enjoying peacetime, the Jews still performed their daily sacrifices and rites of purification with meticulous precision. They did not omit one detail of their devotion to God. As a matter of fact, even when the temple was ultimately captured and the Jews were themselves slain on its altar, they did not cease to observe their laws' holy ordinances. Three months of difficult siege passed before the Romans were able to overthrow one of Jerusalem's towers and get into the temple. The first to climb over the wall was Faustus Cornelius, Sulla's son. Following him were Furius and Fabius with their cohorts. They surrounded the Jews and slaughtered them, some while they were running to hide in the sanctuary and others as they fought in self-defense.

5. It was then that many of the priests calmly continued their sacred ministry and worship, even when they saw their enemies attacking with drawn swords. They were cut down while giving their drink offerings and while burning incense, preferring their duty to God above self-preservation. The majority of the priests were killed by their own countrymen who had supported Hyrcanus, and a great number jumped to their deaths over the precipitous walls. Some, driven mad by the hopelessness of the situation, set fire to the buildings near the wall and were themselves consumed by the flames. The Jews slain numbered 12,000. On the other hand, only a few Romans perished but many were wounded.

6. Of all the distresses suffered by the Jews in the calamitous events of that time, none was as devastating as that of having the most Holy Place—the Holy of Holies—opened up to the eyes of Gentiles for the first time. Pompey himself, together with his entire staff, went into the inner sanctuary of the temple and saw the sacred items it contained. They saw what the law allowed only the high priest to see: the candlestick with its lamps, the table, the pouring vessels, the solid gold censers, a vast mountain of spices, and 2,000 sacred gold talents. Yet Pompey touched neither the money nor the implements of worship. Instead, the next day he commanded the priests who had been loyal to Hyrcanus to clean the sanctuary and to resume their ritual sacrifices. Further, he reinstated Hyrcanus as high priest in return for his enthusiastic support during the siege, and for keeping large numbers of his countrymen from siding with Aristobulus. Hyrcanus's benevolent spirit had made him more of a general than one who commanded by terror. Among the Roman prisoners was Aristobulus's father-in-law, who was also his uncle. Pompey executed those upon whose shoulders lay the central responsibilities for the hostilities. But upon those of his own troops who had fought so bravely he gave glorious rewards for bravery, including Faustus and those who first climbed over the temple wall. Meanwhile, Jerusalem and all of Judea was placed under tribute to Rome.

7. Pompey also took from the Jews those cities they had captured in Coele-Syria, and placed them under a Roman governor appointed to administer the region. He therefore reduced Judea to its historic boundaries. At the request of Demetrius—a Gadarene and one of Pompey's own

freedmen—Pompey rebuilt Gadara, a city that the Jews had destroyed. He also liberated other centrally located cities that had not yet been burned to the ground. Among these were Hippo, Scythopolis, Pella, Samaria, and Marissa, as well as Ashdod, Jamnia, and Arethusa. In the same way, the seaports of Gaza, Joppa, and Dora, as well as that city formerly known as Strato's Tower, were also liberated. The latter town was to be rebuilt with magnificent structures by King Herod, who renamed it Caesarea. All of these towns Pompey restored to their former citizenry and annexed to the province of Syria. Pompey made Scaurus governor of Syria, Judea, Egypt, and all the lands as far as the Euphrates, giving him two legions of fighting men as support. Then Pompey marched quickly through Cilicia on his way to Rome accompanied by his prisoners, Aristobulus and his children. The journey began with two daughters and two sons. But Alexander, one of the sons, made his escape along the way. The youngest son, Antigonus, and his sisters were all transported to Rome.

Chapter 8

Aristobulos's Son Alexander Who Ran from Pompey, Makes an Expedition Against Hyrcanus, but Is Instead Overrun by Gabinius and Surrenders the Fortresses to Him. Following this, Aristobulos Escapes from Rome and Assembles an Army, but Beaten by the Romans, Is Brought Back to Rome. Other Matters Relate to Gabinius, Crassus, and Cassius

1. Meanwhile, Scaurus invaded Arabia but saw his advance halted by the difficult terrain near Petra. He ravaged the countryside surrounding Pella as his army underwent great hardships and was on the verge of starvation. To meet his needs, Hyrcanus sent Antipater with provisions for Scaurus. Scaurus then sent Antipater to king Aretas, since the two were friends, to propose that Aretas pay Scaurus money in exchange for a Roman withdrawal. The king of Arabia accepted the proposal and sent Scaurus 300 talents. Scaurus, therefore, withdrew his troops from the country.

2. During this time, Alexander, the son of Aristobulus who had escaped from Pompey, assembled a considerable force and pestered Hyrcanus with continual raids within Judea. Indeed, Alexander advanced as far as Jerusalem and had the gall to begin rebuilding the wall that Pompey had battered down. He probably would have overthrown the city had not

Gabinius marched against him. This successor to Scaurus as governor of Syria had shown his courage on other occasions. But Alexander, warned of the approaching Gabinius, enlarged his army to 10,000 armed foot soldiers and 1,500 cavalry. He also built defensive fortifications around the strategic cities of Alexandrium, Hyrcanium, and Macherus, adjacent to the mountains of Arabia.

3. Gabinius sent out an advanced division led by Mark Antony, then followed with the main body of his troops. The handpicked troops of Antipater and another force of Jewish soldiers under the command of Malichus and Pitholaus joined the advanced division under Mark Antony and marched with it to meet Alexander. Gabinius soon arrived with his legions of infantry. When Alexander realized that he would not be able to withstand the assault of such a large force, he beat a hasty retreat, but on approaching Jerusalem Alexander was forced into a fight. He proceeded to lose 6,000 men in the battle that ensued. The killed numbered 3,000 and the captured 3,000. Licking his wounds, Alexander fled with the rest of his troops to Alexandrium.

4. When Gabinius reached Alexandrium, he found a great many of Alexander's men camped outside the walls. Before beginning his attack, Gabinius offered the Jews pardon for past offenses if they would come over to his side before the fighting began. When Alexander's troops proudly refused to come to terms, Gabinius killed many of them while the rest took refuge inside the fortress. Mark Antony received many honors during this engagement. He had always displayed great courage, but it was never so in evidence as at Alexandrium. So Gabinius left a force sufficient to capture the fortress and then began a tour of the country restoring order in those towns still standing. He rebuilt those he found in ruins. By his command, the following cities were restored: Scythopolis, Samaria, Anthedon, Apollonia, Jamnia, Raphia, Marissa, Adoreus, Gamala, and Ashdod, along with many others. These places were quickly re-inhabited by their dispossessed citizenry.

5. After supervising these reconstructions, Gabinius returned to Alexandrium and vigorously intensified the siege there. When Alexander realized that he would never succeed in making himself Judea's king, he sent ambassadors to Gabinius asking for forgiveness and surrendering the re-

maining fortresses at Hyrcanium and Macherus. He subsequently delivered up Alexandrium as well. Gabinius proceeded to demolish all three of these fortress cities. The destruction was done at the request of Alexander's mother so that they might not serve as bases of operation in future wars. She had come out to pacify Gabinius out of concern for her husband and children, who were prisoners in Rome. Gabinius then brought Hyrcanus back to Jerusalem and placed him in charge of the temple. Gabinius established a new aristocratic government. He also divided the nation into five city-states. Jerusalem, Gadara, Amathus, Jericho, and Sepphoris in Galilee were to be the capitals of each. The Jews were happy to be released from monarchial rule, and from that time forward were governed by an aristocracy.

6. But Aristobulus was to be the source of renewed trouble. He had succeeded in making his escape from Rome and had put together a large band of revolutionaries — men who had long been his devoted admirers. Aristobulus first took Alexandrium and attempted to construct a wall around it. But when he heard that Gabinius had dispatched a force against him under the joint command of Sisenna, Mark Antony, and Servilius, he retreated to Macherus. He quickly dismissed all of his followers except his 8,000 armed fighting men. Among the latter was Pitholaus, who had been second in command at Jerusalem and who had deserted to Aristobulos with a thousand of his men. The Romans pursued Aristobulus, and battle ensued eventually. Aristobulus's men held their ground for a long while, fighting courageously. But, in time, the Romans overran them. Those killed numbered 5,000 while about 2,000 escaped to a position on a nearby hill. The thousand that remained with Aristobulos broke through the Roman lines and reached Macherus. There, as he camped among the ruins of the demolished fortress, the former king dreamed of raising another army, if only he could evade war for a short time. Accordingly, he proceeded to erect some weak fortifications. But when the Romans attacked, Aristobulus could only hold out for two days before they overran Macherus, and he was taken prisoner. Together with Antigonus — his son with whom he had escaped from Rome — Aristobulos was taken to Gabinius, who sent him back to Rome in chains. The senate imprisoned Aristobulus but sent his children back to their mother in Judea. Gabinius had informed them that he had promised their mother to do so after the surrender of the three fortresses.

7. Gabinius then turned his attention to the Parthians and began a campaign against them. But he was stopped by Ptolemy, who sought to be restored to the Egyptian throne upon his return from the banks of the Euphrates. Hyrcanus and Antipater promised to provide Gabinius with everything he would need for the Egyptian campaign. In addition to supplying money, corn, arms, and auxiliary troops, Antipater persuaded the Jewish guards on the Egyptian border at Pelusium to let Gabinius through in peace. But following Gabinius's departure to Egypt, a general revolt began in Syria. Aristobulus's son, Alexander, had also begun a fresh revolt in Judea, and after mustering a vast army he proceeded to slaughter all the Romans in the country. Gabinius was obviously alarmed at these events. He had already returned from Egypt, having heard of the disturbances in Syria and Judea. He sent Antipater ahead to reason with the revolutionaries and who convinced some of them to return to their homes. But Alexander still had 30,000 men and was itching for a fight. So Gabinius took to the field and met the Jews near Mt. Tabor. A total of 10,000 Jews fell in the battle, and the rest of Alexander's army fled – every man for himself. Gabinius then proceeded to Jerusalem where he reorganized the government in accordance with Antipater's wishes. He subsequently marched against the Nabateans and defeated them. After that engagement, Gabinius privately let two fugitives from Parthia slip away, Mithridates and Orsanes, while getting the word out to his troops that the pair had escaped.

8. The government of Syria subsequently passed from the hands of Gabinius to Crassus. In order to underwrite his campaign against the Parthians, Crassus stripped the temple in Jerusalem of all its gold, including the 2,000 talents in the inner sanctuary that Pompey had left untouched. Crassus then crossed the Euphrates and perished with his entire army. But this is not the proper time to deal with those events.

9. After Crassus's death, the Parthians quickly counterattacked across the Euphrates into Syria, but were repulsed by Cassius, who had escaped into that province. Once he had made Syria secure, Cassius marched into Judea. He captured Tracheae and enslaved 30,000 Jews. He also executed Pitholaus upon the advice of Antipater. Pitholaus had tried to rally those who had once supported Aristobulus. Meanwhile, Antipater had married

a woman named Cypros, from a very prominent Arabian family. They had four sons: Phasael, Herod (who later became king), Joseph, and Pheroras, as well as a daughter named Salome. Antipater, by virtue of his kind favors and great hospitality, had made many friends in high places. Above all, through his marriage to Cypros, he had won the friendship of her relative, the king of Arabia. So it was to the king that Antipater entrusted his children when he went off to war against Aristobulus. Meanwhile, Cassius had forced Alexander to remain peaceful, so he returned to the Euphrates to keep the Parthians from reentering Syria. But we cannot take up that topic here.

Chapter 9

Aristobulos Is Released by Pompey's Friends, as Is His Son Alexander by Scipio. Antipater Cultivates a Friendship with Julius Caesar Following Pompey's Death; Antipater also Performs Great Exploits In the War Assisting Mithridates.

1. It came to pass (in 49 BC) that Julius Caesar became emperor of Rome, sending Pompey and the entire Roman senate in flight across the Ionian Sea. He also released Aristobulus and sent him immediately to Syria with two legions under his command. Caesar hoped Aristobulus would conquer Syria and the area around Judea, but Aristobulus's zeal for the task and Caesar's hopes were soon dashed. Some of Pompey's friends poisoned the Jew. For a long time, Aristobulus's corpse remained unburied in Judea. It lay preserved in honey above ground until Mark Antony finally sent it to the Jews for burial in the royal tombs.

2. Aristobulus's son, Alexander, also perished. He was beheaded by Scipio at Antioch following a trial in which he had been accused of significant grievances against the Romans. Ptolemy, son of Mennaeus, prince of Chalcis in the Lebanon Valley, then took Alexander's brother and sisters under his wing. Ptolemy had sent his son, Philippion, to Ascalon to fetch them. The young man then proceeded to fall in love with one of Aristobulus's daughters, Alexandra, and married her. But Ptolemy also loved the girl, so he murdered Philippion and took his son's widow for himself. His marriage caused Ptolemy to pay more attention to Alexandra's brother and sister.

3. After Pompey's death, Antipater decided to switch sides and to cultivate Caesar's friendship. When Mithridates of Pergamus tried to lead his army into Egypt, (on a mission for Caesar), he was refused passage at the Pelusium border and was forced to remain at Ascalon. Antipater persuaded his Arabian friends to assist Mithridates. He then joined his friend at Ascalon with a force of 3,000 armed Jews. Antipater also encouraged some powerful Syrians, such as Ptolemy—son of Sohemus of Lebanon, and Jambilicus, to bring the men of that country into the war. Mithridates was encouraged by these reinforcements. When refused passage through Pelusium, he attacked and laid siege to the town. Antipater greatly distinguished himself in the battle. His troops breached the part of the wall that they faced. Antipater himself was the first to leap into the city followed by his men.

4. And so Pelusium fell. But the joint forces of Mithridates and Antipater were then stopped by the Egyptian Jews in that area of the land called Onias, after the man who had built the replica of the temple there. Antipater, however, persuaded them to lift their opposition and even to furnish supplies to the invaders. Because of this, they encountered no opposition in Memphis. Many of its citizens even joined Mithridates's army that then marched around the Delta and met the army of Egypt at a place called the "Jew's Camp." In the battle that ensued, Mithridates's right wing got into serious trouble. But Antipater, on the left wing, wheeled his troops about and came along the bank of the Nile to his rescue. Attacking the Egyptians who were pursuing the retreating Mithridates, Antipater's forces killed a host of them and pushed the rest back to their camp that he captured. Antipater lost but 80 of his men. In contrast, Mithridates had lost about 800. Mithridates was unexpectedly saved and later bore unimpeachable testimony to Caesar of Antipater's courage and skill.

5. Caesar's great praise for Antipater's heroic action and his own hopes for future reward spurred Antipater on to undertake even more hazardous tasks. He proved himself to be a valiant fighter on many occasions. The many scars on his body bore witness to his courage in combat. Later, as Caesar had settled matters in Egypt and was returning to Syria, he conferred the privileges of Roman citizenship upon Antipater. This honor gave him freedom from taxation and made him a man to be envied.

Caesar further pleased Antipater by reconfirming Hyrcanus to the high priesthood.

Chapter 10

Caesar Appoints Antipater Procurator of Judea. Antipater Appoints Phasael Governor of Jerusalem and Herod Governor of Galilee. Herod is Called to Answer to the Sanhedrin but Is Acquitted. Sextus Caesar is Assassinated by Bassus and Is Succeeded by Marcus.

1. It was about this time that Antigonus, Aristobulus's son, paid a visit to Caesar that ironically became the occasion of further promotion for Antipater. Antigonus should have restricted his remarks to grieving over the death of his father, who he believed was poisoned because of his differences with Pompey, and to his complaint against Scipio for the cruelty to his brother Alexander. Instead, with Antipater present, he railed on against Hyrcanus and Antipater, and mixed obvious jealousy with his plea for pity. He accused Hyrcanus and Antipater of unjustly banishing himself and his sisters from their native land, and of repeatedly harming the nation. For instance, he charged that they had sent supporting troops into Egypt, not out of patronage to Caesar, but out of fear of the consequences of previous disputes and to gain Caesar's pardon for their former alliance with his enemy, Pompey.

2. Hearing Antigonus's accusations, Antipater immediately stripped off his clothes and exposed his numerous scars. He said that his loyalty to Caesar needed no words. The scars themselves bore mute witness to it. He was astounded at Antigonus's audacity. Here was the son of a Roman enemy and fugitive who had inherited his father's passion for treason and revolution. And this man dared to accuse others to the Roman emperor, looking for favors when he ought to be thankful just to be alive! Indeed, Antipater said, Antigonus's ambitions were not due to his impoverished state but in order that he might sow seeds of treason among the Jews while using the resources of Rome to do it.

3. When Caesar heard all this, he pronounced Hyrcanus to be worthy of the high priesthood and asked Antipater to choose whatever high office he wanted. But Antipater chose to leave the decision of his office to Caesar,

who immediately appointed him procurator of Judea (in 47 BC). Caesar further authorized him to rebuild the walls of Jerusalem. After these appointments, Caesar sent orders to Rome for these honors to be engraved in the Capitol as a memorial to his own justice and to Antipater's courage.

4. After escorting Caesar across Syria, Antipater returned to Judea. His first act was to begin the reconstruction of the walls that had been torn down by Pompey. He then began a tour of the country, snuffing out any potential local disturbances by both threats and reason. Their support of Hyrcanus, he told the people, would ensure prosperity and peace in the future. But if they followed those who for personal profit desired revolution, they would find him to be their slave master rather than their procurator, and find Hyrcanus to be a tyrant rather than a king. Both the Romans and Caesar would be their enemies instead of their rulers and friends, for Rome would not allow its own nominee swept from his office. As he was going around speaking like this, Antipater realized that Hyrcanus was lazy and lacked the energy to lead the country. So he took the affairs of Judea into his own hands. He appointed his eldest son, Phasael, to be governor of Jerusalem and its surrounding territory. The second son, Herod, though a mere boy, was sent to govern Galilee.

5. Young Herod was very energetic, and soon found problems enough with which to deal. After discovering that one Hezekias, the leader of a large band of thieves, was ravaging and pillaging the area of Judea near the Syrian border, he caught him and executed both Hezekias and many of his band of robbers. The Syrians greatly praised this most welcome achievement. Throughout their villages and towns, songs were sung to the praise of Herod, the restorer of the peace and the preserver of their possessions. Further, the exploit brought Herod to the attention of Sextus Caesar, a relative of Julius Caesar and the governor of Syria. Herod's brother, Phasael, was moved by the young man's growing reputation enough to emulate him. Phasael's popularity with the citizens of Jerusalem increased as he managed the affairs of government without abusing his authority or ruling in a disagreeable way. All of this served to elevate Antipater's honor as if he were king – the absolute lord of the land. Notwithstanding this public acclaim, Antipater continued steadfast in his love for, and loyalty to, Hyrcanus.

6. But one's prosperity draws envy like honey draws flies. The fame of Antipater's sons even caused Hyrcanus to be jealous, although he did not show it openly. He was particularly vexed by young Herod's success, and by the news brought by messenger after messenger of each new honor heaped upon the young man. His resentment was further inflamed by those court gossips that had taken offense at the wise deeds of Antipater and his boys. Hyrcanus, they said, had been stripped of all his kingly authority and remained a king in name only while Antipater and his sons ruled Judea. They had ceased to masquerade as procurators and had declared themselves dictators, thrusting Hyrcanus aside. For instance, they argued, Herod had executed a great number of people in opposition to Jewish law and without Hyrcanus's consent. They concluded that if Herod is only a commoner and not a king, then he ought to be brought to trial to answer to Hyrcanus for opposing the country's laws. No one was to be put to death without a trial.

7. Their words gradually inflamed Hyrcanus until finally, exploding in rage, he ordered Herod to stand trial. So Herod, on the advice of his father and as soon as the affairs of Galilee would permit, came to Jerusalem after posting garrisons of troops throughout Galilee. He surrounded himself with a sufficient escort of soldiers so that—while not wanting to appear to intend to depose Hyrcanus by force of arms, neither did he wish to run the risk of falling victim to those who were jealous of him. Sextus Caesar, however, fearing that the young ruler would be isolated and subsequently punished by his enemies, sent express orders to Hyrcanus to clear Herod of the manslaughter charges. Being inclined to do exactly that because of his love for the young man, Hyrcanus acquitted him.

8. But Herod thought that he had escaped punishment against Hyrcanus's will, so he went to Sextus in Damascus and prepared to refuse a second summons to Jerusalem, should one come. Meanwhile, the troublemakers at court continued to pester Hyrcanus, telling him that Herod had gone away angry and was now preparing to make war upon him. The king believed them, but he didn't know what to do for he knew that he was no match for Herod. When Sextus Caesar promoted Herod to be governor of Coele-Syria and Samaria, Herod became doubly formidable, both for the affection in which the nation held him and in his own increased power. Consequently, Hyrcanus became very fearful and imagined that Herod

might send his army against him at any moment.

9. Hyrcanus was not mistaken in his assessment of the situation. In his anger at the public accusations against him, Herod mustered his forces and led them to Jerusalem to overthrow Hyrcanus. He would have carried out his plan had not his father and brother gone out together to intercept him and calm him down. They urged Herod to go no further than to frighten Hyrcanus by his threats, but to spare the king. After all, it was under Hyrcanus that Herod had received his power in the first place. They argued that Herod should not be angry at his trial, but rather thankful for his acquittal. After facing the bleak prospect of condemnation, the young man ought to be grateful that he had escaped with his life. Further, they said, if we believe that the fortunes of war are in the hands of God, an unjust cause in wartime can be a greater disadvantage than the advantage of a great army. Herod should therefore not be so confident of success in battle against Hyrcanus, his king and friend. After all, he had often been Herod's benefactor, they contended, and never his oppressor. They said Hyrcanus's evil counselors were the cause of the trouble , and that they had attacked Herod only with unjust words, not deeds. Herod yielded to the advice of Antipater and Phasael, calculating that he had accomplished all that was needed to ensure his future hopes. His exhibition of power to the nation would be sufficient.

10. Meanwhile, there arose trouble for the Romans in Apamia where a civil war was brewing. Cecilius Bassus, out of devotion to Pompey, assassinated Sextus Caesar and took command of his forces. At this outrage, the rest of Caesar's generals attacked Bassus with everything they had in order to punish him for his crime. Antipater, for the sake of his friends—the assassinated and surviving Caesars, sent reinforcements into Syria under the command of his sons. As the war dragged on, Marcus arrived in Syria to succeed Sextus.

Chapter 11

Herod Is Appointed Procurator of All Syria. Afraid of Herod, Malichus Poisons Antipater, Upon Which the Tribunes Are Prevailed Upon to Assassinate Him.

1. A great war broke out between Roman factions (in 44 BC) on the

sudden and treacherous assassination of Julius Caesar by Cassius and Brutus. Caesar had reigned in Rome for three years and seven months. His murder produced a tremendous upheaval as the leaders of the empire split into warring parties, each joining the side where he supposed his self-interests and ambitions would best be served. Accordingly, Cassius went into Syria to take command of the armies concentrated around Apamia. He succeeded in arranging a truce between Murcus and Bassus and their opposing legions. Then, placing himself at the head of their armies, Cassius went around Syria exacting tribute from its cities, demanding much more than they were able to pay.

2. Cassius demanded that the Jews should pay 700 talents. Antipater, out of fear of Cassius's threats, divided the required sum among his sons and some of his acquaintances in order to expedite its collection. He was even required to force one of his enemies, Malichus, to also raise a sum of money. Herod was the first to bring his share—100 talents—out of Galilee. By his quick action, he appeased Cassius and became a close friend. But the rest of the cities of Judea were tardy in their collections, and the angry Cassius reviled them for it. He made slaves of the citizens of Gophna and Emmaus, along with Lydda and Thamna. He was ready to kill Malichus for his slowness in collecting the tribute money. But Antipater stepped in and prevented Malichus's death and the destruction of other cities. Antipater also received Cassius's favor by his quick collection of 100 talents.

3. However, after Cassius had left the country, Malichus forgot the kindness of Antipater and plotted against the life of the man who had saved his. He was impatient to get rid of the one man who stood in the way of his wicked ambitions. But Antipater, fearing Malichus's strength and cunning, crossed the Jordan to raise an army with which to overthrow the conspiracy against him. Malichus, aware that his plot against Antipater had been discovered, went to Antipater's sons and outwitted them. By a great host of excuses and oaths, Malichus deluded Phasael, Jerusalem's governor, and Herod, who maintained the nation's war weapons, to act as mediators between himself and their father. The outcome of all of this was that Antipater saved Malichus's life once again. He persuaded the Syrian governor, Marcus, to drop his resolution to execute Malichus for being a revolutionary.

4. When the young Augustus Caesar and Mark Antony declared war on Cassius and Brutus, Cassius and Marcus mustered an army in Syria. They saw Herod as a substantial future ally and appointed him procurator of all Syria, giving him an army of both infantry and cavalry. Cassius further promised Herod that on the successful conclusion of the war he would make Antipater's son king of all Judea. But this elevation of the power and hopes of Herod led ultimately to his father's death. Malichus was alarmed by the portent of Herod as king, so he bribed one of the royal cupbearers to serve poison to Antipater at a feast. Finally falling victim to the wickedness of Malichus, Antipater died after leaving the banquet (in 37 BC). Antipater had been a man of great energy in the affairs of Judea. His crowning achievement was the restoration and preservation of Hyrcanus's crown.

5. Malichus was immediately suspected of poisoning Antipater, but when an angry crowd rose up against him he denied the charge and convinced the people of his innocence. He also strengthened his position by raising a force of men for protection. He did not think Herod would sit still for long. Indeed, Herod soon marched against Malichus with his troops, intent upon avenging his father's death. However, Herod's brother Phasael advised him not to take his revenge openly for fear that the people might riot in its wake. So Herod, for the moment, accepted Malichus's pronouncement of his innocence and professed to clear him of any suspicion. He then went ahead with plans for a magnificent state funeral for his father.

6. Herod soon went to Samaria where an uprising was taking place and restored peace to the city. During the feast of Pentecost, Herod returned to Jerusalem with his forces surrounding him. Alarmed by his approach to the city, Malichus urged Hyrcanus to stop Herod's foreign troops from entering Jerusalem during the Jews' time of purification. But Herod saw through the ruse, and despising the man who had instigated it, entered the city by night. So Malichus came to him and wept over Antipater's death. Though angered at Malichus's behavior, Herod managed to maintain his composure and faked his acceptance of Malichus's tears as real. At the same time, Herod sent a letter to Cassius lamenting Antipater's death. Cassius also had reasons for hating Malichus and responded that Herod

should avenge his father's murderer. He gave secret orders to his tribunes to assist Herod in the righteous deed.

7. When Cassius took Laodicea and all the influential men from every part of the country had gathered together to give him gifts and crowns, Herod saw his opportunity for revenge. But Malichus suspected as much. So when Malichus reached Tyre to rescue his son who was a prisoner there he prepared to flee back into Judea. Malichus's desperation caused him to dream up even greater schemes, such as provoking a general revolt against the Romans while Cassius was busy fighting Mark Antony's army. Malichus reasoned that he could easily wrest the throne from Hyrcanus and become king himself.

8. But fate had other ideas. Herod saw through Malichus's intentions and invited him to supper, along with Hyrcanus. He then dispatched one of his servants, ostensibly to oversee the banquet preparation. In reality, the servant was to alert the tribunes to prepare for an ambush. Remembering Cassius's orders, they came out to the seashore in front of Tyre and surrounded Malichus with unsheathed swords. They then ran him through with many thrusts, killing him on the spot. Hyrcanus was so overcome with fear that he fainted and fell to the ground. When he came to some time later, he asked Herod who had killed Malichus. To which one of the tribunes replied, "Cassius commanded it." Hyrcanus said, "Cassius, then, has saved both me and my country, by executing the one who conspired against us both." It is uncertain whether Hyrcanus expressed his true thoughts or whether he said what he did out of fear. Be that as it may, it was according to this plan that Herod got his revenge.

Chapter 12

Phasaelus Defeats Felix; Herod Defeats Antigonus; The Jews Accuse both Herod and Phasaelus, but they both are Acquitted by Mark Antony Who Makes them Tetrarchs.

1. Now when Cassius left Syria another outbreak of hostilities occurred in Jerusalem. Felix, Malichus's brother, led an army against Phasael in an attempt to punish Herod for his relative's death. At the time, Herod was in Damascus with the Roman general, Fabius. As he was leaving for Jerusalem to help Phasael, Herod grew ill and was forced to delay his

journey. Meanwhile, Phasael defeated Felix without Herod's aid. He then rebuked Hyrcanus for his ingratitude and for both assisting Malichus and overlooking Felix on the occasion of the latter man's capture of the many fortresses in the country, including its strongest, Masada.

2. But no fortress would be sufficient to withstand Herod's might. As soon as he was well again, he regained the fortresses and threw Felix, begging for mercy, out of Masada. Herod also drove Marion, the tyrant of Tyre, out of Galilee. Marion had taken three of the fortresses there. The Tyrians Herod captured were all spared. In fact, Herod even sent some of them back to Tyre with presents in order to gain their affection and to increase their hatred of Marion, who had gained his power from Cassius, who had placed local tyrants throughout Syria. Out of his hatred for Herod, Marion had assisted Antigonus, Aristobulos's son. In this aid, he had been helped by Fabius, whom Antigonus had bribed to help him in an attempt to be restored to the throne of Judea. Ptolemy, Antigonus's brother-in-law, supplied all the money for this effort.

3. Herod waged war against all these enemies at the border to Judea and was victorious. He drove Antigonus out of the country, and then returned to Jerusalem where his glorious victory had won the hearts of everyone. Even those who had previously been noncommittal to his leadership now joined with Herod due to his marriage into Hyrcanus's family. He had formerly married a Jewess of noble blood named Doris, by whom he had a son named Antipater. Now, he married Miriamne, the daughter of Alexander, Aristobulos' son, and Hyrcanus's granddaughter. Herod had become one of the king's relatives by marriage.

4. When Caesar and Mark Antony killed Cassius near Philippi, Caesar returned to Italy and Antony went into Asia. Ambassadors from the various countries came to Antony in Bithynia; among them were certain Jewish leaders. These men accused Phasael and Herod of usurping the powers of Judean government and leaving Hyrcanus king in name only. Herod also appeared in Bithynia and by his large bribes kept Antony from granting his adversaries a hearing. So, for the time being, Herod's enemies were dispersed.

5. Later, though, one hundred powerful Jews went to Daphne near An-

tioch to meet with Antony, who had been enslaved by his love for Cleopatra. Among these men who accused Phasael and Herod were the most dignified and eloquent Jews. Messala came to the brothers' defense while Hyrcanus stood by in agreement because of his new marriage connection to Herod. When Antony had heard both sides, he asked Hyrcanus which party was best prepared to govern the country. Hyrcanus replied that Herod and his brother were the best qualified. Antony was delighted at this response, for their father Antipater had once treated him in the most gracious and hospitable way, when Antony had marched into Judea with Gabinius. So, he appointed the brothers tetrarchs and entrusted them with the administration of Judea.

6. When the opposing ambassadors vented their anger at this decision, Antony arrested fifteen of them and made preparations to execute them. The rest he drove away in disgrace. This action by Mark Antony inflamed the already simmering, violent situation in Jerusalem. A second company of 1,000 ambassadors was then sent to Tyre, where Mark Antony had stopped on his journey to Jerusalem. Antony ordered that the governor of Tyre intercept this clamorous company and punish as many of them as he could catch, in support of the tetrarchs whom he had already appointed.

7. Before giving this order, both Herod and Hyrcanus went down to the seashore near Tyre and earnestly contended with the ambassadors not to bring destruction upon themselves or war on their native country. But as the throng only grew more pugnacious, Antony sent out an armed guard that slew a great many of them and wounded even more. Hyrcanus buried the dead ambassadors and also provided physicians for the wounded. But those escaping would not be quiet; they created such a disorder in the city that Antony ordered his prisoners put to death.

Chapter 13

The Parthians Return Antigonus to Judea and Cast Hyrcanus and Phasael into Prison. Herod's Flight and the Taking of Jerusalem, and What Hyrcanus and Phasael Suffered.

1. Two years later, Barzapharnes, the Parthian satrap, together with Pacorus, the Parthian king's son, occupied Syria. Lysanias, who had succeeded his late father Ptolemy—son of Menneus, to the governorship of

Chalcis, promised the satrap 1,000 talents and five hundred women for assistance in restoring Antigonus to the throne of Judea, which would effectively depose Hyrcanus. Lured by this offer, Pacorus advanced along the Mediterranean coast while he ordered Barzapharnes to march through the interior. But the maritime city of Tyre closed its gates to Pacorus, although he had formerly been received at Ptolemais and Sidon. So Pacorus committed a squadron of cavalry to one of the royal cupbearers who shared his name—Pacorus, and gave him orders to march into Judea. He was to both gather intelligence regarding the enemy's position and to give such support to Antigonus as he might need.

2. While these forces were raiding Carmel, many Jews flocked to Antigonus and volunteered for an invasion of Judea. So he sent them ahead to seize a place called Drymus, which means "The Woodland." There, they engaged their enemies who supported Hyrcanus and drove them back toward Jerusalem. As their ranks swelled with new volunteers they proceeded as far as the king's palace. They were met there by a strong force, headed by Hyrcanus and Phasael, and a battle in the marketplace ensued. The Herodians routed their adversaries, incarcerating them in the temple and posting sixty men in adjoining houses to guard them. But others who fought against the brothers attacked this garrison of sixty guards and burned them to death. This action so infuriated Herod that he turned his wrath upon the citizens of Jerusalem and killed many of them. As the battle went on, day after day, small bands sallied out against one another, and there was continual slaughter.

3. When the feast of Pentecost arrived, the temple area and the entire city were full of a great many pilgrims who had come out of the country. The majority of them bore arms. Phasael's men defended the walls; Herod's men the palace. With his small force, Herod found his enemy in disorder and fell upon them in the northern part of the city. He killed a great many and caused the rest to flee for their lives. Some of them were trapped in the city, some in the temple, and the rest in the entrenched camp outside the walls. Meanwhile, Antigonus requested that Pacorus [the cupbearer, not the prince], be admitted to the city to serve as a mediator. Phasael agreed and allowed the Parthian into the city with five hundred cavalry troops, treating him hospitably. Pacorus, who had ostensibly come to try to put an end to the fighting, came to Antigonus's support. He cunningly laid a

trap for Phasael, urging him to go as an ambassador to Barzapharnes to seek an end to the hostilities. Herod argued vehemently against the idea and exhorted his brother to kill Pacorus rather than expose himself to the obvious trap of the naturally dishonest barbarian. But Phasael left the city anyway, accompanied by Hyrcanus. To reduce any further suspicions, Pacorus left some of his cavalry, called "Freemen," with Herod while he escorted Hyrcanus and Phasael on their way.

4. On arriving in Galilee, Hyrcanus and Phasael found the residents there in revolt and up in arms. The satrap Barzapharnes received the two Jewish leaders—though his open hospitality concealed treacherous motives. He presented them with many gifts, but later, as they left the audience he began to spring his trap. They discovered the treachery at a place called Ecdippon, a maritime town about halfway between Tyre and the spit of land at Carmel where they had made camp. They heard there that Antigonus had promised Barzapharnes and the prince 1,000 talents for the kingdom. Phasael and Hyrcanus also found out that most of the five hundred women in the deal were Jewesses, and discovered that the Parthians constantly spied them on at night. Further, the two learned they would have been arrested much earlier had the conspirators not wanted to first capture Herod in Jerusalem. The Parthians feared that the arrest of Hyrcanus and Phasael would put Herod on his guard. But now all that had been suspected was about to be confirmed. They could see their Parthian captors already stationed some distance away.

5. Phasael would not consider forsaking Hyrcanus and making his escape, even though Ophellius had tried to persuade him to do so. This latter gentleman had heard of the plot from Saramalla, then the richest man in Syria. Instead, Phasael went directly to Barzapharnes and candidly rebuked him for his treachery, particularly since his greed was the culprit. Phasael then offered the satrap a larger sum of money for his life than Antigonus had offered for a kingdom. In response to the accusations, Barzapharnes declared himself to be innocent, and denied the charges with many oaths. He then left to join the prince. Upon his departure, the Parthians who were left behind arrested Phasael and Hyrcanus as the prisoners cursed their captors for their lies and their breach of faith.

6. Meanwhile, the cupbearer Pacorus was commanded to go back to

Jerusalem to instigate a plot to trick Herod into being lured out of the city. Herod, who had suspected the barbarians of treachery from the start, now learned that the letters informing him of their conspiracy had fallen into enemy hands. He therefore refused to leave the city, even though Pacorus plausibly argued that he should go out to meet the letter bearers. He said that Herod's enemies had not intercepted the letters, which had not contained any mention of a plot but were only a historical report of Phasael's activities. Herod had heard of his brother's capture from other sources. Alternately, Hyrcanus's daughter, Miriamne—a very wise woman [and Herod's future mother-in-law], begged Herod not to go out or trust himself to the barbarians, who were openly preparing to kill him.

7. While Pacorus and his henchmen were secretly busy attempting to spring their trap—for they knew that any outright attack upon Herod would fail—Herod secretly slipped out of the city at night with some of his closest family members and headed for Idumea. As soon as the Parthians discovered he had gone, they gave pursuit. So Herod ordered his mother and sister, along with his fiancée and her mother and his youngest brother, to continue on their way while he and his men covered their retreat. In each subsequent encounter with the barbarians, Herod's men killed many of them. He then headed for the fortress at Masada on the west coast of the Dead Sea.

8. But Herod found that the Jews had become more troublesome than the Parthians. After he had journeyed about eight miles from Jerusalem, they perpetually harassed him with an almost continual engagement. He was able to finally subdue these Jews after killing a great many of them. Later, he built a city on the spot of his victory over them. He erected a citadel of fortress-like strength, and with it many costly palaces. Herod called it Herodium. While Herod's forces pressed on toward Masada, a new group of men joined him daily. On reaching Thressa in Idumea, he was met by his brother Joseph, who advised Herod to send most of his followers away. Masada, he said, could not accommodate a large crowd of more than 9,000 fighting men. So Herod, accepting this advice, and after giving them provisions for their journey, sent away those men who were more of a burden than a benefit. He retained the best fighting men of their number, and together with his cherished relatives safely reached the fortress at Masada. Then, leaving a contingent of eight hundred men

to guard the women, along with sufficient provisions to withstand a siege, he quickly pushed on to Petra in Arabia.

9. Meanwhile, the Parthians who remained in Jerusalem flung themselves into plundering the homes of those who had fled. They also fell upon the king's palace, sparing nothing but Hyrcanus's money, amounting to no more than three hundred talents. The barbarians found much less treasure than they had expected. Herod, anticipating their treachery, had taken the precaution of carrying his most precious treasures to Idumea, and had advised all those accompanying him to do the same. After their pillaging, the Parthians proceeded to carry their injustices to an extreme. They exposed the entire country to the horrors of an undeclared war, totally destroying the city of Marissa. They then elevated Antigonus to the throne and brought Phasael and Hyrcanus to him in chains to be tortured. Falling down on his knees in front of Antigonus, Hyrcanus had his ears bitten off by the newly crowned king. No longer free from physical defect as required by the Law for the priesthood, Hyrcanus would never again—under any circumstances—be allowed to serve as high priest.

10. Phasael's courage prevented Antigonus from experiencing the pleasure of torturing him. Without the use of either of his hands, or of his sword, Herod's brother dashed his own head against a rock. He showed himself to be a true brother of Herod and, at the same time, showed Hyrcanus to be a most defective relative by dying with the same bravery with which he had lived his life. There is an account of Phasael's death in which he recovers from his self-inflicted blow. Antigonus reportedly sent a physician ostensibly to attend to him, but who instead filled Phasael's head wound with poison, killing him. It matters little which account is true, for Phasael's actions would have been heroic in each. Other reports say that before he died, a woman told him of Herod's escape. He replied, "Now I can die happy because I leave behind me one who will exact vengeance upon my enemies."

11. Such was the death of Phasael. The Parthians were disappointed that they never got the five hundred Jewish women they had most wanted. Nevertheless, they installed Antigonus on the Judean throne in Jerusalem and took Hyrcanus to Parthia in chains.

Chapter 14

When Herod Is Ejected from Arabia, He Heads to Rome Where Antony and Caesar Join in Making Him King of the Jews.

1. Believing that his brother Phasael was still alive, Herod pressed on toward Arabia, where he sought the ransom money from its king that he hoped would tempt the barbarians to spare Phasael's life. He reasoned that if the Arab king had forgotten his friendship with his father, Antipater, or if the king were too greedy to give him the money outright, he would borrow from him the necessary sum, leaving Phasael's son as security. Accordingly, he had with him his seven-year-old nephew. Herod was ready to give three hundred talents for his brother's release, and intended to use the Tyrians to negotiate for him as they had volunteered to do. But fate seems to have outrun Herod's zeal. Phasael was dead, and Herod's filial affection for him was now in vain. Further, he soon found that the Arabians were no longer his friends. Their king, Malichus, sent an immediate message to him ordering that Herod leave his territory. Malichus pretended to have received a formal diplomatic demand from the Parthians expelling Herod from Arabia. In truth, Malichus was determined not to pay back his debts to Antipater, nor be shamed by Herod's need for his father's former wealth. The powerful men in the Arabian court had advised the king on this matter, knowing that he, like themselves, wanted to appropriate for himself the money that Antipater had entrusted to them.

2. When Herod realized that the Arabians were his enemies—ironically for the same reason that he had supposed them to be his friends—he sent them an emotional message in response to their treachery and turned back toward Egypt. That first night he camped in an Egyptian temple and picked up those men he had left in the rear. The next day he marched on to Rhinocorura, a seaside town on the border of Palestine and Egypt. There, he learned of his brother's death and of the manner in which he had died. With his anxiety over his brother's welfare now replaced with an equally heavy burden of grief, Herod resumed his march. Meanwhile, the Arabian king thought better of what he had done and sent messengers to call Herod back. But Herod moved too fast for them, having already reached Pelusium. Although refused passage for his troops by a fleet that lay in the harbor, he petitioned the higher authorities—who, due to their

respect for Herod's fame and dignity, escorted him to Alexandria. When he reached that city, Cleopatra met Herod with a grandiose reception. She had hoped to give Herod command of a military expedition she was preparing, but Herod rejected the queen's proposal and sailed for Rome instead, neither being afraid of the winter storms that threatened the Mediterranean nor of the tumultuous political situation in Italy.

3. At Pamphylia, Herod's ship narrowly avoided disaster by casting the bulk of the ship's cargo overboard. At last, with much difficulty, he arrived safely at Rhodes, a place that had undergone great suffering during the war with Cassius. Ptolemy and Sappinius welcomed Herod, and although he lacked funds Herod underwrote the construction of an immense three-decked vessel in which he and his friends then sailed to Brundusium. He quickly moved on from there to Rome where he had his first meeting with Mark Antony, who had been his father's friend. Herod told Antony of the calamities he and his family had suffered and how he had left his closest relatives besieged in a fortress while he sailed through treacherous waters in the midst of winter to seek Antony's assistance.

4. Moved with compassion at Herod's reversal of fortune, Antony recalled the hospitable treatment he had received from Antipater and, above all, Herod's own heroic qualities. Antony determined then and there to make the man he had previously appointed tetrarch king of the Jews. Besides his admiration of Herod, Antony had as additional inducement his aversion for Antigonus, who he regarded as a rebel and an enemy of Rome. Caesar was even more enthusiastic for Herod's promotion than Antony, as he recalled anew the memory of Antipater's steadfast loyalty and hospitality when they had fought side by side in the Egyptian campaigns. Caesar (Octavius) also saw the same enterprising spirit in Herod, so he convened the Senate. Messalas, seconded by Atratinus, presented Herod to the assembly and spoke of the many merits of Antipater and Herod's own goodwill to the Romans. At the same time, they argued that Antigonus was their enemy, not only because of their earlier quarrel with him but because he had held Rome in contempt by gaining his crown at the hands of the Parthians. Their arguments greatly moved the Senate. Meanwhile, Antony stood and told them that it would be advantageous in the war with Parthia if Herod were crowned king. The senate then unanimously approved Herod's nomination. Following adjournment, Antony

and Caesar left the Senate chambers with Herod between them. Preceding them were the counsels and the other magistrates who were to offer sacrifices and to lay the decree in the Capitol. Antony then gave a banquet for Herod on the first day of his reign (in 37 BC).

Chapter 15

Antigonus Besieges those Holding Masada; Herod Returns from Rome and Marches on Jerusalem Where He Finds Silo Corrupted by Bribes

1. All during this time, Antigonus had maintained his siege of the Masada fortress. Although Herod's besieged friends had plenty of most necessities, water was scarce. To remedy this lack, Herod's brother Joseph planned an escape to Arabia with two hundred of his men, having heard of Malichus's repentance of the shameful way he had treated Herod. On the night he was to take flight a heavy rain fell and refilled all the cisterns, so he canceled his plans. Instead, the garrison began a strategy of sallying out to skirmish with Antigonus's troops. They killed a great many of them using this strategy, some in open combat and some as the result of ambushes. These raids were not altogether successful as they were occasionally beaten back and forced to retreat into the fortress.

2. Meanwhile, the Roman general Ventidius was sent from Syria to check the raids of the Parthians. In his pursuit, he had advanced into Judea, ostensibly to come to the aid of Joseph and his friends. In reality, his hidden mission was to extort money from Antigonus. So Ventidius camped close by Jerusalem. After obtaining the bribe he sought, he left, taking the bulk of his army with him. But Ventidius left a detachment under the command of Silo in order to direct attention away from his extortion plot which would have become obvious had he withdrawn his entire force. Antigonus hoped that the Parthians would come to his aid, and tried to court Silo's favor with more bribes. He tried to forestall any trouble from the Roman commander until he realized his hopes of rescue.

3. But by this time Herod had already sailed from Italy to Ptolemais and had gathered a considerable army of both foreign and native troops. He was advancing through Galilee to attack Antigonus. Dellius, Mark Antony's ambassador, had convinced Ventidius and Silo to cooperate with

Herod in his reinstatement to the throne. But Ventidius was busy quelling local disturbances caused by the Parthian invasion while Silo lingered in Judea, held still by Antigonus's bribes. Herod, on the other hand, had no lack of forces. The numbers of his army swelled daily. With few exceptions, all the men of Galilee joined with him. The most urgent task facing him was Masada and the liberation of his relatives from the siege. But the hostile city of Joppa stood in the way. It was necessary for Herod to take that city first so that no stronghold would be left in his rear when he later turned and marched on Jerusalem. Silo soon proceeded to join Herod, happy to have an excuse to leave Jerusalem. In his departure, Silo was hotly pursued by the Jews. Herod came to his aid by counterattacking the Jews with a small body of his men. He routed them and rescued Silo, who had been under some duress.

4. Then, after taking Joppa, Herod quickly marched to Masada to rescue his relatives. As he marched, many of the local people joined his ranks. Some were drawn by their affection for his late father Antipater, while others were drawn by Herod's own reputation. Some came to repay both Herod and Antipater for former favors from both men. But most of the recruits were attracted by their expectation that Herod's claim to the throne was already assured since he had assembled such an unbeatable army around him. While Antigonus tried to halt Herod's advance, his traps and ambushes caused little or no harm to Herod's forces. So Herod easily rescued his loved ones from Masada, recaptured the fortress at Ressa, and then marched toward Jerusalem. Silo's troops joined him there as did many who had fled from the city after seeing the size of Herod's army.

5. After pitching camp on the west side of the city, Herod's troops were assailed by the arrows and darts of the guards posted on that sector of the city's walls. Others of Antigonus's forces sallied out in combat teams to assault those nearest the city. Herod ordered proclamations to be shouted to those on the walls that he had come for the good of the people and to save the city. He proclaimed that he had no intention of punishing his enemies but that he would grant amnesty to even his most bitter foes. But Antigonus's troops made a great clamor against these exhortations and prohibited anyone from listening to Herod or from going over to his side. Antigonus immediately gave his men orders to force Herod's army back from the walls. With showers of darts from the towers, they put them to

flight.

6. It was in this situation that Silo's conduct exposed his former corruption. He induced a great many of his soldiers to complain loudly of their lack of supplies, to demand money for provisions, and to be marched to suitable winter quarters. Antigonus's troops had previously stripped all sectors of the city and had laid waste to the entire area around it. So Silo attempted to break camp and retreat from the field. But Herod went to Silo's commanders and then spoke to the assembly of Silo's soldiers urging them not to desert him. He argued that he had been commissioned personally by Caesar, Antony, and the entire Senate, and also promised to supply all of their needs immediately. Herod then personally scoured the countryside and brought back such an abundance of supplies that he stripped away all of Silo's excuses for withdrawal. Furthermore, to ensure against future shortages Herod instructed the inhabitants of Samaria—who held him in favor—to bring corn, wine, oil, and cattle down to Jericho. Getting word of this, Antigonus sent orders throughout the country for his men to lay ambushes for these supply convoys. Several groups of armed men gathered in the mountains around Jericho to look for those transporting Herod's provisions.

Meanwhile, Herod was not idle. He took five Roman and five Jewish cohorts intermixed with mercenary troops, plus a small cavalry unit, and proceeded to Jericho. He found the city deserted and its citadel occupied by five hundred men with their wives and children. He took them captive, but subsequently released them. The Romans then plundered Jericho, finding every sort of treasure in the abandoned houses there. So Herod left a garrison at Jericho and then dismissed the remainder of his troops to winter quarters in those areas that had come to his side: Idumea, Galilee, and Samaria. Meanwhile, Antigonus tried to ingratiate himself to Mark Antony by bribing Silo to shelter a portion of his troops in Lydda on the western border of Judea.

Chapter 16

Herod Takes Sepphoris and Subdues the Robbers Hiding in Caves. After that, He Avenges Himself upon Macheras and Goes to Antony Who was Laying Siege to Samosata.

1. So the Romans rested from war and lived off the fat of the land. But

Herod, never idle, sent 2,000 foot troops together with four hundred cavalrymen under his brother Joseph and occupied Idumea in order to keep that territory from going over to Antigonus. He then personally moved his mother and the other relatives he had rescued from Masada to Samaria. Having settled them safely there, Herod set out to take the remaining fortresses in Galilee and to drive the forces of Antigonus out of that region.

2. He pushed on to Sepphoris through a very heavy snowstorm and took the city without a fight. Its guards had fled the place before Herod arrived. His men, who had been severely tested by the storm en route, were refreshed by the abundant supplies in Sepphoris and soon mounted a campaign against the highwaymen who hid themselves in local caves. These robber bands had infested a large area and were as much a problem for the inhabitants as an all-out war might be. Herod sent an advance assault team of three battalions of infantry troops and a squadron of cavalry to the village of Arbela. But the enemy was not demoralized by Herod's approach. These seasoned veterans combined their experience at fighting with the boldness of thieves. Meeting Herod in the field they proceeded to collapse his left flank with their right. But Herod instantly wheeled his right flank around, which he was leading personally, and came to their relief. He checked the retreat of his men on the left flank, then fell upon their charging pursuers, breaking both their assault and their courage. Overpowered by Herod's frontal attacks, the robbers gave ground then ran away.

3. But Herod pursued them, and maintaining contact, killed many as they fled to the Jordan. He slew a large number, and the rest scattered across the river. So Galilee was freed from the terror they had endured, except for a remnant of the thieves still holed up in caves. It was to take some time to smoke them all out. After the rout of the highwaymen, Herod distributed to his men the fruit of their labors – 150 drachmas each. Commanders received even more. He then dismissed his army to their winter quarters. Herod instructed his youngest brother, Pheroras, to take charge of the commissary for them and to fortify Alexandrium. Both of these orders were carried out.

4. At this time, Mark Antony had taken up residence near Athens. His

general Ventidius called both Silo and Herod to enter the war against the Parthians after they had settled matters in Judea. So Herod gave Silo leave to go to Ventidius while he went back to clear out the caves where some of the robbers still hid themselves. These caves opened onto high cliffs in the mountains and were inaccessible except by some steep and extremely narrow paths leading up to them. The cliffs at the mouths of these caves dropped off sharply into deep ravines and dry stream-beds far below. Consequently, the king was perplexed by the difficult nature of the terrain. But after a while, he came up with a very hazardous plan of attack. Using ropes, some of his bravest warriors were lowered to the mouths of the caves in cradles. These able-bodied men were able then to attack and kill the highwaymen and their families, setting fire to those who offered any resistance.

But Herod wanted to save some of them, so he had a proclamation read, requesting that they come forth peacefully to meet with him. Not one of them surrendered willingly, and of those who were forced to come out, many preferred deaths to captivity. It was then that one old man, the father of seven children, was asked by his sons and their mother to be allowed to leave peacefully. Instead, the old man ordered them to come out of the cave one by one as he stood by the entrance and killed each son as they came forward. Herod was watching all this from a nearby vantage point and was profoundly affected by what he saw. He reached out his right hand to the old man, begging him to spare his children. But the old man wouldn't listen to Herod and even rebuked the king, calling him a lowborn upstart. He then concluded the slaughter of his sons by slaying his wife. After having flung their corpses into the deep ravine, he threw himself down after them.

5. So Herod cleared the caves of their occupants. Then, leaving Ptolemy in charge of a force capable—in his reckoning—of preventing further trouble, the king returned to Samaria. Herod took with him a force of 3,000 heavy infantry and six hundred cavalry troops to attack Antigonus. But upon Herod's departure, the men who had previously fomented rebellion in Galilee saw their opportunity. They made a surprise attack upon General Ptolemy and killed him. After that, they raped and pillaged the country before seeking refuge in marshes and other concealed hiding places. When news of the revolt reached Herod, he wheeled his men around and returned to Galilee, killing many of the rebels. He besieged

and destroyed their fortresses and imposed a fine of 100 talents upon those towns involved in the sedition.

6. By this time, the Parthians had been driven out of the country, and Pacorus was dead. So Ventidius, under instructions from Mark Antony, sent a detachment of 1,000 cavalry and two legions of foot troops to support Herod against Antigonus. The officer placed in charge of the force was Macheras. Antigonus wrote to General Macheras requesting that he come to his own assistance rather than to Herod's. He cited Herod's brutality and the many instances where Herod had harmed the kingdom. Furthermore, he offered Macheras a large sum of money to switch sides. Macheras was not prepared for this outright offer to betray the trust of his superior officers. Besides, Herod had offered him a larger sum. So he pretended to be friendly toward Antigonus and went out to spy on his position and strength even though Herod had tried to persuade him from doing so. But Antigonus had rightly perceived Macheras's intentions and refused to let him enter the city, and by his guards on the walls, repulsed him as an enemy. Finally, Macheras retired from Jerusalem in shame to join Herod in Emmaus. But his fury at being embarrassed was such that he killed all the Jews he met on his march. He didn't even spare the Herodians, but treated everyone as if they were Antigonus's allies.

7. Herod was indignant at this cruelty. He made ready to fight Macheras as his enemy but soon restrained his anger and instead set out to meet with Antony to accuse Macheras of his atrocities. Macheras, meanwhile, had reflected upon what he had done and set our after Herod, he earnestly appealed for forgiveness and reconciliation. But Herod continued on to meet with Antony. His intelligence had reported that the Roman was besieging Samosata, a large city near the Euphrates, with a sizable army. On hearing the news, Herod stepped up his pace as he saw this as an opportunity to display his courage and strengthen his ties with Mark Antony. Indeed, his arrival at Samosata soon put an end to the siege. He killed scores of the barbarians and took an abundance of plunder. The result was that Antony, who had long admired Herod's courage, now held him in even higher regard. Antony heaped many more honors upon Herod and reassured him of his hopes of gaining the kingdom of Judea, even as King Antiochus was forced to surrender Samosata.

Chapter 17

Herod's Brother, Joseph, Dies following Herod's Premonitions. Herod's Life Is Marvelously Preserved Twice. He Decapitates Joseph's Murderer and Gives the Head to His other Brother, Pheroras. Shortly thereafter, Herod Besieges Jerusalem and Marries Miriamne.

1. Meanwhile, Herod had suffered reversals in Judea. He had left his brother, Joseph, in charge of the kingdom with express orders not to take any actions against Antigonus until his return. Herod saw that he could not trust Macheras in such a position because of his previous failures. But as soon as Joseph received word that his brother was at a safe distance away, he disregarded his orders and marched toward Jericho with five cohorts that Macheras had sent to him. His object was to harvest the mid-summer corn crop there. On the way to Jericho, he was attacked in the difficult terrain of the mountain passes. Although he fought bravely in the engagement, Joseph was killed, and all of his five cohorts were wiped out, comprised as they were of raw recruits from Syria. There were no seasoned veterans among them who could have supported these who were unskilled in armed combat.

2. The victory itself did not satisfy Antigonus's anger. He was in such a lather of rage that he viciously desecrated Joseph's body. When all the bodies of the dead had been rounded up, Antigonus cut off Joseph's head even though Joseph's brother Pheroras had offered 50 talents for the return of his sibling's whole body. The victory also led to an uprising in Galilee by Antigonus's partisans. They seized Herod's main supporters there, dragged them to the lake and drowned them. There was also trouble in Idumea where Macheras was building a wall around the fortress called Gittha. But of all this Herod was still uninformed. After the fall of Samosata when Antony had appointed Sossius governor of Syria with orders to support Herod against Antigonus, Antony left for Egypt. Sossius, therefore, sent an advance party of two legions into Judea to assist Herod and soon followed them with the rest of his army.

3. While Herod was at Daphne, near Antioch, he had been forewarned in dreams of his brother's death. Just as he leapt from his bed in horror from the nightmare, messengers arrived with news of the catastrophe.

After mourning the loss of his brother for a short time, Herod regained his composure and quickly marched back to Judea to meet his enemy. He pushed his troops in a forced march that tested their endurance. When Herod came to Libanus, he received a reinforcement of eight hundred mountain men and a Roman legion. Strengthened, Herod moved on to Galilee before daybreak. Enemy forces met him there, but he soon drove them back to the positions they had just left. Herod then set up an immediate siege of their fortress, but was forced to fall back to take shelter in the neighboring towns by a violent storm. A few days later he was joined by the second of Antony's legions. Seeing Herod's increased strength, the enemy grew fearful and evacuated the fortress, stealing away in the night.

4. After this victory, Herod marched through Jericho, making haste to avenge his brother's murderers. In Jericho, he had a miraculous and providential escape that gave him the reputation of being loved and protected by God. That evening, a large group of magistrates had dined with him. After the feast was over and Herod and all the guests had left, the building collapsed. Judging this escape to be both an omen of the danger he was to face in battle and of the divine preservation that would accompany him, Herod moved his troops into the field before daybreak. Soon, some 6,000 enemy soldiers rushed down out of the hills to attack his advance party. They did not dare enter into hand-to-hand combat with the Roman troops, but threw stones and javelins at them from a distance, wounding many. Herod himself was wounded in the side by a dart.

5. Antigonus, wanting his forces to appear more numerous and more courageous than Herod's, sent an army into Samaria under the command of Pappus, one of his henchmen. Pappus was to encounter Macheras there. But in the interim, Herod was ravaging the territory held by his enemies. He destroyed five small towns and killed 2,000 men within them before returning to his bivouac. He was then headquartered in a village named Cana, a place north of Jerusalem near the border of Samaria and Judea.

6. Multitudes of Jews joined Herod each day, both from Jericho and other parts of the country. Some came out of their hatred of Antigonus, and some in awe of Herod's successes. But most came out of a gut-level desire for change. Herod welcomed them all. The forces of Pappus, which were dismayed neither by the number nor the zeal of their adversaries, eagerly

advanced to engage them in hand-to-hand combat. Pappas's army made a brief fight of it, but Herod's rage at the murder of his brother and his zealous desire for revenge soon prevailed. He quickly overcame Pappas's troops and then, directing his men against the units that still offered resistance, turned the battle into a rout. A bloody scene ensued as the enemy was driven back into the village where they had camped. Herod maintained contact with their retreat and massacred a vast number of them. Chasing his foes into the village, he found every house packed with soldiers, and every rooftop filled with men who attacked him from above. When they had slain all those enemy soldiers who remained outside, Herod's men tore the houses to pieces and dragged those inside into the streets. Many perished as roofs collapsed on their heads while those who tried to escape ran into the drawn swords of Herod's men. Heaps of corpses lined the streets. The victorious army could not pass through them.

    The loss overcame the rest of Pappas's army, which tried to rally after the battle. Seeing the village strewn with the bodies of their fallen comrades, they dispersed and fled. Encouraged by his victory, Herod wanted to move immediately on Jerusalem but was stopped by a violent winter storm. The weather was to be a major impediment to his eventual success over Antigonus, who was by this time considering withdrawal from the holy city.

7. That evening, Herod dismissed his junior officers to refresh themselves from the fatigue of battle. He went off to take a bath, still hot and sweating underneath his armor. Herod was to bathe like a common soldier, with only one slave attending him. As he was entering the bathhouse, one of the enemy bearing a drawn sword came face-to-face with the unarmed Herod. A second man then appeared, then a third, and then several more. These were men who had run away from the battle and had, in fear for their lives, hidden themselves fully armed in the bathhouse. When they saw the king, they were panic-stricken and ran trembling past him, rushing to the exit even though Herod was unarmed. Just by chance, no one else was there to lay a hand on them and so they all escaped. Herod was relieved to have escaped unharmed.

8. Antigonus's general, Pappas, had been killed in the fighting. On the following day, Herod cut off Pappas's head and sent it to his brother, Pheroras, in retribution for the murder of their brother, as it was Pappas

who had killed Joseph. Now, as spring was arriving, Herod marched toward Jerusalem and positioned his army beneath its walls. Three years had passed since he had been declared king in Rome. He pitched his camp opposite the temple, for the city was most vulnerable to attack there—it was through that sector that Pompey had captured the city. He then gave each unit its separate orders. They cut down all the trees surrounding the city and constructed three lines of earth-works on which Herod ordered three towers built. Herod then left all this labor to his lieutenants while he took off for Samaria to bring back as his wife Miriamne, the daughter of Alexander, son of Aristobulus. She had been engaged to him before, as we reported earlier. He showed his enormous contempt for his enemy by inserting his wedding ceremony as an intermission to the siege.

9. Once he had married Miriamne, Herod returned to Jerusalem with an even greater force. Sossius joined him with a large army of both foot and mounted troops. The general had sent them ahead of him by an interior route while he later arrived in Jerusalem via Phoenicia. The full strength of their united armies was eleven battalions of infantry troops and 6,000 cavalrymen, not including the Syrian mercenaries who made up a large part of the total force. The two generals made camp near the north wall. Herod relied upon the decree of the Roman senate that had pronounced him king, as Sossius relied upon Mark Antony, by whose orders he was sent to Herod's side.

Chapter 18

Herod and Sossius Take Jerusalem by Force; Antigonus Dies; and a Look at Cleopatra's Covetous Temper.

1. The great number of Jews in the city was divided into several factions. The weakest among them crowded around the temple area, declaring that given the perilous times the most blessed and religious among them were to be the ones who died first. The more bold and daring men organized themselves into various groups and went about robbing places outside the city, seizing what little food was left nearby either for the men or their horses. Some of the more disciplined men, seasoned war veterans, were to defend the city during the siege, driving away those who built the earth-works. They also continually devised ways to stymie the enemy's siege engines, and enjoyed some success using the underground passageways

that honeycombed the earth beneath Jerusalem.

2. The king set up ambushes to stop those raiders who ventured out of the city to bring back food. Meanwhile, he brought in provisions from great distances away to supply his own army's needs. The military experience of the Romans gave them a distinct advantage over the Jews in the city, although the latter were courageous to the point of foolhardiness. The Jews, however, knew that to meet the Romans face-to-face in close combat was to invite certain death. Instead, they used the underground passageways to attack suddenly in the midst of their enemies. Before one defensive wall was battered down by the Romans the Jews had built its replacement. In short, the Jews lacked neither brains nor energy in their determination to hold out to the bitter end. In fact, notwithstanding the superior strength of the siege forces, the Jews held out for five months until some of Herod's hand picked men were able to scale the wall and leap into the city, followed by Sossius's centurions.

The Romans first seized the ground that surrounded the temple. When the troops poured in, it was a scene of wholesale massacre, as the length of their siege infuriated the Romans and because the Jews in Herod's camp were determined to leave none of their adversaries alive. Scores were butchered in alleyways, or crowded together in houses while trying to seek refuge in the temple sanctuary. Mercy was shown to neither infants nor to the aged nor to defenseless women. Even though the king sent messengers everywhere trying to stop the slaughter, no one could be persuaded to withhold his sword. Like madmen, they continued to slay people of all ages. This carnage prevailed as Antigonus, without regard for his former or present fortune, came down from the citadel and fell at Sossius's feet. Sossius had no pity on the beggar's misfortune, calling him the feminine rendering of his name, "Antigona." But he did not treat Antigonus as a woman, nor did he let him go free. He was put in chains and kept under strict guard.

3. With the victory over his enemies complete, Herod's next task was to control his foreign mercenaries. Many of these Gentiles were eager to see the temple and the sacred contents of the holy sanctuary. The king tried to restrain them by exhortation, and then by threatening, and ultimately by force of arms. If these people were to set eyes on any sacred, unapproachable objects, the victory would become a defeat. He also had the

task of restraining the pillaging of the city. Herod forcefully argued with Sossius that if the Roman troops despoiled the city of both its money and its men, his kingdom would be that of an empty desert. He told Sossius that even a worldwide empire was too small a realm if purchased with the slaughter of so many of Jerusalem's citizens. But Sossius justified the pillaging as a means of repaying the soldiers for their labors during the siege. Hearing this, Herod offered to pay the men out of his own resources. So, having redeemed what remained of his country, Herod kept his promise and paid each soldier liberally—with proportionately more going to officers, and with a gift to Sossius of truly royal proportions. They all went away wealthy. Sossius then left Jerusalem after dedicating a royal crown to God, and taking with him Antigonus to be delivered to Antony in chains. The prisoner, holding on to but slim hopes of continuing to live, was to die by the ax, justly ending his infamous and cowardly career.

4. Herod then proceeded to separate the population into two classes. He gave great honors to those who had associated themselves most closely with him, but executed those who had backed Antigonus. Finding his available funds now running low, Herod converted all his valuables and jewelry into cash and sent it all to Antony and his staff. But even this gift was not able to ensure against future problems. Antony had fallen irresponsibly in love with Cleopatra, becoming totally enslaved by his passion for her. Meanwhile, Cleopatra had murdered her entire family one after another, until not a single one of her relatives remained alive. She now thirsted after the blood of those unrelated to her. Cleopatra laid before Mark Antony slanderous charges against the Syrian high governmental officials and persuaded him to execute them, believing that she would then have no trouble absconding with their possessions. Now, her ambitions had reached into Judea and Arabia. She secretly plotted to have Herod and Malichus, kings of those nations, slain by Antony's hand.

5. While Antony complied with some of her orders, Cleopatra's latest ambitions brought him to his senses. He thought it abominable to take the lives of innocent men, particularly when they were kings of such eminence. Nevertheless, he threw aside the friendship he previously had with them and proceeded to take away large portions of their territories. For instance, he took the palm grove in Jericho where the balsam trees grow and presented it as a gift to Cleopatra. He also gave her all the towns to

the south of the river Eleutherus with the exceptions of Tyre and Sidon. Now mistress of all this land, she accompanied Antony, who had begun a campaign against the Parthians, as far as the Euphrates. Traveling through Apamia and Damascus, Cleopatra entered Judea. There, with many magnificent gifts, Herod appeased her acrimony against him and agreed to lease back the lands taken from him at the annual rental of 200 talents. He then accompanied her to Pelusium, treating her with the utmost respect. Antony soon returned from Parthia bringing with him Tigranes' son, Artabazes, in chains. He presented the Parthian to his beloved Cleopatra along with his money and all his spoils of war.

Chapter 19

Cleopatra Persuades Antony to Make War on Arabia and how He Was Victorious. Also, a Great Earthquake Strikes.

1. When war broke out in Actium between Mark Antony and Caesar Octavius, Herod prepared to join forces with his friend Antony. He had extinguished the rebellion in Judea and captured the fortress at Hyrcania formerly held by Antigonus's sister. But his plans to aid Antony were precluded by Cleopatra's guile. As we have already seen, she had plotted against both Herod and the Arabian king, Malichus. She now convinced Antony to pit Herod against the Arabs in a war. If Herod was successful, she hoped to become the mistress of Arabia. If Herod lost, she would rule Judea. Either way she intended to use one king to overthrow the other.

2. Her plot, however, was to work to Herod's ultimate advantage. Gathering a large army of cavalry, Herod charged the Arabians near Diospolis. He emerged victoriously, though met with stiff resistance. The defeat shocked the Arabs who subsequently assembled a vast army at Kanatha in Coele-Syria to await the next Jewish attack. When Herod's forces arrived he had thought it best to proceed cautiously, and so had given orders to fortify the army's camp. But his men did not comply with his orders. Instead, encouraged by their first victory and itching to fight, they rushed headlong at the Arabs. Their first charge was successful. The Arabs ran away with the Jews closely on their heels. But Athenion, one of Cleopatra's generals, had laid a trap for Herod's men during their pursuit. Athenion had always been Herod's enemy and now made a surprise counterattack with the natives of Kanatha. Emboldened by this new development,

the Arabs turned around, regrouped in the difficult, rocky terrain and put Herod's army to flight, slaughtering many of them. Those who escaped fled to Ormiza. But the Arabs surrounded their camp and captured everyone in it.

3. Herod arrived with fresh reinforcements a short time later, but it was too late. The catastrophe had been due to the disobedience of his divisional officers. Had they not ordered the attack Athenion would have had no opportunity to spring his trap. But Herod was to have his revenge on the Arabs. He began a series of successful raids on their territory that made the Arabs forget the joy of their single victory. But while he was avenging himself on his enemies, another disaster struck. It occurred in the seventh year of his reign (in 30 BC), when the war at Actium was at its peak.

In early spring, a tremendous earthquake annihilated innumerable cattle and killed 30,000 residents of Judea. The army, which was operating away from the cities, escaped harm. But the news of the tragic earthquake gave the Arabs fresh confidence. As rumors often do, the magnitude of the tragedy was greatly exaggerated among them. They imagined that all of Judea lay in ruins and that all they had to do was to walk in and take control. After executing the ambassadors that the Jews had sent to them, offering them up as sacrifices, the Arabs marched immediately into Judea. The Jewish army was greatly alarmed at the invasion, being already demoralized by the calamities that had fallen upon them. So Herod called them together and tried, by the following speech, to encourage them to defend the country.

4. "The despondency that has come over you is most unreasonable. To be dismayed by a heavenly chastisement is only natural, but to be terrified in the face of an attack by a human enemy is cowardly. Speaking for myself, I am far from being intimidated by the invasion following the earthquake. I believe that the catastrophe was a trap that God has set up to decoy the Arabs into attacking so that we may have our vengeance upon them. They have not come because of the confidence in their weapons or their strength. They have attacked because they are relying on unpredictable catastrophes to aid them. But any hope based on someone else's circumstances and not on one's own strength is unfounded. What's more, mankind's fortune is never set permanently in one direction or another. Rather, it changes and swings from one side to the other. You know this

from examples in your own experience. We won the first battle with the Arabs, but then they defeated us. Now, they are in all probability expecting a victory and will be defeated. Overconfidence throws men off their guard, while fear promotes caution. So, your fearfulness gives me great reassurance. When you rashly cast aside my advice and rushed out to attack our foes, you gave Athenion the opportunity to spring his trap. But now your hesitance and apparent despair is to me a pledge and assurance of victory. But appropriate as your feelings are as you await a battle, once you are in action your confidence must rise, and you must teach those scoundrels that no disaster, whether of God or man, will ever weaken the courage of Jews, so long as breath remains in their bodies. Not one Jew will allow his belongings to pass to the hands of some Arab, who in the past has narrowly escaped becoming his slave.

"Do not let the shaking of the inert earth bother you or imagine that an earthquake is an omen of upcoming disasters. These chance events to which nature is subject have physical causes, and beyond the immediate injury they bring, have no further consequences. A sign or a premonition may precede a plague or famine or earthquake, but these catastrophes are limited only to their immediate effects. Let me ask you, can war with our enemies, even if it prove to be violent, do us any more harm than the earthquake? No!

"Our enemies, on the other hand, must deal with a great omen of their impending doom. The omen is not a natural omen, nor one which comes from the actions of others. But it is this: Contrary to mankind's common law, they have brutally murdered our ambassadors. They did so as if they were sacrifices to God in order to assure their victory in this war. But they will not escape God's mighty eye, nor his invincible right hand. They will soon answer for their atrocities if we retain but a small portion of the courage of our forefathers, and we now rise to inflict vengeance for their violation of the sacred covenant. Let each of us go to battle not to defend our wives, or children, or country, but to avenge our ambassadors. These dead ones will conduct this war of ours much better than those of us who are alive. If you obey my orders, I will lead you into the battle. Be sure of this, your courage is irresistible if you do not harm yourselves by some rash action."

5. Herod's speech served to elevate the morale of his army. Seeing their renewed courage, Herod offered sacrifices to God and then proceeded to

march his troops across the Jordan. He camped close to the enemy's position near Philadelphia. Anxious to force a fight, Herod began skirmishes for the possession of the fort that lay between the opposing lines. The enemy had sent out a party to occupy the post, but the unit sent forward by the king quickly forced their retreat and secured the fort for the Jews. Herod then began to march his troops out each day, forming them into full battle array, inviting the Arabs to fight. But none would come out of their camp. They were terribly afraid, and their general, Elthymus, was paralyzed with fear in the presence of his troops. So Herod attacked them and tore their fortification to pieces. Therefore, the Arabs were forced to come out and engage the Jews. But they came out in disorder, their infantry mixed up with their cavalrymen. Even though they outnumbered the Jews, the Arabs were less than eager to fight. Nevertheless, they were obliged to expose themselves to danger even if victory seemed hopeless.

6. As long as their lines held together, the Arabian losses were minor. But as soon as they began to give ground and to run, a great many were killed by the Jews, while their own countrymen trampled many others to death. About 5,000 fell dead in the rout while those remaining escaped immediate death by crowding inside the remains of their entrenched fortification. Herod surrounded them and attacked. With the assault likely to finish them off, the Arabs had another problem to confront. Their water supply had given out, and their thirst forced them to seek terms of surrender. The king mocked their ambassadors, and even though they offered him 500 talents for their redemption, he pressed the attack even more. Parched with thirst, the Arabs soon came out by the scores and willingly surrendered to the Jews. In five days time, 4,000 were taken prisoner. On the sixth day, those Arabs remaining in the fortification came out to fight desperately. Herod's men took them on, killing some 7,000 more. Herod punished Arabia severely, and broke its people's spirit. Herod had gained such a reputation in his victory over them that the Arabs chose him to be their protector.

Chapter 20

Herod Is Confirmed king by Caesar and Cultivates a Friendship with the Emperor through Magnificent Gifts, While Caesar Responds to His Kindness by Restoring that Part of His Kingdom that Cleopatra and Taken, and Giving Him Zenodorus's

Country As Well.

1. With this crisis behind him, Herod immediately became concerned about the security of his position as king of Judea. He was Mark Antony's friend, and Octavius Caesar had defeated Antony at Actium. Herod thought that the best course of action was to confront the danger head on and so sailed to Rhodes where the emperor was staying and presented himself to Caesar. He dressed and acted with the humility of a commoner, appearing without a crown but retaining the spirit of a king. He spoke with candor and presented the truth, withholding nothing.

"Caesar," he said, "Antony made me to be king of the Jews, and I admit that I have used that authority to be of service to him in every possible way. Nor will I conceal the fact that, had I not been preoccupied with fighting the Arabs, you would have certainly found me fighting at Antony's side in Actium. I sent him all the auxiliary troops I could spare, besides many thousand measures of corn. Even after his defeat at Actium, I did not desert my patron. While no longer useful as an ally, I became his best advisor. I told him that there was but one way of recovering what he had lost, and that was to execute Cleopatra. But he would not kill her. I promised him that upon her death I would give him money and walls for his protection. I would equip him with an army and even come out myself as his brother-in-arms in his war against you. But his love for Cleopatra made him deaf, as also did God, who has given the government to you. I, therefore, share Antony's defeat and lay down my crown. I have come before you resting my hope of safety on my integrity. I pray that the question will not be whose friend have I been, but how loyal a friend have I been."

2. Caesar then replied, "No. Be assured that you shall be both safe and king and that you shall reign with more confidence than ever before. Your loyalty as a friend speaks well of your ability to rule over many subjects. I ask that you endeavor to be as loyal to we who have been more fortunate, since, speaking for myself, I will depend upon your fine spirit for my success. As it turns out, Antony did well to obey Cleopatra rather than you. For through his folly we have gained you as a friend. But it seems that you have already been of service to me. Quintas Didius informs me that you sent a force to assist him against the gladiators that Cleopatra called to fight for Antony. I, therefore, confirm your kingdom to you by decree. I

shall also try to render you additional kindnesses in the future so that you may not feel the loss of Antony."

3. Having so graciously spoken to the king, Octavius placed the crown back on Herod's head and publicly proclaimed his kingship by decree, in which he commended Herod with even more honor than before. After showering Caesar with gifts in supplication for his next request, Herod asked the emperor to pardon Alexander, one of Antony's friends who had come to the emperor to plead for mercy. But Caesar's resentment was too strong and with many vehement accusations against the alleged crimes of Alexander, he denied Herod's request. Later, when Caesar passed through Syria on his way to Egypt, Herod entertained him with all the means at his disposal. Then—after first riding with the emperor as he inspected his army near Ptolemais—Herod held a banquet for him and all of his friends. The king then provided a great feast for the enjoyment of the rest of Caesar's army. He also provided them with an abundant supply of water for their march through the arid desert to Pelusium. In a word, Caesar's army lacked no necessity. Observing Herod's liberality, it was Caesar's opinion and also the opinion of his soldiers that Herod's realm was much too small for the magnificent way in which he had treated them. Accordingly, when Caesar had reached Egypt following the deaths of Antony and Cleopatra, he not only conferred new honors upon Herod, but also gave him the territories that Cleopatra had appropriated for herself. He also gave Herod Gadara, Hippos, Samaria, and the coastal towns of Gaza, Anthedon, Joppa, and Strato's Tower (later known as Caesarea). Caesar also gave Herod Cleopatra's four hundred Galatian bodyguards. Moving the emperor to such generosity in these presents was the equal generosity of their recipient.

4. After the first games in Actium (in 28 BC), Caesar added to Herod's dominion the country called Trachonitis and its adjacent districts of Batanea and Auranitis. One Zenodorus, who held a leasehold interest in a district named Lysanias, occasioned this grant. Zenodorus was continually sending bandits from Trachonitis to harass the residents of Damascus. These citizens asked for protection from Varro, the governor of Syria, requesting that he report their problems to Caesar. On learning of the matter, Caesar sent back orders to exterminate the thieves. So Varro led his army against them and rid the district of these pests—at the same time relieving Zeno-

dorus of his leasehold. The leasehold was the territory that Caesar gave to Herod to prevent it from ever again falling into the hands of bandits as a base of operations against Damascus. Ten years after his first visit (in 20 BC), Caesar returned to the province and made Herod procurator of all Syria. The other rulers of that area could thereafter do nothing without Herod's concurrence. Later, when Zenodorus died, Caesar gave Herod all the territory between Trachonitis and Galilee. But of more importance to Herod than all these lands was the fact that only Agrippa was more beloved by Caesar than he and that he stood only after Caesar in Agrippa's affection. Herod was, therefore, elevated to the heights of prosperity. His soul, moreover, rose to even greater heights as he spent the majority of his energies and largess on the promotion of piety in the kingdom.

Chapter 21

The Jerusalem Temple Is Restored and other Cities and Buildings Are Built by Herod. He Shows His Magnificence to Foreigners as Fortune Smiled Upon Him in Every Respect.

1. In the fifteenth year of his reign, Herod restored the temple. By constructing new foundation walls, he enlarged its surrounding enclosed area to twice its former size. He invested incalculable sums of money in the project, arriving at unsurpassed magnificence. The colonnades around the temple and a fortress that rose above it to the north provided clearly seen evidence. The colonnades were constructed on a new foundation while the fortress, restored at tremendous expense, was like a royal palace. He called it Antonia in honor of Antony. He also built a palace for himself in the upper city, consisting of two very large and beautiful buildings, the size of which dwarfed the temple. Herod named these two wings of the palace for his friends. The one building he named Caesareum and the other Agrippeum.

2. But Herod was not content to honor the names of his patrons by mere buildings. His largess extended to the construction of entire cities. In Samaria, he built a city surrounded by walls two and one-half miles long on each side. He brought in 6,000 inhabitants and gave them all parcels of the richly productive 4,000 acres. In the center of the settlement, he erected a huge temple in the middle of one hundred twenty two acres of sacred land. He called the city Sebaste after Sebastus, or Augustus. He

then gave its settlers a special constitution and government.

3. Later on, when Caesar was to give him additional territory, Herod built another temple. This one was of white marble, and located near the source of the Jordan at a place called Panium, (later named Caesarea Philippi). At this spot, Mt. Hermon's summit is at an exceedingly lofty height. At the mountain's base is the mouth of an immense, dark cavern. Inside the cavern, a deep pit yawns, containing a great quantity of still water. Sounding lines have never been able to plumb its depths, as no length of line is sufficient to find its bottom. Some think that the Jordan wells up from this deep pool and flows out of it. We will speak more of this matter later.

4. At Jericho, between the fortress of Cypros—constructed in honor of Herod's mother—and the former palace, the king erected new buildings for guests and travelers that were far more spacious and well appointed than what had been there. He also named all these for his friends. In a word, there was no suitable spot in the entire kingdom that was not used to honor Caesar. Then, when Herod had filled his own land with temples, he let his magnanimity overflow into the province, where he erected many cities as monuments to Caesar, calling them Caesarea.

5. Herod had taken notice of a seaside town named Strato's Tower. Though old and decaying, it was beautifully situated and well suited for the king's extravagance. He had the city entirely rebuilt in white stone and adorned it with several splendid palaces, in which he truly demonstrated the magnitude of his largess. The entire coast between Dora and Joppa was without a suitable harbor, and Strato's Tower was right in the middle of it. Ships bound from Egypt to Phoenicia had to anchor in the open sea and were subject to high winds from the southwest. Even a moderate breeze from that quarter would send huge waves crashing upon the rocky cliffs of the shoreline. The backwash created by these waves reached far out to sea. But Herod overcame nature and by great expenditure and industry constructed a harbor larger than the Piraeus [at Athens]. Furthermore, he dredged roads and deep anchorages within it for the ships taking refuge there.

6. Given the difficult nature of the site, Herod's plans to build the harbor

seemed impossible. But his accomplishment was such that not only did the resulting sea walls and harbor provide a secure and firm anchorage, but its beauty made it seem as if the task were easy. After he had determined what size the harbor was to be, Herod had gigantic blocks of stone lowered into one hundred twenty feet of water. Most of these blocks measured fifty feet in length by nine feet in depth and ten or more feet in width. Some were even larger. On top of this underwater foundation, he built a sea wall extending above the surface that was two hundred feet wide. The first one hundred feet of its width were to break the surge, and the inside portion served to support a stone wall encircling the harbor. From this wall rose massive towers, the highest and most beautiful of which was named Drusium, after Caesar's stepson.

7. Numerous long notches in the lee side of the wall provided dock space for ships entering the harbor. A circular quay in front of these slips served as a wide promenade for embarking or disembarking passengers. The entrance to the port faced to the north as the north wind on that portion of the coast is the most gentle and reliable. At this mouth to the harbor stood gigantic statues—or Colossi—three on each side, which rested on columns. The columns on the coastward side were supported by a massive tower while those to seaward stood on two upright blocks of stone joined together. The height of these two stones exceeded that of the tower on the other side. White stone houses spaced equidistantly apart existed along the harbor and all the city's streets, leading down to the port. On a prominent spot overlooking the harbor stood a striking temple dedicated to Caesar. It was of beautiful proportions and contained a gigantic statue of the emperor modeled after the Olympic Zeus. Another huge statue of Rome rivaled that of Hera at Argos. Herod dedicated the city to the province, and the harbor to the sailors who would put in there. He gave Caesar the glory for the entire project, and so named the place Caesarea.

8. The rest of the buildings of Caesarea—the amphitheater, the theater, and the marketplace, were all designed and built in keeping with the city's grandeur. He further ordered that there be games held every fifth year, to be called "Caesar's Games." So Herod inaugurated them in the 190th Olympiad, and offered valuable prizes, not only for the winners, but those in second and third places would also share in the royal bounty. He also rebuilt the coastal city of Anthedon that had been destroyed in

the wars, renaming it Agrippium. So great was Herod's affection for his friend Agrippa that he even had his name engraved on one of the gates he built at the temple in Jerusalem.

9. No man ever loved his father more than did Herod. As a memorial to Antipater, he established a city on the finest plain in the country, a place full of rivers and trees, calling it Antipatris. He also built walls around a fortress above Jericho, a work of both beauty and strength, and dedicated it to his mother, naming it Cypros. He also erected a tower in Jerusalem in memory of his brother Phasael. We shall describe its appearance and beautiful proportions in a later chapter. Herod also honored Phasael by naming a city after him, which he had built in a valley north of Jericho.

10. While Herod was immortalizing the names of his family and friends, he did not forget to leave memorials to himself. He constructed a fortress on the Arabian border and named it Herodium. Herod erected a huge earthwork in the shape of a woman's breast eight miles from Jerusalem, to which he gave the same name, but embellished it more elaborately. He crowned its crest with a ring of circular towers, and filled its center with lush palaces. Their magnificence was not limited to their interior space but was profusely lavished on outer walls, battlements, and roofs. At great expense, Herod also brought in a vast supply of water from a great distance away. The sum of two hundred steps of white marble provided easy access to the artificial mound's heights. Around the base of the mound, he built other palaces for his furniture and his friends. The resources of the place were like those of a city while its size was only that of a simple palace.

11. After establishing all this, Herod continued to display his generosity to a number of cities outside Judea. He built gymnasiums for Tripolis, Damascus, and Ptolemais; a wall around Byblus; halls, porticoes, temples, and marketplaces for Berytos and Tyre, and theaters for Sidon and Damascus. He built aqueducts for the coastal town of Laodicea. For Ascalon he financed baths, lush fountains, and colonnades around a courtyard, famous for their architecture and workmanship. To other communities he dedicated groves and meadows. Many cities, as if they were a part of Herod's kingdom, received land grants. Others, like Cos, were given sums of money sufficient to endow the perpetual salaries of coaches and physi-

cal trainers. Herod gave corn to whoever asked for it. He made numerous contributions to Rhodes for shipbuilding. When their Pythian temple burned down, Herod gave the funds to rebuild it on an even more majestic scale. Do I need to mention his gifts to the people of Lycia or Samos, or his generosity that met the needs of every district of Ionia? Are not the Athenians, the Lacedemonians, the citizens of Nicopolis and Pergamum in Mysia full of Herod's gifts? And how about that wide boulevard in Syrian Antioch, formerly bypassed because of its mud? Didn't Herod pave its three-mile length with polished marble and then to protect it when it rains, cover its entire length with a colonnade?

12. One might argue that the individual communities receiving Herod's gifts were his sole beneficiaries. But his generosity to the people of Elis was a gift not only to Greece but to the whole world, wherever the glory of the Olympic games reaches. When the king realized that the games were in decline for lack of funds and that this priceless relic of ancient Greece was disappearing, he accepted the leadership of their quadrennial celebration. His leadership began when he happened to be in Rome, but he also endowed the games in perpetuity in memory of the time he served them there. For me to list all of the debts or taxes that he paid for others would take an eternity. In that way he lightened the tax burden of the citizens of Phasaelis, Batanea, and various small towns in Cilicia. Herod often had to be careful that his generosity not he mistaken by some jealous governor as a power play. He frequently conferred greater benefits upon some cities than they were receiving from their own rulers.

13. Herod's brain matched his brawn. He was an excellent hunter. The king was always in the lead due to his skill in horsemanship. On one hunt, he brought down forty wild beasts in a single day. The country not only breeds boars, but even more stags and wild donkeys. Herod was also an unbeatable warrior. In practice, spectators were often shocked at the precision with which he threw the javelin, and the unerring skill with which his bow sent arrows to their target. But besides this excellence of both body and soul, he was blessed by good fortune. He rarely met with defeat in wartime, and when he did, it was not due to his own failure but to either treachery or to the recklessness of his soldiers.

Chapter 22

The Murders of Aristobulos and Hyrcanus, the High Priests, and that of Queen Miriamne.

1. Even though fortune blessed Herod with public prosperity, it had vengeance upon him with trouble and discord at home. The problems started with his wife, Miriamne, of whom he was very fond. After Herod's ascension to the throne, he had divorced Doris, that native of Jerusalem whom he had married while still a commoner. He then married Miriamne, Alexander's daughter and Aristobulos's granddaughter. But Miriamne brought trouble into Herod's household. The problems had begun in the early days of their marriage, but increased markedly following the king's return from Rome. First of all, in the interests of his children by Queen Miriamne, Herod banished Antipater, his son by Doris, from the capital. Antipater was allowed to visit Jerusalem only during the festivals. Next, Herod executed Hyrcanus, Miriamne's grandfather, who had returned from Parthia to join Herod's court. The king had suspected the old man of treason. Barzaphernes took Hyrcanus prisoner when he overran Syria, but the actions of some of his countrymen who lived north of the Euphrates subsequently freed him. These men advised Hyrcanus not to return to Judea. Had he listened to them he would have avoided his tragic demise. But Hyrcanus was lured to his death by his granddaughter's marriage. He placed reliance for his safety upon that but was drawn principally by a great love for his native land. Herod's resentment against Hyrcanus was not because the old man had made a claim to the throne of Judea. He had not. But Herod knew that Hyrcanus was the rightful heir to it.

2. Herod had five children by Miriamne, two daughters and three sons. The youngest boy died while in school in Rome. The two older boys received a royal education, both because of their mother's royal heritage and because they had been born after Herod's ascent to the throne. But stronger than both of these reasons was Herod's love for Miriamne, the flame of which grew brighter and stronger with every day that passed. His love blinded Herod to the trouble that Miriamne was causing for him. Indeed, her hatred for the king was every bit as strong as his love for her. What he had done to her family gave her just reason to be angry. Further, Herod's love for her gave her the boldness to speak candidly to him, and she open-

ly rebuked the king for the fate of her grandfather Hyrcanus, and that of her brother Aristobulus. For even though this young man was but a lad of seventeen, Herod had Aristobulus murdered shortly after conferring the high priesthood upon him. It occurred during a festival in the city. As the young high priest approached the altar clad in his sacred garments, the great crowd that looked on burst into tears as if they were one man. Herod had Aristobulus escorted to Jericho by night, where, in accordance with the king's instructions, he was drowned in a swimming pool.

3. So it was on these grounds that Miriamne screamed at Herod. She also violently abused Herod's mother and sister. The king's love for her paralyzed him. But his mother and sister knew how to get the king's attention. In their hatred for Miriamne, they brought a charge of adultery against the queen. They invented numerous stories in order to convince the king, including one that accused her of sending her picture to Antony in Egypt. They said that her lust had driven her to expose herself, although at a distance, to a man mad in his desire for sex and powerful enough to use violence to get it. The charge hit Herod like a thunderbolt. His love for Miriamne only intensified his jealousy. He recalled Cleopatra's shrewdness and how because of her, king Lysanias and king Malichus of Arabia met their ends. He, therefore, reasoned that not only was his marriage in danger, but his very life as well.

4. On the evening before a trip abroad, Herod entrusted Miriamne to his sister's husband, Joseph, as a man he knew he could trust because of his marriage relationship. He left instructions that if Antony were to kill him, Joseph was to execute the queen. Unfortunately, Joseph was to tell Miriamne of the king's secret instructions. He didn't tell her out of any malicious intent, but merely to convince her of the king's love for her, saying Herod had not wanted even death to separate them. When Herod returned, he confirmed his love for Miriamne with many oaths assuring her that he had never had the same love for any other woman. To which the queen replied, "A fine demonstration of your love! — to give Joseph orders to put me to death!"

5. Herod was beside himself the instant the words of his secret came out of her mouth. He screamed that Joseph would never have told her of his orders had he not seduced her. In a rage of anger, the king leapt out of

his bed and stormed wildly about the palace. His sister, Salome, seizing the opportunity to slander the queen, confirmed Herod's suspicion of Joseph. Mad with jealousy, Herod gave orders to execute both Joseph and Miriamne immediately. As soon as his rage had worn off, Herod was filled with remorse for what he had done. But his love for Miriamne was rekindled when his anger had subsided. Indeed, so consumed was he with love for her that he would not think of her as dead and in his affliction still spoke to her as if she was alive. Ultimately, the passage of time brought the reality of his loss home to Herod. Miriamne was dead, and Herod's grief in losing her was as profound as had been his love for her.

Chapter 23

Miriamne's Sons Are Slandered as Antipater Is Preferred; They Are Accused Before Caesar; Herod Is Reconciled to them.

1. Miriamne's sons, Alexander and Aristobulus, inherited their mother's resentment of the king, which only intensified due to their father's abominable crime against her. Even in their early days of schooling in Rome they had begun to see him as their enemy. Such hatred grew as they returned to Judea and as they advanced in years. When they were of marriageable age, Aristobulos married his aunt's daughter, Salome, the woman who had accused his mother. The other boy, Alexander, married the daughter of Archelaus, king of Cappadocia, and they both began to vent their anger openly. The boys' boldness opened the door for others to slander them, and from that time on certain men at court began to speak overtly to Herod accusing both of his sons of conspiracy against him. Alexander, courting the influence of his father-in-law, Archelaus, was preparing to journey to Rome to accuse Herod before Caesar. Meanwhile, the king was under a constant barrage of scandalous news. Drunk with this gossip, the king recalled Antipater, his son by Doris, as a defensive measure against Aristobulus and Alexander. Herod then proceeded to honor Antipater, elevating him above his other sons.

2. This new move by the king was intolerable to Miriamne's sons. At the news of the promotion of this son of a common woman, these men, prideful of their own royal birth, could not restrain their indignation. They began to display their anger openly at the slightest provocation. It got them nowhere. Every day their fortunes declined while Antipater

steadily gained favor because of his exceptional abilities, his vociferous praise of Herod, and his vilification of his two half-brothers. He concocted various disparaging rumors against Aristobulus and Alexander, some of which he began himself, while others he left to his associates to circulate. Ultimately, he had completely destroyed any hopes that Miriamne's sons had of rising to the throne. By now, Antipater was almost assured to be the heir to the crown, both by his father and his own public appearances. When Herod sent him as an ambassador to Caesar, Antipater went as a prince, with all the pomp and circumstance of royalty, save the crown. Soon, his influence was such that his mother, Doris, was brought back into Miriamne's bed. Using his two weapons of flattery and slander, Antipater had cunningly put the thought of executing his other two sons into the king's mind.

3. Herod went so far as to drag Alexander off to Caesar in Rome, accusing his son of attempting to poison him. The young man, wisely understanding that he was in the presence of a judge who was more experienced than Antipater and wiser than Herod, was careful not to accuse his father of any wrongdoing. Instead, he focused his reasoned arguments against the slander directed against him. He also demonstrated the innocence of his brother—who faced the same dangers and railed against the cunning deceptions of Antipater—as he exposed the wrongs perpetrated upon them. Alexander matched his obviously clear conscience with his eloquence. He was an extremely capable public speaker. Alexander concluded his speech with the charge that if his father believed they wanted to kill him, it was in his power as king to execute them. His listeners were reduced to tears, and so deeply touched was Caesar, that he pronounced the sons' innocence while requiring that their father be reconciled to them immediately. Caesar also made the further stipulations that Alexander and Aristobulus give their father complete obedience, while giving Herod the power to leave the kingdom to whomever of his sons he wished.

4. During his journey home from Rome, Herod seems to have forgiven the charges against his sons, even though he had not relinquished his suspicions. Antipater, the author of all the hatred against the sons of Miriamne, accompanied him. But out of reverence for Caesar, who had required their reconciliation, Antipater now held his tongue. Sailing along the coast of Cilicia, Herod and Antipater anchored at Eleusa, where King

Archelaus received them warmly. He was delighted at his son-in-law's acquittal and pleased at the reconciliation. Archelaus had previously written his friends in Rome asking them to assist Alexander in the trial. He accompanied his guests as far as Zephyrium, giving them presents valued at more than 30 talents.

5. After Herod's return to Jerusalem, he assembled the people and presented his three sons to them. He then apologized for his absence from the capital and thanked both God and Caesar for settling the disputes of his disrupted household. He said that he had achieved the reconciliation of his sons, a gift far greater than the kingdom itself.

"I intend to bind us together even more closely," Herod said as he addressed the multitude. "For Caesar has given me the power to be lord of the realm and to leave the reins of government to whomever I choose. And so, to repay Caesar's kindness, and for my own advantage, I hereby declare that all three of my sons shall be kings. I ask both you and God to endorse my decision. They are each entitled to be my successors, Antipater because of his age, and the other two because of their royal birth. After all, my kingdom is sufficiently large to accommodate even more kings.

"Now I ask you to affirm these who Caesar has united and who their father has appointed. May the honors bestowed upon them be neither undeserved nor uneven, but apportioned to each according to his rank. If you pay undue respect, you do not bring joy to the one honored beyond that which his age allows as much as you bring sorrow to the one slighted. I will personally select the advisers and those who will attend to each son and shall hold them responsible for maintaining the peace. I am well aware that dissensions and quarrels between rulers are produced by the influence of malicious advisers while courts filled with upright and honest counselors maintain natural affections.

"But for the present, the sons and their advisers and also the officers of the army must look to me alone for their hopes. For I am not now bestowing the title to the kingdom, but merely giving them royal honors. My sons will enjoy the pleasures of high office, but upon me will rest its responsibilities, whether I like it or not. I want each one of you to take into account my age, the way I have conducted my life, and the piety I have exercised. I am not so old that you may expect me to die soon. Nor have I endangered my health in the enjoyment of a luxurious debauchery that can shorten the lives of even younger men. I have served God well, and that raises my expectations of a long life.

"Therefore, whoever tries to manipulate my sons in order to cause my downfall shall be punished by me for my sons' benefit. There is no jealousy toward my sons that seeks to limit the honors paid to them. But I understand that too much flattery causes youthful men to act rashly. Everyone who has contact with my sons should remember this: I will duly reward those who act in an honorable way. But treason will not be rewarded even by the one the seditious man seeks to honor. If you remember this then you will all have my best interests and the best interests of my sons at heart. For it is advantageous to my sons that I should govern and advantageous to me that they live in harmony with one another.

"As for you, my good children, remember the natural ties of your brotherhood. Even those in a family of wild beasts hold natural affection for one another. Next, remember Caesar, who is the cause of our reconciliation. Finally, remember me as one who requests that you do that which I have the power to command of you. Continue to be brothers! I present to you the robes and honors of royalty and pray that God will uphold my decision, if only you are united to one another."

When the king had said all these things, he tenderly embraced all his sons and then dismissed the people. Some of the people cheered Herod and hoped that all would be as he had spoken. But others who longed for change in the government pretended that they didn't even hear him.

Chapter 24

Antipater and Doris's Maliciousness; Alexander Is Uneasy Regarding Glaphyra. Herod Pardons Pheroras whom He Suspected, as well as Salome, who He Knew Was Causing Trouble. Herod's Eunuchs Are Tortured, And Alexander Is Placed In Chains.

1. The tension between Herod's sons continued as they parted company. In fact, they separated more suspicious of one another than ever before. Alexander and Aristobulus were greatly distressed that Herod had confirmed Antipater with the rights of the first-born. Likewise, Antipater resented the rank given to his half-brothers though it was lesser than his own. But this eldest son was subtle and politically adroit. He knew how to hold his tongue and shrewdly concealed the hatred he had for them. On the other hand, the sons of Miriamne, prideful of their noble birth,

let their lips speak what was on their minds. Men who tried to stir up their anger also troubled them. An even larger number penetrated their inner circle of friends in order to spy on their activities. Every word that Alexander said was soon brought to Antipater, and then on to Herod after being modified to suit Antipater's designs. The young man could not make the mildest remark without being incriminated by the slanderous distortion of his words. If Alexander became somewhat freer with his remarks, the most insignificant matter was blown way out of proportion. Antipater also continually used his spies to provoke Alexander, hoping that the young man would say something to put Antipater's lies on a more solid foundation of truth. If only one thing that Alexander said could be proven true, that would imply that everything else reported to the king was true as well.

Meanwhile, Antipater's friends were either a secretive lot by nature or had been bribed to conceal their thoughts, for none of Antipater's secrets leaked out through their lips. One would not be wrong to conclude that Antipater lived a wicked life. He either bribed Alexander's friends with money or won them over with seductive flattery that Antipater always found effective. By these means, he converted them to traitors and spies who reported everything his younger brothers said and did. In all these subterfuges, Antipater acted with the utmost shrewdness. He always played the role of a devoted brother to Alexander and Aristobulus while at the same time hiring men to report what they did to the king. When someone said anything against Alexander, Antipater would speak up and deny the allegations. But afterward, in private, he would proceed to affirm them. By this strategy, he greatly aroused the king's animosity toward Alexander. His purpose was to make everything look like Alexander was plotting his father's death. Antipater's defense of his brother served to lend further credence to his lies.

2. Antipater's methods also engendered Herod's fury. The king's affection for his youngest sons diminished daily while his love for Antipater increased. The people in his court shared he king's growing estrangement from Miriamne's sons. Some distanced themselves from the brothers of their own accord while others were under orders to remain aloof, such as the king's closest friend, Ptolemy, Herod's brothers, and all the rest of his family. Antipater became all-powerful. Even more distressing to Alexander was that Antipater's mother, Doris, also held unlimited power.

She, too, plotted against him and Aristobulus and was much more severe to them than a normal stepmother. Doris hated the former queen's sons more than might be more common in such relationships. So everyone sought Antipater's favor, believing his expectations for power were more secure than his brothers'. After all, the king had commanded everyone, even his closest friends and associates, to not go near or to pay any attention whatever to Alexander, Aristobulus, and their friends.

Moreover, Herod's power extended beyond Judea to his friends abroad. Caesar had given Herod power above that of any other ruler. That is, he was able to reach out into any other jurisdiction and bring back to Judea any fugitive who had fled from his realm. Herod had never even mentioned to Alexander and Aristobulus the slanderous accusations leveled against them. They were, therefore, ignorant of the charges and could not, for that reason, take precautions against further abuse. Neither had the king spoken out in public against them. They did sense, however, a certain coldness in his attitude toward them and noticed his overreaction to small matters that troubled him. Antipater had also caused their uncle, Pheroras, and their aunt, Salome, to become their enemies. He was always speaking with her with great emotion as a husband would speak to his wife, and arousing her animosity.

Alexander's wife, Glaphyra, also exacerbated Salome's hostility. She boasted of her royal lineage as the daughter of Archelaus, king of Cappadocia. She claimed to be superior to all others in the kingdom. Glaphyra was descended on her father's side from Temenus and on her mother's side from Darius, the son of Hystaspes. Glaphyra also poked fun at the common ancestry of Herod's sister and wives, saying that the latter were chosen only for their beauty and not for their breeding. Herod had many wives, since Jewish custom permitted polygamy and the king enthusiastically adopted the practice. But his wives hated Glaphyra because of her boasting and her insults. Because of her they also hated Alexander.

3. Aristobulus also alienated his mother-in-law, Salome, who was already enraged at Glaphyra's insults. Aristobulos was perpetually mocking his wife for her heritage as a simple commoner, complaining that while Alexander had married a princess, he had married an ordinary lowborn. In tears, Salome's daughter had reported this to her mother. She added that Alexander and Aristobulus had bragged that when they took the throne they would make the mothers of their other brothers work as weavers

with their servant girls, and make their sons village clerks. Such work, they laughed, would be appropriate to their education. Hearing the slander, Salome burned with anger, and reported the whole affair to Herod. Since she was accusing her own son-in-law, her words were to carry significant weight. Meanwhile, more slander reached the king's ears that further infuriated him. He heard that Alexander and Aristobulus were constantly speaking of their mother, Miriamne, cursing him while they grieved for her. Herod also heard that they detested his distribution of Miriamne's clothes among his new wives. They had threatened to one day strip his wives of their royal robes and clothe them in rags.

4. These tales all served to increase Herod's fear of his high-spirited sons while he still held out hope for their restoration. He sent for the young men just prior to setting sail for Rome. He first threatened them as their king but spent most of his time giving them fatherly advice and urging them to love their brothers. He told them that he would forgive their previous offenses if they would behave themselves in the future. But the brothers refuted the charges, claiming that they were lies. They assured their father that their actions would speak louder than the words of their accusers and urged the king to shut the mouths of the rumormongers by refusing to listen to them. They said that there would be no end to those who wanted to slander them as long as someone was there to listen.

5. Herod was reassured by their words, and the sons' fears of immediate danger eased. But a foreboding for the future remained, as they now had been made aware of the animosity of Salome and their uncle, Pheroras. Both of them were powerful and dangerous enemies. This animosity was especially true of Pheroras, who shared in the king's entire royal honor, save for his crown. He had a personal income of 100 talents per year over and above that received from his ownership of all the land east of the Jordan River. The land had been a gift from Herod, who had obtained permission from Caesar to make Pheroras a tetrarch. Herod had also given his brother a royal wife—his own wife's sister. After her death, Herod wanted to give Pheroras the hand of his eldest daughter, Salampsio, daughter of Miriamne, together with a dowry of 300 talents. But Pheroras refused the marriage, preferring to chase after one of his maidservants instead. Herod was very angry at this rebuff and subsequently gave his daughter to a nephew, Joseph, who was killed by the Parthians. Later Herod was to

forgive Pheroras for being infatuated with his slave girl.

6. Many years earlier, when Miriamne was still alive, Pheroras had been accused of plotting to poison Herod. There were so many that came forward to accuse Pheroras that Herod, even though he loved his brother, began to believe what was rumored and feared for his life. The king tortured many of those suspected in the plot, finally arresting Pheroras's own friends. No one confessed that such a plot existed. But Herod learned that Pheroras had planned to flee to Parthia, taking his beloved mistress with him, and that his accomplice in the matter was Costobarus, Salome's husband. The king had given the man his sister's hand in marriage after her former husband had been executed on charges of adultery. Salome herself did not escape vilification. Pheroras, her brother, had accused her of agreeing to marry Silleus—the procurator of Obodas, king of Arabia—and Herod's most bitter enemy. While she was convicted of this and other charges against her, the king pardoned Salome. Herod also pardoned Pheroras of the charges against him.

7. The storm now bursting upon Herod's household hovered over Alexander's head. The king employed three eunuchs in his service that he held in high regard and to whom he had entrusted three crucial offices. One was his wine steward, one his chef, and the third attended to his bedchamber and slept nearby. Alexander offered the eunuchs large bribes to side with him against Herod. When the king got wind of the plot, he proceeded to torture the three to get at the truth. They confessed to their seditious conversations with Alexander and explained how he had deluded them. He had told the three not to pin their future hopes on an old man who dyed his hair in order to fool people into thinking he was younger. Rather, they should side with the one who would be Herod's successor in the kingdom, whether the king liked it or not. For Alexander had told them that he would soon avenge himself upon his enemies, making his friends happy and blessed, chief among them being the three eunuchs. Alexander had further told them that the ranking members of the court and the generals and high-ranking officers of the army came to him for secret talks.

8. Their confession so terrified Herod that he did not dare to make it public. Instead, he sent spies out both night and day to gather data. If Herod suspected anyone of treason, he was executed immediately. The palace

was in the grip of anarchy. Everyone invented accusations against their enemies and tried to turn the wrath of the throne against them. Lies were given instant credibility and reprisals came sooner than the accusations that caused them. Accusers suddenly became the accused, and were led off to die accompanied by those whom they had maligned. The danger in which the king saw himself made for brief examinations. Herod was so bitter that he could look at no one without suspicion. Even his friends received barbarous treatment at his hands. Many were forbidden to come into his presence. Those who he did not have the power to punish physically were to feel the lash of Herod's tongue.

To add to Alexander's troubles, Antipater gathered a group of his friends and began to disparage the young man in every conceivable way. The king was so terrorized by all of these slanders and lies that he thought he saw Alexander coming at him with a drawn sword in his hand. As a consequence, he had Alexander arrested immediately and put in chains. Many of Alexander's friends would die in the tortures they suffered. Most were silent or said nothing beyond what they knew to be the truth. But some were forced to utter lies in the face of the pain inflicted by the king's torturers. They falsely confessed that Alexander and his brother Aristobulus had plotted against the king, waiting for an opportunity to kill him while he was hunting, and then to flee to Rome. The king readily believed these confessions even they were unbelievable and given solely under the great duress of torture. But the king found comfort in his rationalization that he had not arrested his son for an unwarranted or unjust reason.

Chapter 25

Archelaus Secures a Reconciliation between Alexander, Pheroras, and Herod.

1. Alexander realized that it would be impossible to persuade his father of his innocence, so he resolved to face the crisis boldly. He wrote four books answering his enemies. While he confessed to being part of a plot, he argued that they were in it with him, particularly Pheroras and Salome. Salome, he charged, had once forced him to sleep with her against his will. These books that pressed other shocking charges against some of the most powerful men of the realm were then given to Herod. It was about that time that Archelaus, concerned for the welfare of his daughter and son-in-law, made a hasty trip to Judea. He came prudently as a counselor, and

succeeded in diverting Herod's threatened plans by his own wily scheme.

"Where is that scoundrel of a son-in-law?" Archelaus cried out when arriving in Herod's presence. "When I lay eyes on this man who plotted to murder his father, I'll tear him to pieces with my own hands! I'll do the same to my daughter! Even though she has had no part in the scheme, as his wife, she, too, has become polluted. I am astonished at your patience, who as the victim of this treason, has let Alexander live this long. I have hurried here from Cappadocia expecting to find that he had already been put to death. I came to question you about my daughter, who, because of your exalted position, I had given Alexander to be his wife. If your fatherly affections weaken your will to punish the son who plotted to kill you, let us trade places in order to inflict judgment upon our children."

2. When Archelaus had finished this boisterous speech, he saw that Herod's anger had subsided. The king then handed Alexander's books to Archelaus and proceeded to go over them page by page with him, frequently stopping. The Cappadocian saw this as an opportunity to continue his gambit, and little by little managed to shift blame onto the men whose names appeared in the books, particularly Pheroras. When he saw that Herod believed him, Archelaus said, "We must judiciously conclude whether these wicked men plotted against the young man or whether the young man plotted against you. I can arrive at no motivation that would cause Alexander to commit this awful crime. After all, he has already enjoyed the honor of royalty and the expectation of at least a portion of the throne. Unless someone was behind the scenes urging him on, taking advantage of his youthful naiveté. For by such people not only are young men led astray but old men as well. Great families and even kingdoms are occasionally overthrown by such as these."

3. Herod agreed with Archelaus and gradually his anger against Alexander lessened, refocusing instead upon Pheroras, the central subject of Alexander's books. Pheroras sensed this quick shift in the king's sympathies and the profound influence Herod's friend Archelaus had upon him. He perceived that there was no honorable way to save himself, so the king's brother resorted to audacity. He abandoned Alexander and threw himself at the mercy of Archelaus. The latter responded that it would be impossible to support anyone under such grave charges. They proved he had plot-

ted against the king and was the cause of Alexander's predicament, unless he was prepared to renounce his treason and his denials of it, confess to the accusations, and seek forgiveness from his brother, who loved him. If Pheroras were to do all of that, Archelaus told him; he would assist him in every possible way.

4. Pheroras took Archelaus's advice. He dressed completely in black in order to arouse Herod's deepest compassion. Then, with tears streaming down his face, Pheroras threw himself at Herod's feet and begged forgiveness for what he had done, confessing that he had acted wickedly and that he was guilty of all the charges against him. Pheroras mourned over his mind's madness, and said that his attraction to a woman had been the cause of his derangement. So Archelaus, who had moved Pheroras to both accuse and bear witness against himself, now proceeded to plead for mercy for him seeking to appease Herod's wrath. He cited parallel examples from cases involving his own family. He said that he had suffered worse treatment from one of his brothers but had taken the course of brotherly love rather than revenge. Kingdoms were like overweight people, Archelaus explained. There was always a part of the body that suffered under the heaviness of the body's weight. But an amputation was not the answer. Rather, a gentle cure was more effective and kept the body whole.

5. Archelaus's arguments succeeded in assuaging Herod's anger against Pheroras. Meanwhile, the Cappadocian monarch continued to feign anger against his son-in-law saying he would order his daughter to divorce Alexander and would take her back to her own country with him. Ultimately, he brought Herod back to arguing on his son's behalf and contending against any divorce. Sounding sincere, Archelaus told the king that he had his permission to marry her to anyone in the kingdom, with the exception of Alexander, for he wanted to maintain the marriage link to Herod and Judea. The king responded that if Archelaus would consent to allow the marriage to continue he would be returning his son to him since the couple had children and since the young man was so deeply devoted to his wife. If Glaphyra were to remain Alexander's wife, her very presence would make him ashamed of his past mistakes. But if torn from him, her absence would drive him to desperation. Recklessness is best squelched in the context of a loving home life. So, reluctantly, Archelaus agreed and was reconciled to the young man and reconciled Herod to

him also. He added, however, that Alexander should be sent to Rome to meet with Caesar because he had already sent a full report of the matter to the emperor.

6. Archelaus's ruse worked, and he rescued his son-in-law from danger. Upon everyone's reconciliation, a time of great feasting and enjoyment followed. When Archelaus left Herod, he gave him 70 talents, a golden throne set with precious stones, some eunuchs, and a concubine named Pannychis. The king then gave favors to his friends in keeping with their rank. At the king's command, Archelaus also received many magnificent presents from all of the king's family. Herod and his officials then escorted Archelaus as far as Antioch.

Chapter 26

How Eurycles Slandered Miriamne's Sons, and how Euaratus's Apology had no Effect.

1. Shortly afterward, a man arrived in Judea whose wiles were vastly superior to those of Archelaus. He not only destroyed the reconciliation that the king of Cappadocia had negotiated for Alexander, but he also became the reason for the latter's ultimate downfall. He was a Lacedemonian named Eurycles. He came to Judea looking for money because Greece could no longer supply his demand for extravagant luxuries. Eurycles brought magnificent gifts to Herod as bait to catch bigger fish. Herod quickly repaid Eurycles for his apparent generosity but at a very high interest rate. Eurycles gave no gift unless it returned a profit in the blood of the people of the kingdom. So, using flattery, subtlety, and false compliments, Eurycles imposed his will upon Herod. He quickly uncovered the king's weaknesses and began to say and do everything he could to please him. He soon became one of Herod's closest friends. Indeed, both the king and all of his courtesans held the Spartan in high regard because of his country and manner.

2. As soon as Eurycles learned all the gossip about the royal family's troubles, of the quarrel between Herod's sons, and of the king's feelings toward each of them, he began his ruse. Though Eurycles first accepted Antipater's hospitality, he won Alexander's friendship by claiming to be an old friend of his father-in-law, Archelaus. That recommendation quickly brought

Eurycles the friendship of Alexander, who immediately introduced the Spartan to his brother Aristobulus. Playing all these parts, as if in a theatrical production, Eurycles slipped surreptitiously into the friendship of each brother. He played different roles with each, but principally that of Antipater's right-hand man, and then to Alexander the role of Antipater's traitor. He told Antipater, on the one hand, that it was disgraceful that he, the eldest son and heir, should overlook those who might stand in his way to the throne. To Alexander, on the other hand, he claimed it shameful that this son of a princess and the husband of another should allow the son of a common woman to lay claim to the crown, especially when Alexander had the formidable support of his father-in-law behind him. Eurycles' fictitious claim to be a friend of Archelaus gave weight to his counsel. So, without withholding anything, Alexander poured out upon the Spartan all of his grievances against Antipater, adding that it would not be surprising if Herod deprived both him and Aristobulos of their rightful claims inasmuch as the king had already murdered their mother. Eurycles pretended to have pity on the young man and consoled him. He then lured Aristobulus into making the same complaints against Herod.

Eurycles soon went to Antipater and told them all that the brothers had said in confidence. He added another little creative detail. He told Antipater that his half-brothers were plotting against him waiting for their opportunity to attack with drawn swords. Antipater paid richly for this information to Eurycles, who then hurried off to Herod to sing the praises of Antipater. Eurycles had, for a price, brought about the ultimate executions of Alexander and Aristobulus, and now went to their father to accuse them.

Arriving in Herod's presence, Eurycles told him that he had come to save his life in return for all that the king had done for him; the light of life for his kind hospitality. He told the king that a sword had long been sharpened, and Alexander's right hand was ready to wield it against him. But, Eurycles said, he had slowed down the plot by pretending to assist the young man in his cause. Alexander said, he continued, that Herod had not been content to reign in a kingdom that belonged to others. He had not only murdered their mother and squandered her realm, but was now proceeding to pass the crown to a bastard, offering their grandfather's throne to that pest Antipater. But he—Eurycles said of Alexander—stood ready to avenge the ghosts of Hyrcanus and Miriamne, for it was not right to usurp the power of the king without bloodshed.

Eurycles then spoke of the provocations to which Alexander said he had been subjected, adding his own comment in advance that every word spoken by the youth was slanderous. When the subject of noble birth had been broached, Alexander said that his father had rudely insulted him by commenting, "No one is of noble birth but Alexander here, who even scorns the low birth of his father!" While hunting, Alexander said that he offended others even if he said nothing, yet thought sarcastic if he opened his mouth to congratulate someone. In fact, Alexander had found his father exceedingly harsh toward him and his brother. Herod had loved only Antipater. He would, therefore, willingly forfeit his life should his plot fail. On the other hand, if Herod died, he had men who would protect him.

First was his father-in-law, Archelaus, to whom he could easily escape. Next was Caesar, who remained ignorant of Herod's true character. This time, when the opportunity came to stand before the emperor, Alexander said that he would not be overawed by his father's presence nor would he confine his remarks to his personal grievances. Rather, he would tell the whole world of his nation's sufferings, how they had been bled to death by excessive taxation. Then he would go on to describe the luxury and wickedness purchased with the blood of his countrymen. According to Eurycles, Alexander had said that he would also describe to Caesar the evil characters of the men who had become rich at his and Aristobulus's expense, and how Herod lavished money on particular cities out of selfish motives. Eurycles continued. He reported to Herod that Alexander said that he would request Caesar to make an inquiry into the deaths of his (great) grandfather Hyrcanus and his mother Miriamne, and make all the abominations of Herod's reign a matter of public record. Under such conditions, Alexander said an accusation of patricide against him would never stick.

3. When Eurycles had finished his ominous tirade against Alexander, he proceeded extolling the great virtues of Antipater. He described Herod's eldest son as the only one who loved his father and because of that love had been able to frustrate the plot against him. Herod, who had barely been able to control his temper during the accusations against Alexander, now burst into irrepressible rage. Antipater seized the opportunity to call in other men who accused his brothers of holding clandestine meetings with Jucundus and Tyrannus, former commanders of the king's cavalry.

Herod had demoted these men for various offenses. This accusation sent Herod's blood to the boiling point. He immediately ordered that the two men be sent to the torturers, but they made no confession of wrongdoing. But then a letter was produced, purportedly written by Alexander to the commander of Alexandrium asking him to let them seek refuge in his fortress following Herod's death, and requesting from him weapons and other provisions. Alexander swore that the letter was a forgery skillfully created by Diophantus, one of the king's secretaries who had a clever knack of imitating the handwriting of others. This man had been recently put to death after being discovered to be the author of one of many forgeries. Herod subsequently tortured the commander of Alexandrium, but failed to get any information either to confirm or to deny the letter's origin.

4. Herod knew that the evidence against his sons was weak, but he nevertheless ordered that a close watch be placed upon them, otherwise allowing them to remain free. The king then claimed Eurycles, who had been the arch-villain in this plan and the instigator of the forgery, to be his savior and benefactor, and gave him a reward of 50 talents. Before Herod determined whether of not the story was true, the rascal slipped away to Cappadocia where he extorted even more money from Archelaus, claiming to have brought reconciliation between Herod and Alexander. After that, Eurycles crossed over to Greece where he spent the money he had received on his wicked missions on even more wicked purposes. He was ultimately tried before Caesar and convicted of provoking rebellion in Achaia then of looting its cities, and was sentenced to perpetual exile. Fortune repaid Eurycles for his role in the betrayal of Alexander and Aristobulus.

5. In contrast to the wickedness of the Spartan, the name of Euaratus of Cos is worth mentioning. One of Alexander's closest friends, he visited Judea at the same time as Eurycles. Being questioned by the king regarding the allegations of Eurycles, Euaratus affirmed on oath that he had heard nothing of the kind from the young man. Unfortunately, his testimony was on no avail to the unfortunate Alexander. Herod was predisposed to listen only to slander and lies. Only those who shared his gullibility and anger found honor in Herod's eyes.

Chapter 27

Caesar Orders Herod to Accuse His Sons at Berytus. They Are Condemned in Absentia, Taken to Sebaste, and Strangled.

1. Salome provided further motivation for Herod's cruelty toward his sons. It seems that Aristobulus, wanting to involve his former mother-in-law and aunt in his own difficulties, warned her to watch out for her personal safety. He told Salome that the king was planning to kill her as a result of a previous accusation against her. In her desire to marry the king's enemy, Silleus the Arab, she had apparently revealed many of Herod's secrets to him. Aristobulus unwittingly touched off the storm that was to prove the final undoing of the young mens' fortunes. Salome ran to the king and advised him of the warning she had received. On hearing her, Herod's patience ended. He had the brothers slapped in irons and separated from one another. He then sent Volumnius, the military tribune, and one of his friends, Olympus, to Caesar with the case against his sons in writing. Setting sail for Rome, they delivered the king's letters to Caesar, who, while deeply troubled for the young men, did not think it right to usurp their father's authority over them. Caesar replied accordingly, giving Herod full freedom of action. He did add his recommendation that the king hold a formal inquiry into the plot before a joint council of his own relatives and the provincial governors. Convicted by such a hearing, his sons should be put to death. But if they had only planned to flee the country, a lesser sentence imposed.

2. Herod took Caesar's advice and left for Berytus (Beirut) where Caesar had designated the court assembled. Following written instructions from Caesar, Roman officers were to preside, specifically the Syrian governor Sentius Saturninus, together with his legates, Pedanius and others. Volumnius the procurator also sat on the bench. Next were the king's relatives and friends including Salome and Pheroras, and then all the aristocracy of Syria, except King Archelaus—Herod distrusted Alexander's father-in-law. The king did not allow his sons to appear at the hearings, a precaution he thought wise. It would assure that no undue sympathy would arise for them. Had they been permitted to speak, Alexander would have had no difficulty in rebutting the charges against them. So the brothers were held in custody in Platane, a village in Sidon south of Berytus.

3. The king stood up in court and denounced his sons as if they were in the courtroom with him. As he had little evidence, the case he made against them was weak. So he focused on the many provocations, indignities, insults, and offensive remarks leveled against him, which, he told the court, were crueler than death itself. When no one rose to contradict him, Herod asked the court for pity for himself for even if he won the victory over his sons his winning would be like losing. He then asked the judges to express their opinions. Saturninus was the first to speak. The Syrian governor argued that the young men were guilty but were not worthy of death. He said, as one who had three children present in court that it was not right for him to condemn the sons of another man. Saturninus's two legates agreed with him as did some others.

Volumnius was the first to speak out in favor of the death sentence. All who then spoke followed his lead in condemning the young men to death. Some wanted to flatter Herod with their vote to execute while the motivation of others was their hatred of the Judean king. But no one voted the death penalty because they held any animosity against the prisoners. From that time on, all Syria and Judea waited anxiously to see what would happen next. No one thought that Herod's cruelty would extend to murdering his own children. Meanwhile, the king transported his sons to Tyre. Then, sailing for Caesarea, he pondered the best way to execute them.

4. An old soldier in the king's army named Tero had a son who was Alexander's close friend. Tero himself had a deep personal affection for both of the princes. Because of this, he lost control of his tongue, letting his anger get the best of him. First, he began shouting that justice had been trampled under foot and that the truth had been murdered. He screamed that the laws of nature were turned upside-down and that the whole world was full of evil, along with anything else that might come to a man who no longer cared whether he lived or died. Ultimately, Tero spoke up in the king's presence, saying, "My opinion is that you are the most despicable of men! You give credence to the most wicked scoundrels and by their words condemn those nearest and dearest to you. The very ones you have frequently sentenced to death, Salome and Pheroras, have now made you believe their slander against your own children. They are cutting off your rightful heirs, leaving only Antipater, their choice to be the most pliable

of puppets. Be aware that the death of Antipater's brothers will one day arouse the hatred of the army against you. For there is not one man under arms who does not pity the young men and many of the officers are already freely expressing their anger." Tero finished his speech by naming the officers to whom he had alluded. The king promptly arrested them all, including Tero and his son.

5. Upon hearing all of this, a court barber named Trypho, as if demon possessed, sprang forward and accused himself. He said to the king, "This same Tero tried to persuade me to cut your throat with my razor as I was trimming your beard, promising me a large reward from Alexander." When Herod heard these words, he had Tero, his son, and the barber all delivered over to the torturers. The father and sons denied the charges and the barber would add nothing to what he had already said, so Herod gave orders to rack Tero more severely. But Tero's son, out of pity for his father, promised to tell the king everything if he would but spare his father further torture. The king agreed, and the son told him that his father had been persuaded by Alexander to assassinate him. Some say he concocted this in order to spare his father further torment while others say that all of it was true.

6. Herod then gathered a large crowd of people and formally accused Tero and the others involved requesting the people's assistance in executing them. So they were all put to death including Trypho the barber, clubbed and stoned on the spot. The king then sent his sons to Sebaste (Samaria), a city near Caesarea, and ordered their immediate deaths by strangulation. He then commanded that their bodies be brought to the fortress at Alexandrium for burial next to Alexander, their maternal grandfather. So ended the lives of Alexander and Aristobulus.

Chapter 28

The People Hate Antipater; Herod Engages His Slain Sons' Sons to Girls In the Family, but Antipater Makes Him Change His Mind. Of Herod's Marriage and Children.

1. Now that Antipater had the indisputable right of succession to the throne, he became the object of the nation's utter hatred. Everyone knew that it was he who had concocted all of the slanders against his broth-

ers. He also became alarmed at the sight of his victims' children growing into adulthood. For Alexander had two sons by Glaphyra: Tigranes and Alexander. By his marriage to Salome's daughter, Bernice, he had three children. They were Herod, Agrippa, and Aristobulus, along with two sisters, Herodias and Miriamne. Following Alexander's execution, Herod had sent Glaphyra back to her father in Cappadocia with her dowry. He then gave Aristobulus's widow, Bernice, to be the wife of Theudion, Antipater's maternal uncle. The match was arranged by Antipater to appease his enemy, Salome. Antipater also sought to win Pheroras's favor through many gifts and flatteries and also with Caesar's friends by sending substantial monetary gifts to Rome. Antipater also saturated Saturninus and all his staff in Syria with presents. The more that Antipater gave away, the more hatred arose against him. His gifts were seen as motivated by fear rather than generosity. The result was that the recipients were not swayed to befriend him, and those who got nothing hated Antipater all the more. But the presents became greater with each day that passed, as, contrary to what Antipater had expected, Herod was paying close attention to the orphans, showing remorse for the murder of his sons by having compassion upon their children.

2. One day, Herod called all of his relatives and friends together. Setting the young children before them, he said with tears in his eyes, "An unfortunate fate has taken from me the fathers of these children. But natural pity for orphans has commended them to my care. While I have been the most unsuccessful father, I will try to make a much better grandfather, and then following my death, leave these in the protection of those dearest to me. Therefore, I betroth your daughter, Pheroras, to the eldest of the sons of Alexander, in order that this alliance will oblige you to watch over him. To Antipater's son I betroth Aristobulus's daughter, Miriamne, that you may become a father to this orphan girl. Her sister Herodias will become the wife of my own Herod, for he is the grandson of the high priest on his mother's side. Let, therefore, my wishes be granted, and let no one be called my friend who seeks to frustrate them. I pray that God will bless these marriages so that the kingdom and my descendants may prosper. May He look down with more peace upon these children than He gave to their fathers."

3. Having spoken these words, Herod wept and joined hands with the

children. Then, after fondly embracing one after the other, he dismissed the gathering. Immediately Antipater's blood ran cold. Disappointment was written all over his face. He figured that the honor that Herod had just given these children had signaled his own ruin. Alexander thought that his claim to the throne was again endangered, particularly if Alexander's children were to have not only the support of their grandfather, King Archelaus, but also of the tetrarch, Pheroras. He also considered how the nation hated him and how they pitied the orphans. Antipater thought of the affection the Jews had held his brothers while they were alive, and their fond memories of them now—that because of him they were dead. He then resolved by all means not to let any of these marriages take place.

4. Antipater was afraid to approach the matter with his father using his usual craftiness, knowing of Herod's harshness and how the least suspicion of his motives might arouse the king's anger. So he boldly entered the king's presence and asked him outright neither to deprive him of what he promised nor to leave him with just the title of king while others enjoyed the power. He argued that he would never have control if Alexander's son, with Archelaus his grandfather, also had Pheroras as his father-in-law. He, therefore, begged the king, since the palace contained such a large family, to change the intended marriages. The king, in fact, had nine living wives, seven of whom had children. Antipater himself was Doris's son; Herod II of Miriamne, the high priest's daughter; Antipater and Archelaus were sons of Malthace, the Samaritan; Olympias, also a daughter of Malthace, had married the king's late nephew, Joseph. By Cleopatra of Jerusalem, the king had Herod and Philip, and by Pallas, Phasael. Besides these he had other daughters, Roxana and Salome, one by Phaedra and the other by Elpis. Two of Herod's wives, one a cousin and the other a niece, were barren. Additionally, there were two daughters, Salampsio and Cypros, by Miriamne, sisters of Alexander and Aristobulus. In view of Herod's large family, Antipater asked him to alter the plans for their future marriages.

5. When the king realized Antipater's attitude toward the orphans, he became very angry. The thought crossed his mind that perhaps the man standing before him now was responsible for the murder of his sons. So Herod immediately began a long, angry speech, after which he dismissed Antipater from his presence. Later, though, after having been seduced once again by Antipater's cunning, the king did change his plans. He gave

Aristobulus's daughter to Antipater himself and Pheroras's daughter to Antipater's son.

6. One can arrive at an understanding of just how powerful was Antipater's guile when compared to Salome's failure in a similar situation. Although she was Herod's sister and had the added benefit of Empress Livia (Julia) pleading her case with the king for permission to marry the Arab, Silleus, she lost. In fact, Herod swore that if she did not give up the idea, he would regard her as his bitterest enemy. Ultimately, against his will, Herod married Salome to one of his friends named Alexas and one of her daughters to Alexas's son. The other daughter, Bernice—Aristobulus's widow—he married to Theudion, Antipater's uncle on his mother's side. Of Herod's daughters by Miriamne, he gave Cypros to Salome's son Antipater, and Salampsio to Phasael, son of his brother Pheroras.

Chapter 29

Antipater Becomes Intolerable. He Is Sent to Rome Carrying Herod's Will with Him. Pheroras Dies at Home.

1. After Antipater had rearranged the marriages to his own advantage and had cut off the expectations of Alexander and Aristobulos's orphaned children, he became insufferable. His wickedness became inflamed by his assurance of power. Unable to avoid the hatred he inspired in the people, he found security through terror and intimidation. Pheroras assisted him, viewing Antipater's ascent to the crown as a foregone conclusion.

Meanwhile, a circle of women at court was causing new problems for the king. Pheroras's wife, in union with her mother and sister and Antipater's mother, Doris, displayed a continual arrogance in the palace. Pheroras's wife even managed to insult the king's daughters, Salome and Roxana. As a consequence, she became a special object of Herod's impassioned hatred. But even with the king's animosity toward these domineering women, they kept at it. Salome, the king's sister, provided their sole opposition. She informed the king of their meetings and warned him of the danger inherent in their union. Upon hearing these charges made against them and Herod's indignant response to them, the women stopped meeting publicly and ceased all social gatherings. Instead, they pretended to quarrel with one another whenever the king was within earshot. Antipater joined in their little game by publicly opposing Pheroras.

But the women continued to hold secret meetings and private parties after nightfall. The knowledge that they were being watched only served to draw them closer together. Salome, however, knew their every move and reported it all to Herod.

2. But the king continued to be inflamed with anger against the women—in particular Pheroras's wife, the central figure of Salome's charges. Herod, therefore, formed a council of his friends and relatives and accused this woman of many wrongdoings, particularly of insults toward his daughters, of giving the Pharisees the funds that enabled them to oppose him, and of alienating his brother, Pheroras, by addicting him to drugs. Ultimately, Herod addressed Pheroras and gave him this choice: choose his brother or his wife, one or the other. Herod was perplexed when Pheroras claimed he would sooner die than forsake his wife. So the king turned to Antipater and ordered him to have nothing further to do with Pheroras's wife or any of her cronies. Antipater did not openly defy Herod's command, but continued to meet secretly with the women under cover of darkness. But figuring that Salome might be spying on him even then, Antipater planned a visit to Rome with the help of some friends in Italy. A letter was received from them urging that Antipater be sent at the earliest opportunity to Caesar's court. Without a moment's delay, Herod sent his son off with a princely entourage and a large sum of money. He also sent with him a copy of his will naming Antipater as the heir to the throne and appointing Herod, the king's son by Miriamne—daughter of Simon the high priest—as Antipater's successor.

3. Silleus the Arab, Caesar's orders notwithstanding, also sailed to Rome in order to oppose Antipater with all of his energies. At issue were the charges of Nicolas of Damascus, who accused Silleus of treasonous activities against the empire. There was also a grave matter between Silleus and his own king, Aretas. For Silleus had killed a number of Aretas's friends, including Sohemus, one of the most powerful men in Petra. By a large bribe Silleus obtained the services of Caesar's treasurer, Fhabatus, to assist him against Herod. But Herod gave Phabatus an even larger sum of money to join with him against Silleus. Herod wanted to extract from Silleus the fine previously imposed by Caesar for his unlawful behavior. But Silleus refused to pay anything. Furthermore, he accused Phabatus to Caesar, charging that the emperor's secretary was not loyal to Caesar, but

to Herod. Phabatus was indignant at the assertion. Still on Herod's payroll, however, he betrayed Silleus by informing the king that the Arab had bribed one of Herod's bodyguards, a man named Corinthus, warning him to be on his guard against an attack. Herod took his advice, knowing that although Corinthus had grown up in Judea he was an Arab by birth. He immediately had Corinthus arrested along with two other Arabs found with him. One was a friend of Silleus and the other a tribal chief. The two men were tortured and confessed that they had been offered a bribe by Corinthus to kill Herod. So, after further examination by Saturninus, governor of Syria, they were shipped off to Rome to stand trial.

4. Meanwhile, Herod never relaxed his efforts to require that Pheroras divorce his wife. Even though he had more than sufficient cause to hate the woman, he could imagine no way to punish her. Finally, filled with indignation, he banished both her and his brother from the kingdom. Pheroras accepted the indignity patiently. He swore that only Herod's death would provide an end to his exile. He would never return to his brother as long as the king lived. Nor, he swore further, would he return to visit Herod even during his illness even though urgently pressed to do so. At one juncture, when Herod had thought he was dying, he wanted to leave Pheroras with certain instructions. But the king recovered and shortly afterward Pheroras himself became ill. Herod, showing brotherly compassion, went to his bedside to take care of him. But his visit accomplished little, for Pheroras died a few days later. A report was circulated that Herod had poisoned him, even though the king had shown only love for his brother as he lay on his deathbed. Be that as it may, Herod had the corpse transported to Jerusalem. He gave orders for a period of national mourning and honored Pheroras with a majestic state funeral. So the curtain fell on one of the murderers of Alexander and Aristobulus.

Chapter 30

Herod Discovers Antipater's Foul Play in Pheroras's Death. Herod Casts Doris and Her Collaborators, as well as Miriamne, out of the Palace. He Writes His Son, Herod, out of His Will.

1. But punishment was about to descend upon the real murderer, Antipater, and was to have its genesis in the death of Pheroras. Some of Pheroras's freedmen came dejectedly to the king and told him that his brother

had died of poison. They said that he had become ill immediately after eating some strange food prepared for him by his wife. They then added the fact that just two days before, her mother and sister had imported a woman from Arabia, an expert in drugs, to create a love potion for Pheroras. Instead, she had given him a deadly poison at the instigation of Silleus, who knew the woman.

2. Herod was beside himself with suspicions. He had the maidservants and some of the freedwomen above them tortured. In pain torture one victim cried out in agony, "May the God who governs the earth and heaven punish the source of our miseries, Antipater's mother, Doris!" Clutching at this clue, Herod pushed his investigation of the matter further. The woman who had screamed out then further disclosed the intimacy that had existed between Pheroras and Antipater's mother and the other women of his family, and of the secret meetings that they held together. She added that Pheroras and Antipater, following an interview with the king, spent the entire night drinking with these women while not allowing a single servant, neither male nor female, to be present. The woman who had spoken was a freedwoman.

3. Herod then had each of the slave girls tortured separately. Their additional evidence agreed with what he had already discovered. They added that it was by mutual agreement that Antipater had gone to Rome and Pheroras to Perea for they would often say, "After Alexander and Aristobulus, we and our wives will he Herod's next victims. Since he spared neither Miriamne nor her children, neither will he spare any of us. So it will be better for us to flee as far as possible from the ferocious beast." Antipater, they continued, often complained to his mother that while he was already gray-haired his father seemed to get younger every day. He said that he would probably die first without ever having the opportunity to be king. Even if he did outlive his father, his enjoyment of his office would be extremely short. Then he would complain of those "Hydra-head" sons of Alexander and Aristobulus and how fast they were coming of age. Herod had even robbed Antipater of the hope of being succeeded as king by one of his own children, having nominated instead Miriamne's son, Herod. Furthermore, the slave girls continued, Antipater said that in appointing Herod, the king had only disclosed his own senility if he truly thought that his appointment would stand. For Antipater would see to it that

none of his family would be left alive. Never had a father hated his children more than the king, they reported Antipater saying, but he hated his brothers even more. Only the other day the king had given Antipater 100 talents to break off all contact with Pheroras. When Pheroras had asked, "Why? What harm are we to him?" Antipater had replied, "It would be better if Herod would take everything we own and leave us naked yet still alive. But it is impossible to escape the bloodthirsty beast, who won't even let us show our affection for one another. Now we must meet secretly, but we shall one day meet openly, if we have but the courage and enough armed men."

4. To all of this the tortured women added that Pheroras had planned to escape with all of them to Petra. Herod was confirmed in his belief of all their admissions because of the detail of the 100 talents. No one but Antipater knew about that. The first to encounter the king's wrath was Antipater's mother, Doris. He stripped her of all the jewelry and finery that he had given her at great expense and threw her out of the palace for the second time. He then made his peace with the maidservants and ladies of Pheroras's court, treating them with special kindness following their torture. But Herod's fear was now magnified and flared up at the slightest provocation. Because of it, many innocent people were delivered to the torturers out of the king's fear that a single culprit might escape.

5. Herod now focused on Antipater the Samaritan, his son Antipater's steward. After torturing the Samaritan, the king learned that Antipater had imported a deadly poison from Egypt, intended for Herod. Antipater had used his friend Antiphilus as his smuggler. From Antiphilus the potion was passed on to Theudion, Antipater's uncle, who delivered it to Pheroras. Antipater had commissioned the king's brother to kill Herod while he, himself, was in Rome and out of suspicion. Pheroras had then entrusted the poison to his wife. So the king sent for her and ordered her to produce the poison immediately. She left as if to get it but instead jumped off the roof of the palace in order to escape conviction and the king's tortures. By God's providence, whose vengeance was pursuing Antipater, she didn't land on her head but on another part of her body, and survived her suicide attempt. She was brought back to the king, knocked senseless by the fall. But after being brought around, Herod asked her why she had jumped, swearing that if she told him the truth he would not

punish her. But if she lied, he swore he would tear her body into so many pieces that there would not be one part remaining fit for burial.

6. The woman hesitated for a second and then replied, "Why should I keep these secrets any longer now that Pheroras is dead? Merely to save Antipater, who is the ruin of us all? Hear me, O king! God, who cannot be deceived, is my witness to the truth of what I am about to say. When you sat weeping as Pheroras was dying, he called me over and said, "Wife, I have been mistaken about my brother's feeling toward me. I hated him and yet he has such affection for me. I plotted to kill the very one who is so stricken with grief as I lay dying. I am now receiving the retribution of my wickedness. I want you to bring the poison that Antipater left with us and that you are keeping to kill Herod, and promptly throw it in the fire so that I may not be subject to vengeance in the afterlife." So I brought it to him as he had commanded and emptied most of it into the fire before his eyes. But I kept a little of it for myself because of the uncertainty of the future and out of my fear of you."

7. When she had said this, she produced the box that still contained a small portion of the poison. The king then tortured Antiphilus's mother and brother. They both confirmed that Antiphilus had brought the box from Egypt, and further, that he had initially procured the poison from another brother, a physician in Alexandria. It was as if the ghosts of Alexander and Aristobulus were going throughout the palace uncovering and disclosing all its secrets and bringing to judgment those who were the least suspect. Even Miriamne, the high priest's daughter, was found to be involved in the plot. The king, therefore, avenged her insolence upon her son, Herod. The king crossed his grandson, Herod, out of his will, the one whom he had appointed the successor to Antipater.

Chapter 31

Antipater Is Convicted by Bathyllus, but still Returns Home from Rome Unaware. Herod Brings Him to Trial.

1. Antipater's freedman Bathyllus now furnished evidence ultimately corroborating Antipater's part in the plot. This man came forward with yet another vial of poison containing the venom of asps and other snakes. This potion was to be used by Pheroras and his wife in case the first poi-

son failed to have the desired effect. He also submitted, in addition to proof of Antipater's intended murder of his father, letters by which Antipater had contrived to harm his brothers, Archelaus and Philip. These two sons of the king, still only growing boys and full of youthful exuberance, were being educated in Rome. Antipater was anxious to get rid of these potential heirs who threatened to spoil his hopes for the throne. So he forged injurious letters in the names of their friends in Rome. He bribed some to say that the young princes were constantly ridiculing their father, publicly deploring the deaths of Alexander and Aristobulus and angrily decrying their summons back to Judea. Their father had indeed called them home, and it was this fact that most troubled Antipater.

2. Even before he had sailed to Rome, while still in Judea, Antipater had used bribes to procure fictitious letters from Rome abusing the brothers. Then, to avoid suspicion, he would appear before their father to make excuses for the boys. He argued that some of the things said in the letters were false while other statements were merely indicative of youthful indiscretions. Now in Rome, Antipater tried to conceal the vast sums that he was spending to bribe men to write the letters. He began purchasing expensive clothing, embroidered carpets, silver and gold cups, and a great many other costly items. His accounts showed expenses of 200 talents, most of which he used in his lawsuit against Silleus. Now, all these petty misdemeanors were being exposed in the light of his discovered felonies. For the examination by the torturers had uncovered his attempt to murder his father, and the letters showed his desire to murder his brothers also. But not one of those who traveled to Rome from Judea bothered to inform Antipater of his turn of fortune at home. Indeed, so intensely did they hate him that seven months went by between his conviction and his return while he remained ignorant of his troubles. Perhaps the ghosts of his murdered brothers held the tongues of those who might have said something. Be that as it may, he wrote to his friends from Rome of his soon return, and of the honors paid to him by Caesar upon his departure.

3. The king was anxious to get his hands on his traitorous son, but at the same time he feared that Antipater might be forewarned and on guard. Herod, therefore, subtly hid his real attitude and sent a letter couched in affectionate terms and bidding him return with all haste. The king, knowing that Antipater was not ignorant of his mother's dismissal from

court, added that if he came quickly he would set aside any complaints he had against her. Antipater had received a previous letter at Tarentum informing him of Pheroras's death. He had displayed the utmost grief at the news, which was applauded by some as a nephew's sadness at the loss of his uncle. But Antipater's true grief lay in the fact that his plot had failed. He did not weep for Pheroras but his accomplice. Moreover, he was concerned about any remaining trail of evidence. Had the poison been discovered? But when he received the current letter from his father while in Cilicia, he pressed on toward home.

After Antipater had sailed from Celenderis, he began to think of his mother's disgraceful misfortunes and there arose in his heart a deep foreboding of the future. His closest friends advised Antipater not to go any further until he had determined the circumstances surrounding his mother's dismissal. They feared that his arrival might only add to the charges against her. But those who were more anxious to see their native land than concerned for his safety urged him to push on and neither arouse his father's suspicions by his procrastination nor allow an opportunity for his enemies to slander him. "Even if it is true," they argued, "that there are accusations against you, they are because you have been out of the country. They would never have arisen had you been there. It is absurd to deprive oneself of certain happiness because of some uncertain suspicions. Run to your father and rescue the kingdom that is teetering on the brink of ruin because of him." Antipater followed this last advice, for Providence was moving him along to his own destruction. So, sailing southeastward over the Mediterranean they landed at the port of Augustus in Caesarea.

4. Once ashore, Antipater was surprised to be surrounded by a profoundly ominous feeling of isolation. Everyone avoided him; no one came near him. Although always hated, such hatred was now shown without fear of reprisal. Additionally, their fear of Herod kept many away from him. Every city was, by now, full of the scandal. Only Antipater was ignorant of his standing with the king. No man had ever had a more extravagant send-off to Rome, only to return to such a dishonorable reception. Realizing for the first time the disaster that had come upon him back home, Antipater tried his best to maintain an outward composure. Though his heart trembled in fright, he cunningly put on a bold face. There was no longer any possibility of retreat from the dangers that encircled him.

However, he still did not have any definite news from the royal palace largely because of the king's threats against informers. So Antipater still held out a small hope. Perhaps nothing had been discovered. Even if it had he still might be able to make things right by his impudence and guile, weapons that had always delivered him in the past.

5. Armed only with these weapons, Antipater entered the palace alone—the guards had stopped his friends at the outer gate. At the time of his arrival, it just so happened that Herod had another visitor, Syria's governor, Quintilius Varus. Antipater went at once into his father's presence. Seeking courage in feigned boldness he approached Herod as if to kiss him. Herod held out his hands and turned his head to the side screaming, "See! His desire to embrace me confirms that he wants to kill me, even in the midst of all the accusations against him! Go to Hades you wicked wretch! Don't dare to touch me until you have cleared yourself of the charges against you. I will set up court for you to that end. Varus, who has arrived at a most opportune time, shall be your judge. Go and prepare your defense for tomorrow's trial. I'll give you that much time to prepare all of your defensive bag of tricks." Antipater left the room. He was so confused he couldn't say anything in response to his father's words. His mother and his wife then came to him and made Antipater aware of all the evidence against him. So he was then able to collect himself and begin to prepare his defense.

Chapter 32

Antipater Is Accused Before Varus and, by Strong Evidence, Is Convicted of Plotting Against His Father. Herod Delays His Punishment until He Recovers, then Alters His Will once more.

1. The next day, the king assembled a council of his friends and relatives and invited Antipater's friends to attend as well. Herod presided, with Varus, and ordered all of the witnesses to be brought in. Among the witnesses were some of Antipater's mother's recently arrested domestic servants, caught in the act of delivering a letter from Doris to her son. The letter read, "Since your father has discovered everything, do not go near him unless you have gotten assistance from Caesar." When these and other witnesses had been brought in, Antipater entered and fell on his face at his father's feet He cried out, "Father, I plead with you not to judge me

in advance but to lend an unprejudiced ear to what I have to say in my defense. If you permit me, I will establish my innocence."

2. Herod shouted at his son to be silent, and then addressed Varus: "I am certain that you, Varus, and any other upright judge will convict Antipater of his vile crimes. But I also fear that you will judge me worthy of punishment for raising such as he. But I should draw pity for being the most devoted father to such abominable wretches. My late sons, whom I saw fit to prepare for the throne while they were quite young, were educated in expensive Roman schools, and also brought into Caesar's friendship. These who were envied by other kings were also found to be plotting against me. They have died mainly to secure Antipater's hopes. For, since he was a young man when designated as heir, my main goal was to keep him safe from danger. But this depraved monster, after feasting on my patience, has turned my abundant gifts against me. He thought that I had lived too long. My advancing years haunted him, and he came to believe that the only way he would ever become king was to murder his father. I have been justly repaid for bringing Antipater back to this court. For he was a nobody, a commoner, whom I made successor to the realm while casting aside my sons born of royal stock. I admit to you, Varus, my great foolishness. I alone provoked my late sons by cutting off their expectations in favor of Antipater. What kindness did I show them that equaled what I have shown this rascal? In a way, I gave up my royal power while yet alive. For I publicly proclaimed his succession to the throne as part of my will. I gave him an annual income of 50 talents besides the vast sums he received from my personal revenues. When he sailed for Rome, I gave him 300 talents and recommended him to Caesar as his father's only deliverer. What crimes of my other sons maybe compared to those of Antipater? What evidence against them was as strong as that which will now demonstrate Antipater's guilt?

"But once again this 'kinsman-killer' has taken the liberty of opening his mouth hoping once again to hide the truth beneath his cunning tricks. Be on your guard, Varus. For I know this beast and foresee his very plausible arguments and the hypocritical mourning that are sure to proceed from his mouth. Antipater is the very man who warned me to watch out for Alexander and not to place my life in the hands of all men! This is the man who saw me to my bedroom, making sure that no assassins lay hidden to trap me! Antipater is the man who made sure I got my sleep, who kept me free from the fear of danger, who comforted me in my depression after the deaths of my two sons, and who reported to me of the love my surviving brothers had for me! This was my protector, my bodyguard! When I recall, Varus, Antipater's consistent shrewdness and hypocrisy, it's hard to believe that I'm still alive. It is a wonder that I have escaped the deep pit of his schemes. But since fate is determined to destroy my household after raising up against me one after the other of those who are nearest to my heart, I may weep over my misfortune. Yet, even though my spirit grieves under the heaviness of this sad burden, not one person who thirsts after my blood shall escape, not even if the evidence involves every one of my children."

3. At this point, Herod's emotion choked off his words, so he signaled to his friend, Nicolas, to cite the evidence against Antipater. Meanwhile, Antipater had looked up from his prone position at his father's feet and exclaimed, "Father, you have made my defense for me! How could I be guilty of parricide if, as you admit, I served as your protector? You have called my sonly affection enormous lies and hypocrisy! How could I, who was so crafty in everything else, have been so stupid not to understand that, while it is difficult to hide from men a plot of such large proportions, it is impossible to hide it from the all-seeing, all-present, Judge of heaven. Was I ignorant of my brothers' death, whom God relentlessly sought out and punished for their wicked schemes against you? Furthermore, what possible motive could I have for plotting against you? Hopes of the throne? But I already reigned! Concern that you hated me? But was I not loved by you? Did I have any other reason to fear you? No. For by protecting you I inspired fear in others. Did I lack money? No one had more money to spend than I. Even if I had been the most unworthy of men, with the heart of a cruel monster, would I not have been restored, father, by your generous benevolence to me? For as you have just said, you recalled me into your court and preferred me over the rest of your sons. You proclaimed me to be king while you still lived and by your many favors made me the envy of all others.

"It was that fateful journey to Rome! It afforded the perfect opportunity for envy. It provided a lengthy time for those who would plot against me! Yet, I took that journey for you, father, to fight your battles and to prevent Silleus from treating your old age with contempt. Rome stands witness to my sonly devotion. Even Caesar, the lord of the universe, has often called me "Philopater" — lover of my father. Read these letters from him, father. These are far more believable than the slanderous accusations raised here. These letters alone will vindicate me. They are proof of the love I have for you. Remember how reluctantly I made the trip to Rome, knowing the hatred for me that lurked within Judea. It was you, father, who unwittingly brought about my ruin, by giving my jealous enemies the opportunity to spread their lies. But I am here to face my accusers. Here I am, the 'parricide' who has traversed both land and sea and has escaped the dangers of them both. But this trial has trapped me, for it seems, father, that I stand condemned already, both before God and you. Since already condemned, I beg you not to rely upon the admissions of others made under the pain of torture. Let the same fire be lit to torment me! Let the instruments of torture rack my being! Don't listen to the cries for mercy from this polluted body. If I am truly a parricide, I should not die without being tortured first."

Antipater's emotional speech, interspersed with wailing and weeping, moved everyone, even Varus, to compassion for him. Herod alone did not cry but was furious at his son's tactics, knowing full well that the evidence against him was true.

4. Now Nicolas stood up and as directed by the king, addressed the assembly. He began by fully exposing Antipater's dishonesty, eroding the pity that his speech had provoked. He then introduced a terrible indictment against the king's son, laying at his feet all the wickedness in the kingdom, in particular the execution of his brothers. Nicolas demonstrated that their deaths were due to Antipater's slanderous tongue. He argued that Antipater had even plotted against those who had survived the brothers presuming them to be jockeying for power. He asked the court, "Would a man who was plotting to poison his father think it wrong to harm his brothers?" Moving along to the evidence regarding the plot to poison Herod, Nicolas brought out, in order, all of the facts obtained from the confessions. On the subject of Pheroras, Nicolas became furious. The idea that Antipater could persuade even him to murder his brother, and, therefore, by infecting his closest loved ones, fill the whole palace with his filth. When Nicolas had made many more accusations against Antipater, and had backed them up with proof, he sat down.

5. Varus then called on Antipater to speak in his own defense. But the

prostrate Antipater remained silent. He would only say, "As God is my witness, I am innocent." The governor then called for the poison and had a prisoner brought in who was under the sentence of death. When the man was forced to drink the poison, he dropped dead immediately. Then, after a private meeting with Herod, Varus wrote up his report of the trial for Caesar, leaving Jerusalem the following day. The king then arrested Antipater and sent messengers to the emperor to inform him of the tragedy.

6. It was later discovered that Antipater had plotted against Salome. A domestic servant of Antiphilus arrived with letters from Rome that were from Acme, a maidservant of Caesar's wife, Julia. Acme had written to Herod to tell him of some letters from Salome she had found among Julia's personal papers. She said she had forwarded them to the king out of her goodwill for him. The letters, which contained scathing denunciations of the king and brutal accusations against him, were forgeries of Antipater, who had also bribed Acme to send them to Herod. This evidence was ratified by another letter from Acme to Antipater that stated, "As you wished; I have written to your father and forwarded the letters, and feel certain that he will not spare his sister once he has read them. Be sure to remember what you promised me when all of this is over."

7. When this letter was discovered along with the letters forged against Salome, the thought crossed Herod's mind that perhaps all the letters against Alexander were forged. He was also deeply disturbed that he had almost executed his sister because of Antipater's maneuvering. He, therefore, determined to wait no longer to punish Antipater for his crimes. But as he proceeded to carry out justice against his son, Herod was struck down by a serious illness. However, he did manage to send Caesar all the facts about Acme and the fraud that had been perpetrated against Salome. He also sent for his will and changed it. He now appointed Antipas to be king, passing over his older sons, Archelaus and Philip, who had also suffered under Antipater's slanders. He bequeathed to Augustus Caesar, in addition to non-monetary items, the sum of 1,000 talents. Then he left the empress, their children, friends, and freedmen about 500 talents. To the other members of his own family Herod left huge parcels of land and large sums of money. He honored his sister Salome with the finest gifts of all. These are the alterations Herod made to his will.

Chapter 33

The Golden Eagle Is Cut to Pieces. About to Die, Herod still Practices Barbarity. He Attempts Suicide. He Commands Antipater's Execution. Herod Survives Him Five Days and then Dies.

1. Herod's illness grew steadily worse, aggravated by old age and his bouts with depression. He was now nearly 70 years old, and troubles with his children had broken his spirit. Even when he had enjoyed good health he took no pleasure in life. The knowledge that Antipater was still alive further exacerbated his disease. The king had resolved that his son's execution not be private, but in a formal, public manner following his own recovery.

2. Now, added to the weight of Herod's other troubles was a rebellion of some of his subjects. Two rabbis in Jerusalem were known as the finest experts in the laws of their country, and who, as a consequence, were held in extremely high regard throughout Judea. Their names were Judas ben Sepphoris, and Matthias ben Margalus. Many young men attended their lectures on the law. Day by day, their students grew into quite an army. Hearing that the king was slipping away under his disease and depression, these rabbis hinted to their friends that perhaps the time was right to come to God's defense and pull down the icons erected in defiance of their fathers' laws, for it was against the law to place anything in the temple—images or faces or the like—which represented any living creature. Ignoring this law, Herod had put up a golden eagle over the great gate of the temple. The rabbis urged their disciples to cut it down, and told them that if they got into trouble for doing so, it was a glorious thing to die protecting the laws of one's country. They said that the souls of those who died in this way obtained immortality and eternal happiness. Only uncouth and vulgar men, not schooled in philosophy, preferred death on a sick-bed to that of a virtuous hero.

3. While these two men were making speeches to their disciples, a rumor spread that the king was dying. The news aroused the young men to boldness. One day at noon several of them lowered themselves down from the top of the temple with ropes into the crowd that was milling about in the temple area. They then hacked down the golden eagle with axes. Their act was immediately reported to the captain of the temple who rushed to

the scene with a large group of soldiers. They arrested about forty of the young men and took them to the king. Herod first asked them whether they had cut down the golden eagle. They admitted that they had. Then he asked, "Who ordered you to do it?" They replied, "The law of our fathers." He then asked why they seemed so joyful, seeing that execution was their fate. They responded, "Because we will enjoy greater happiness when we are dead."

4. Their answers so infuriated the king that he momentarily forgot his illness and had himself carried outside to speak to the people. There he made a strong and lengthy denunciation of the young men as being guilty of sacrilege. He said that they had tried to cover their blasphemy by a pretended zeal for the law while they truly had a somewhat more ambitious motivation. Their punishment would be most severe. The people were afraid of wholesale prosecutions and urged the king to confine his anger to the ringleaders and those who had already been arrested, letting all others go free. Herod reluctantly agreed. Those who had let themselves down from the roof were burned alive together with the rabbis. The remainder of those arrested were delivered to the executioners.

5. From this point on Herod's disease spread to his entire body. A variety of symptoms appeared throughout his system. He had only a slight fever, but his skin itched unbearably over its entire surface. He had continual pain in his intestines, tumors on his swollen feet, and an inflamed abdomen. His private parts grew gangrenous and produced worms. Additionally, he had great difficulty breathing when lying down and had spasms and cramps in his arms and legs. Herod's conditions led some prophets to say that his maladies were in punishment for his treatment of the rabbis. Yet, Herod continued to struggle against his many illnesses. He still had the desire to live and this hope for recovery caused him to try several cures. For instance, he crossed the Jordan to bathe in the warm thermal springs at Callirrhoe, the waters of which descend southward into the Dead Sea but are sweet and pure enough to drink. While there, the king's physicians decided to bathe his entire body in warm oil and lowered him into a large vat filled with the liquid. As he sank into it, Herod fainted, and his eyes rolled back into his head as if he were dying. Such a cry went up from those in attendance that he woke up. But his hopes for recovery soon faded, and he gave orders to distribute 50 drachmas to each Judean

soldier and even larger sums of money to their officers and his friends.

6. Herod then began the journey back to Jerusalem. Stopping in Jericho, he teetered on the brink of death yet proceeded to devise a horrible plan. He assembled all of Judea's most distinguished men out of every town from one end of the country to the other and then ordered them locked inside a building called the Hippodrome. He then called for his sister, Salome, and her husband, Alexas, and said to them: "I know that the Jews will celebrate my death by holding a great feast. But I still can be mourned vicariously and have a magnificent funeral if you will but follow my instructions. Immediately upon my death, I want you to order the soldiers to surround these I have jailed in the Hippodrome and massacre them all. Then all Judea, every family, will weep at my death whether they want to or not."

7. While he was issuing these orders, a letter arrived from Herod's ambassadors in Rome informing him that the maidservant named Acme was executed at Caesar's order and that Antipater was condemned to die. They wrote, however, that Caesar would permit Herod to merely banish his son if that would be preferable. The news revived Herod somewhat, and he regained his desire to live. But soon, being subdued by his pain, lack of nourishment, and a convulsive cough, Herod tried to take his own life. He took an apple and asked for a knife, for his custom was to pare and slice apples before he ate them. After looking around for someone who might try to stop him, he lifted the knife up in his right hand as if to stab himself. His first cousin, Achiabus, rushed toward the king and caught his hand before it could strike. Instantly, a loud mournful cry arose in the palace as if the king had died. Antipater heard the sound and took heart. He joyfully called his jailers and attempted to bribe them to release his chains and let him go. But the head jailer not only refused Antipater's money, he ran and told the king of his son's plan. Hearing the jailer's report, Herod roared louder than it seemed possible for such a sick man. He immediately sent some of his guards into the prison and had Antipater executed. He then gave the order to have his son buried at Hyrcanium, and changed his will again. He made Archelaus, his eldest son and Antipas's brother, heir to the throne. Antipas was designated to be tetrarch.

8. Herod lived for five days following Antipater's execution. He died after

a reign of 34 years following Antigonus's death. But it had been 37 years since the Romans had appointed him king of Judea. He had been exceedingly fortunate in many respects. Only a commoner, he ascended to the throne and after keeping it for such a long time passed it on to his sons. But Herod was most unfortunate in his family affairs.

Before the soldiers in Jericho learned of the king's death, Salome and her husband went to the Hippodrome and released the prisoners whom Herod had ordered massacred. She told the guards that the king had changed his mind and wanted every man sent back to his own home. After the men had left, Salome told the soldiers of the king's death, summoning them and the rest of the citizens to assemble in Jericho's amphitheater. There, Ptolemy, who was entrusted with the king's signet ring, came forward and spoke to the throng, pronouncing a benediction on the dead Herod. After comforting them, he read a letter that the king had written to the soldiers urging them to be loyal to his successor. Finishing the letter, Ptolemy opened and read Herod's will. Philip was to rule Trachonitis and its neighboring districts while Antipas was to be tetrarch, as we mentioned earlier. Archelaus was to be crowned king, and was commanded to carry Herod's ring to Caesar together with all the documents of his reign. The documents were sealed up because Caesar was executor of Herod's will. Caesar was to confirm and ratify the dispositions of this will and the king's earlier bequests.

9. The citizens instantly greeted Archelaus with loud acclamations. Everyone congratulated him upon his ascension. The soldiers and all the people paraded before him in groups promising their loyalty and praying that God would bless his administration. Preparations then began for Herod's funeral. Archelaus omitted no detail that would contribute to its magnificence. He brought out all the royal jewels to accompany the procession honoring the deceased. The funeral bier, made of solid gold embroidered with precious stones, was covered in purple cloth adorned in multicolored embroidery work. On this lay Herod's body clothed in a purple robe. He had a coronet on his head covered by a golden crown, and the scepter was in his right hand. Surrounding the bier were Herod's sons and a large group of his relatives. Following them were the royal guards and the regiment of Thracians, the Germans and the Gauls, all dressed in full battle array. The remainder of the army led the procession, armed and marching in step with their captains and officers. After them, 500 of

Herod's domestic servants and freedmen followed, carrying sweet spices. The body was transported a distance of twenty-five miles to Herodium. Herod was thus buried in accordance with his instructions. So ended the reign of Herod the Great.

The Jewish Wars
Book II
From Herod's Death until Nero Sent Vespasian to Suppress the Jewish Uprising. Containing the Interval of 69 years

Chapter 1
Archelaus Feasts the People in Herod's Honor. A Great Tumult Is Raised Among the People, Archelaus Sends Troops and Kills about 3,000.

1. New disturbances now came in the wake of Archelaus's need to travel to Rome. He had mourned for his father, Herod, for seven days and had given a very expensive funeral feast in his honor. [This custom of the funeral feast caused poverty among many Jews because if one does not give it for a deceased loved one, he is not considered a holy person.] Archelaus put on a white robe and went up to the temple. There, the people cheered wildly for him. Seated on an elevated, golden throne, he spoke kindly to the many gathered there. He thanked the crowd for the zeal they had shown at his father's funeral and for their submission to him, as if he was already king. But he told the assembly that he would not at present take either the authority vested in a king, nor the titles involved, until Caesar had confirmed his succession. After all, only Caesar was capable of establishing his father's will. Archelaus said that when the soldiers stationed at Jericho tried to set the crown on his head, he would not accept it. He said he would apologize not only to the soldiers, but also to all the people for

their enthusiasm and good will toward him. When Rome had given him the full title to the throne, he promised to rule in all ways better than his father had.

2. The throng was pleased at all this, but then asked what he specifically meant by his words. Questions began to arise from those present. Some made noise about lower taxes. Others asked for his views on easing the duty on foreign commodities. Some requested that Archelaus free those held in prisons. To all of these subjects, Archelaus answered the crowd to their satisfaction in order to procure their good will. Following the meeting, Archelaus offered appropriate sacrifices and feasted with all of his friends. Once the time of mourning for Herod ended, many of those who desired innovations came in large numbers and began to mourn over their circumstances. They mourned over those whom Herod had put to death in the matter of the golden eagle that had stood guard over the temple gate. And this was no private grieving. Their cries were loud and their grief solemn, such that the weeping was heard all over the city for those who had died trying to uphold Hebrew law and for the sanctity of the temple. They loudly demanded that punishment should be inflicted on those who had killed these men and who had been subsequently honored by Herod. But primarily, they sought the termination of the high priest, and that someone of more piety and purity should be installed in the office.

3. Archelaus became provoked at these demands, but retrained himself from taking action due to his upcoming trip to Rome. He feared that attacking the crowd physically might detain him in Jerusalem. So Archelaus attempted to silence the innovators by persuasion rather than by force. He sent his general to them privately, who exhorted the crowd to be quiet. But the rebels threw stones at him and drove him into the temple even before he could speak. Other envoys sent by Archelaus were treated in the same ill-mannered way. As the crowd's numbers were considerable, it ultimately looked like they would not settle down. Indeed, the Feast of Unleavened Bread, also known to the Jews as Passover, was underway. In the feast, a huge number of sacrifices were made and innumerable people from the countryside streamed into Jerusalem to worship. Some of these stood in the temple bemoaning the rabbis that had been put to death. They supported themselves in the sedition by begging. At this, Archelaus

feared a riot and secretly sent a tribune with a cohort of soldiers to try to silence them before the entire multitude became infected with their insubordination. He gave orders to use whatever force was necessary to quiet the crowd. Seeing the Roman soldiers, the entire multitude became angry and threw stones at them, killing many of the soldiers. The wounded tribune barely escaped. The multitude then continued to offer sacrifices as if nothing had happened. At this point, because it looked to Archelaus that the crowd could not be restrained without bloodshed, he ordered his entire army to attack. Many infantry advanced through the city and cavalry came by way of the plain. They attacked the crowd as they were offering their sacrifices and killed about 3,000 of them. The rest of the throng was dispersed into the surrounding mountains. Archelaus's heralds followed the soldiers into the city, commanding everyone to return to his own home. So the multitude left the festival and went home.

Chapter 2

Archelaus, Accompanied by Many Relatives, Goes to Rome; There, He Is Accused by Antipater Before Caesar. Nicolaus Successfully Comes to His Defense.

1. Archelaus then headed for the (Mediterranean) sea with his mother and friends Poplas, Ptolemy, and Nicolaus. He left his brother Philip in the palace to administer local affairs in his absence. Salome and her sons also joined Archelaus, as well as the former king's nephews and son-in-law. Ostensibly, they all were travelling to support Archelaus' ascent to the throne of Israel, but secretly wanted to accuse him to Caesar of breaking the law in his prior actions at the temple.

2. But as they approached Caesarea, Sabinus, the Syrian procurator, met them. He was secretly on his way to Judea to manage Herod's personal estate, but was detained on his journey by Varus, Syria's governor. Archelaus's emissary, Ptolemy, had asked Varus to meet him. Meanwhile, Sabinus, in order to please Varus, hadn't gone up to the citadels in Jerusalem to prohibit Archelaus from getting at Herod's treasures. He promised that he would wait until Caesar was notified as to what was going on. So Sabinus holed up in Caesarea, and as soon as those who opposed him had gone, Varus went to Antioch while Archelaus sailed for Rome. He then left straight away for Jerusalem to take control of the palace. Sabinus sent

for the governors of the citadels and king Herod's private stewards, trying to locate Herod's treasures. He tried to take possession of the strongholds, but their commanders, remembering Archelaus's orders, held Sabinus off, arguing that the citadels belonged to Caesar.

3. Meanwhile, Antipas also sailed for Rome to present his own claim to the throne. He was to demand that his father Herod's first will—wherein he was named successor—was prior to the second will, and therefore, the valid document. Salome had vowed to back Antipas, as did several of Archelaus's own relatives, who had sailed with him. Antipas took his mother along, as well as Ptolemy, Nicolaus's brother. Ptolemy was thought to carry high authority because of Herod's complete confidence in him. He had been one of Herod's most trusted associates. But Antipas relied most heavily on Ireneus, a gifted public speaker, who had advised Antipas to reject the advice of some that he yield to Archelaus, for the reasons that Archelaus was his older brother and because the second will gave the throne to him. Archelaus's relatives also hated him and began showing their support for Antipas as soon as they had arrived in Rome. Nevertheless, all of these would prefer living under the rule of law as opposed to the rule of a king, and under subjection to Rome. If that didn't come to pass, then they preferred Antipas to Archelaus as their king.

4. Sabinus had sent with them (to Rome) letters accusing Archelaus while praising Antipas. Salome also wrote down the crimes of which they accused Archelaus and gave Sabinus's notes to Caesar. On the other hand, Archelaus also wrote down the reasons for his claim to the throne, and through Ptolemy, sent Caesar his father Herod's ring and personal letters. Caesar then gathered all of these allegations and weighed both the problems in Judea and the revenues generated there. He also considered the many offspring that Herod had sired and read all the letters and accounts from Varus and Sabinus. Caesar then assembled several high-ranking Romans including Agrippa's son, Caius, a man whom Caesar had also adopted as his own son. Caesar then brought in the various parties from Jerusalem and asked them to speak.

5. So Antipater, Salome's son, who of all Antipas's supporters was the best speaker, stood up to accuse Archelaus, making the following speech:

"Archelaus has argued for his succession to the throne, but he has long been exercising the authority of a king without Caesar's permission. In doing so, he has insulted Caesar to speak now in this assembly, since after Herod's death he has already seated himself on the throne without waiting for Caesar's approval. He has illegally bribed several people to place a crown on his head! He has also given commands only a king can give. He has changed the disposition of the army. He has granted high offices to some. He has also agreed to many requests of the people as their king. He has also freed many important prisoners put there by his father. Now, after all these things, he has the gall to ask you for the shadow of royal authority, the substance of which he has already seized! In doing so, Archelaus has lowered Caesar's power over real events to power only over mere words. Archelaus has also reproached Caesar further by only pretending to mourn for his father. Yes, he put on a sad face during the day, while drinking and partying late into the night. This behavior has led to many recent disturbances by those people who were ignorant of his duplicity."

The thrust of Antipater's whole speech was to elevate Antipas's crime in killing many of the throng gathered around the temple. He said that the crowd had come only for the festival but were wantonly murdered right in the middle of their sacrificial offerings. Antipater said that there were so many dead bodies piled upon each other in the temple that it looked as if it was an undefended surprise attack by a foreign invader, though they could not have slaughtered and then stacked that many corpses. He then added that Archelaus's father had foreseen this kind of barbarity in his son—Archelaus—and had subsequently refused to give him any hope of ascending to the throne. Antipas continued:

"But when Herod's mind began to fail, and he lost his rational capabilities and wasn't cognizant of this son's evil character, he wrote this second will appointing him king. At the time of his second will, Herod had no complaints against him who he had named in the first will. When Herod wrote that first will, he was sound in both body and mind. In order to make Archelaus king, one would need to think that Herod's judgment was better when he was ill than it was before his illness. Else, one would need to think that Archelaus has not forfeited his claim to the throne by his illegal actions. What sort of king will this man Archelaus make after Caesar has approved him, when he has murdered so many people even before he became king?"

6. After Antipater had said these things, he brought forth many of Archelaus's relatives as witnesses in order to prove every aspect of his accusations. After they had testified, Nicolaus rose to speak for Archelaus. He said that the temple massacre was unavoidable. Those slain were not only Archelaus's enemies, but also Caesar's enemies. He noted that those who have accused Archelaus actually encouraged him to commit more acts of which he could be accused. Nicolaus also insisted in upholding Herod's second will because Herod had made clear that the appointment of the new king should rest entirely with Caesar. He argued that the man, Herod, showed the prudence to yield the succession to the Judean throne to him who ruled over the entire world—Caesar. Herod cannot be presumed to err in judgment about the man whom he approved to be next in line when he knew the final decision would rest with a man who knew Archelaus well.

7. When Nicolaus had made his arguments, Archelaus silently fell on his knees before Caesar. Caesar then lifted him up and declared that Archelaus was worthy of succeeding his father. Yet, Caesar did not fully confirm him. So Caesar dismissed all in attendance so that he might think over the matter and the allegations that he had heard. He also considered whether either man named in Herod's two wills were fit to hold the office. Should the government in Judea be parsed out among all of Herod's sons, because so many stood in need of the revenues generated from Judean taxes?

Chapter 3

The Jews Fight a Battle Against Sabinus's Soldiers and the Temple Area Suffers Wide Destruction

1. Before Caesar had made his decision, Archelaus's mother, Malthace, became ill and died. Letters had also been received from Varus that spoke of a new Jewish revolt in Judea. Varus had foreseen these events, and following Archelaus's departure for Rome, had gone to Jerusalem to quash those who were leading the sedition. After it became obvious that the entire country was becoming unsettled, Varus left in the city a legion that had been with him in Syria, and then went to Antioch. After his departure, Sabinus arrived and compelled the citadel's forces to surrender. Sabinus then made an urgent search for the king's treasuries, using his own servants and the legion Varus left behind. He armed all of these men to help him in his covetous search.

When the feast of Pentecost [fifty days or seven weeks following the Passover] was held, the people gathered, but it was not for the worship of God, their normal custom. Because of the indignation at the events that had come to pass, a large multitude had come together from Galilee, Idumea, Jericho, and Perea, across the Jordan River. Men from Judea, who came eagerly and angrily with their teeth bared, joined them, more than doubling their numbers. They set up camp in three areas, distributing themselves among the three camps. One camp was north of the temple, another to the south next to the Hippodrome. The third encampment was to the west in the palace. Thus, they encircled the Romans and began a siege.

2. Sabinus was frightened of the multitude, both as to its size and its vehemence. He sent frequent messages to Varus asking for his quick assistance lest his legion be decimated. Sabinus then climbed to the highest point of the citadel, the Phasaelus Tower. The tower was named for Herod's brother who was killed by the Parthians. Sabinus then signaled the legion below to attack the encircling foe. He was terrified and refused to climb down to be with his men. Having their orders, the Roman soldiers charged the temple and flew into the Jewish encampment there. The Romans soon overwhelmed the Jews who proved no match for their military prowess. But several Jews climbed up to the top of the cloisters and began throwing their spears down on the Romans below. A number of Roman soldiers died in the assault from above. The survivors found hand-to-hand combat with the Jews difficult.

3. The Romans, because of this situation of great distress, set fire to the cloisters, the covered walkways around the temple that were of immense size and beauty. Those Jews who had climbed up above the Romans were caught in the flames and many were burned alive. The Romans who had reached their elevated position also succeeded in annihilating many. Some Jews hurled themselves backward of the high walls, and some cheated the fire by falling on their own swords. But many Jews climbed down and were slaughtered giving battle to the Roman soldiers. Other terrorized Jews succeeded in running away. The Romans then found God's treasure and stole about 400 gold talents, which was in addition to that which Sabinus had carried off.

4. The destruction of the Jewish forces and the temple surroundings enraged many other warlike Jews, encouraging them to gather in opposition to the Roman forces. They encircled the palace and threatened death to all of its occupants unless they fled. They promised that Sabinus would be given safe transport provided he accompanied his legion in retreat. Many of the king's retinue deserted the Romans and joined the Jews. But the men from Sebaste, who were the most warlike, stayed with the Romans. These included Rufus, Gratus, and their subordinate commanders. [Gratus commanded the king's infantry, and Rufus his cavalry.] Each of these men was strong and wise, and even if their troops had not followed them, they alone would have altered the outcome of the battle. But the Jews continued their siege. They attempted to destroy the fortress walls, crying out to Sabinus and those with him to leave Judea and allow the Jews to reestablish the freedom enjoyed by their ancestors. Sabinus was all too happy to remove himself from danger, but he didn't trust the Jews. This thought, together with his hope that Varus would arrive with support, caused Sabinus to continue to hold out against the Jewish siege.

Chapter 4

Herod's Veteran Soldiers Rise Up. Judas Leads a Pack of Thieves. Simon and Athrongeus Crown Themselves King

1. Many more outbreaks of war occurred widely in Judea at this time. The unrest provided an opportunity for many to proclaim themselves kings. In fact, in Idumea, 2,000 of Herod's soldiers joined ranks and, taking up arms, made war against those who had been Herod's supporters. Achaibus, first cousin of the king, joined in battle against these men. He chose for himself strongly fortified positions and tried to avoid fighting in the plains. In Sepphoris, a town in Galilee, a man named Judas, son of the highwayman Hezekias—who had been defeated by Herod—assembled a large army, and, breaking into the king's armory, provided his force with arms. He then attacked those who had so earnestly fought for power.

2. One of the king's servants, one Simon of Perea, a tall and fine looking man, crowned himself king. He gathered a group of bandits and set fire to the royal palace and other buildings in Jericho. He procured much loot by snatching it from the buildings he set on fire. Soon, all of the buildings in Jericho were reduced to rubble. The Trachonite archers, a very tough

outfit from Sebaste led by Gratus, a captain of the king's infantry, fought against Simon. Many of Gratus's soldiers were killed in the battle. As Simon was galloping through a narrow valley attempting to escape, Gratus cut him down by breaking his neck with a thrust of his sword. Also burned in the vast destruction were the royal estates near the Jordan River at Betharampha, as other seditions arose in Perea.

3. Athrongeus, a shepherd, also attempted to set himself up as king. He was an extremely strong and brave man who did not fear death. Athrongeus had four brothers of the same makeup. He gave his brothers each an armed troop of men and appointed them as commanders. In battle, Athrongeus rode like he was already king and only did those things that would befit his royal dignity. He had placed a crown upon his own head and ran through Judea raising havoc with his brothers. He killed both Romans and Herod's men, as well as every Jew he ran across, as if by doing so he could obtain material wealth. In one instance he tried to surround an entire cadre of Roman soldiers at Emmaus who were transporting corn and armaments to their legion. Athrongeus slew their centurion, Arius, and forty of his best men in a hail of darts and arrows. The rest of the Romans were spared a like fate by the arrival of Gratus and his troops from Sebaste. Ultimately, Athrongeus and his brothers were captured and their piracy ended. Archelaus took the older brother, while Gratus and Ptolemus took two other brothers. The fourth also surrendered to Archelaus after receiving assurance that he would not be executed.

Chapter 5

Varus Quells the Disturbances in Judea and Crucifies About 2,000 Rebels.

1. When general Varus received the letters from Sabinus and his own subordinates, he feared for the legion he had left in Ptolemais. So he hurried to relieve them, taking along two other legions with four troops of cavalry. Varus also ordered the auxiliary units of several city leaders to meet him in Ptolemais. Additionally, he took 1,500 armed men from the citizens of Berytus as he marched through their city. The auxiliary units arrived in Ptolemais together with the Arabians. Arestas, out of his hatred for Herod, came with a substantial armed force. Meanwhile, Varus sent part of his army to Galilee near Ptolemais. Leading them was Caius, one

of his friends. Attacking the city of Sepphoris, Caius put its defenders in full retreat, burnt the city and enslaved all of its surviving inhabitants.

Meanwhile, Varus and his army marched to Samaria. He didn't attack the city itself, as no uprising had occurred there, but encamped near the village named Arius. That town was Ptolemy's and because of that the Arabians had plundered it. Their anger at Herod extended to Herod's friends as well. Varus then moved out to the fortified village of Sampho that the Arabians had also pillaged. They had taken the city's public funds and left the place bloody and burning. Resistance against the Arabians was impossible. Emmaus was also torched following its evacuation. But this arson was done on Varus's command, enraged as he was with the massacre of people in those areas surrounding Arius.

2. Varus then went on to Jerusalem. When the Jews encamped around the city caught sight his forces they dispersed immediately, fleeing throughout Judea. But other Jews were happy to see Varus. They reported to him that they had had no part in the uprising there, in fact did not cause any trouble at all, but had only allowed those going to the feast to enter the city. Because of this one action, they had suffered the Jewish siege alongside the Romans and had never assisted the revolutionaries. Before all of this, Varus met Archelaus's first cousin, Joseph, along with Roman generals Gratus and Rufus. The latter led the troops of Sebaste as well as those of the king. Varus had also met with members of the typically armed Roman legion. But Sabinus kept a low profile, and fled from Jerusalem to the shore of the Mediterranean. Varus then sent a portion of his troops into Judea to attack the troublemakers. They captured many of these, jailing those who had only played small parts in the rebellion, but crucifying about 2,000 of the more guilty.

3. Varus then received intelligence that 10,000 armed men remained in Idumea. He also began to understand that the Arabians had not proven to be worthy allies. They had waged war only to suit their own passions and lusts for money. They had done things Varus never intended—and those things only because of their hatred for Herod. So he sent them all back to Arabia, and then focused his army against the revolt in Idumea. Before actual fighting began there, and upon Achiabus's advice, the rebels surrendered. Varus forgave the majority of any offense, but their leaders were carted off to Caesar for interrogation. Caesar ultimately forgave them,

but then gave the order to execute some of Herod's own family members, because they had waged war against some of their own kinsmen. Thus, when Varus had settled the situation in Jerusalem, he returned to Antioch, leaving his former legion to garrison the city.

Chapter 6

In Rome, the Jews Forcefully Complain of Archelaus and Ask for Roman Governors. Caesar Listened, then Allocated Herod's Kingdom Among His Sons According to His Own Pleasure.

1. The Jews again brought accusations against Archelaus while he was in Rome. These came from those representatives who arrived in Rome with Varus's permission to plead for Judean freedom. Over fifty of them were present, representing over 8,000 Jews in Rome. Then Caesar gathered a council of high-ranking Roman citizens in Apollo's temple, the expensive palace he had built and furnished. A great many Jews stood on one side with their representatives, while Archelaus and his friends stood on the other side. The Jews could not situate themselves on the same side with Archelaus because of their hatred and envy. At the same time, they didn't want Caesar to see them standing with those who accused Archelaus. Also present was Archelaus's brother Philip, whom Varus had sent to Rome for two reasons: first, that he might be of some assistance to his brother, and second, that should Caesar distribute the kingdom among the brothers Philip might receive his share of it.

2. When those accusing Archelaus' were permitted to speak, they first told Caesar of Herod's lawless record. They said that Herod really wasn't a king but the worst of tyrants, such was the suffering that he caused. So many were put to death by Herod, they complained, that those who remained alive were so miserable as to admit the dead were happier than they. They said that Herod had imposed physical torture upon his own people, even upon entire communities, many of which were in Judea. He shed Jewish blood merely to satisfy foreigners. Further, Herod had impoverished his people, taking from them their joy and the ancient laws by which they had lived. In sum, the Jews had suffered more under Herod's short reign than in all the years since the return from Babylon during Artaxerxes's

rule. The nation had reached such a low estate, enduring such hardships, that they had submitted to Herod's son Archelaus as king, though it also meant odious enslavement.

The people had joined in mourning Herod's death and had wished Archelaus success as the new king. But Archelaus thought that unless he continued in his father's hateful legacy he might not be thought of as Herod's actual son. Accordingly, he began his rule by murdering 3,000 Jews. It's almost as if he wanted to fill the temple with human corpses, not unlike the many bloody animal sacrifices during the feast. So the people remaining in misery began to think about the calamities they had suffered, and decided that rather than take the lashes across their backs as before, they would face the whipping head on. So they prayed that Rome would have compassion upon those remaining in Judea and not expose them once again to that that before had ripped them to shreds. They would rather join with the Syrians in running Judea's government by their own officials. Then it would become clear that even those tyrants and rebels who love fighting would acquiesce to a more tolerant government. With this statement, the Jews finished making their plea.

Nicolaus then rose to refute the Jewish accusations against Herod and Archelaus. He argued that the Jews had brought all this trouble upon themselves as unruly subjects, unwilling to be obedient to the established power. He then censured Archelaus' family members who had sided with the Jews against their brother.

3. So, after he had heard each position, Caesar recessed the gathering. After a few days, he divided Herod's kingdom into two parts. Half he gave to Archelaus as ethnarch, promising that he would be given the title "king" if he were found worthy of it. The remaining half Caesar divided into two tetrarchies under Herod's other sons. Philip would rule one, and the other ruled by Antipas. Perca and Galilee fell under Antipas, bringing with them an annual revenue of 200 gold talents. Philip would be tetrarch of the surrounding Batanea, Trachonitis, and Auranitis, plus portions of Zeno's estate around Jamnia, with an annual income of 100 talents. Caesar then placed Idumea and all of Judea plus Samaria under Archelaus' ethnarchy. Samaria received a reduction in taxes of twenty five percent inasmuch as it had taken no part in the revolt by the other Jews. Additionally, Archelaus was to rule Strato's Tower and Sebaste, Joppa, and Jerusalem. The Grecian cities of Gaza, Gadara, and Hippos were lopped

off of Herod's former kingdom and given to Syria. Archelaus's total income from his ethnarchy was to be 400 talents.

Caesar then made Salome mistress of Jamnia, Ashdod, and Phasaelis, in addition to what Herod's will had bequeathed her. Caesar also gave her the royal palace of Ascalon, from which she was to receive the annual sum of 60 talents. But Caesar placed all of her holdings under Archelaus's ethnarchy. The remainder of Herod's offspring was allowed to keep what Herod had given them in his two wills. But in addition, Caesar gave Herod's unmarried daughters 500,000 silver drachmas, and betrothed them to Pheroras's sons. After all this, Caesar distributed among Herod's children the sum that Herod had willed to them—1,000 gold talents. Caesar only kept a few of Herod's minor presents to himself, in honor of the deceased.

Chapter 7

The History of the Pseudo-Alexander. Archelaus Is Banished, and Glaphyra Dies After Both of Their Tragedies had been Predicted in Dreams.

1. While all of this was taking place, a man who though born a Jew was raised by a Roman freedman began to pretend that he was Alexander, the son whom Herod executed. The man resembled Alexander in his physical appearance. He went secretly to Rome with a Jewish associate who was wise in the ways of the kingdom. This assistant told the pseudo-Alexander to say that those sent to assassinate him and Aristobulus instead took pity on them and hid them. The would-be murderers then put bodies of similar appearance in their place. These two men fooled the Jews in Crete into giving them a small fortune with which they sailed to Melos in luxury. Once there, the pretender was also thought to be the true Alexander and was given even more money. He even convinced his benefactors to sail to Rome with him. They landed at Dicearchia [Puteoli] and received even more valuable gifts from the Jews living there. Treated as if he were royalty by his father's friends, the man looked so much like Alexander that even those who had known the real Alexander vowed that he was one and the same. The whole Jewish population of Rome ran out to see this man. A crowd beyond number lined the streets where the people who had arrived from Melos carried him in a sedan chair. The people continued to treat him royally at their own expense.

2. But Caesar knew precisely what Alexander looked like because Herod had once accused him in Caesar's presence. He knew that this man was not Alexander even before he saw him. Nevertheless, Caesar understood that the fame and adulation surrounding the imposter carried weight with the masses. So, he sent Celadus, who had known Alexander well, ordering him to bring the young man to him. When Caesar laid eyes on the man, he noticed a difference in his features. His body was well built and muscular like that of a slave. He then was certain that the whole thing was a hoax and was provoked to anger by the man's impudence. When questioned about Aristobulus, the pseudo-Alexander said that Aristobulus had been left on Cyprus on purpose in fear of further attempts on his life. He reasoned that it would be more difficult to plot against them and overpower them if they separated. Then Caesar took the man aside and spoke privately to him.

"I will give you your life if you tell me whom it was who persuaded you to come up with such fabrications," he said. So the young man agreed, and following Caesar, pointed out the Jew who had used him for personal gain. He had accumulated more wealth in the cities than Alexander owned during his entire life. Caesar laughed at the hoax and sentenced the young man to the galleys since he was obviously quite strong. But he ordered the execution of the hoax's mastermind. The people of Melos had been fully punished for their part in it by the great expense the fiasco had cost them.

3. So Archelaus began his reign as ethnarch and began to mistreat not only the Jews but also the Samaritans, due to his previous quarrels with them. Because of this, both parties sent emissaries to Caesar to speak against him. Subsequently, in the ninth year of his rule, Archelaus was banished to Vienna, a city in Gaul, and his wealth was transferred to Caesar. But it is reported that before Caesar sent for him, Archelaus had a dream in which he saw nine ears of corn, lush and full, being eaten by oxen. He sent for his seers and a few Chaldeans and asked what they thought the dream meant. One prophet had one interpretation, while another had a different interpretation. Simon, a man of the Essene sect, thought that the ears of corn represented years, while the oxen represented a cessation of years. Therefore, Simon said Archelaus would reign as many years as there were ears of corn, and after he had passed through certain alter-

ations of fortune, he would die. Archelaus was called to stand trial before Caesar five days after hearing Simon's interpretation.

4. I believe Glaphyra's vision is also worth recording here. Glaphyra was the daughter of Archelaus, king of Cappadocia. She had been married to tetrarch Archelaus's brother, Alexander, who has been the subject of our recent discussion. Alexander was, of course, king Herod's son and also his murder victim. After Alexander's death, Glaphyra married Juba, king of Libya, but on his death she returned home to Judea and lived with her father as a widow. It was at that time that the ethnarch, Archelaus, met her and fell so deeply in love with her that he divorced his wife, Miriamne, and married Glaphyra. So when Glaphyra came into Judea and had lived there a short time, she had a vision of her former husband, Alexander, standing next to her. He said, "Your marriage to Libya's king should have been enough for you, but you weren't content with marrying again. Now you have returned to my family, you impudent woman, and have chosen to wed my brother. But I shall soon have you again, whether you want me or not." Glaphyra only lived two more days after seeing this vision.

Chapter 8

Archelaus's Ethnarchy Is Reduced to a Roman Province. The Galilean Judas's Sedition. The Various Jewish Sects—the Essenes, Pharisees, and Sadducees.

1. Following his exile, Archelaus's former ethnarchy was reduced to a province. Coponious, a Roman of the equestrian order, was made procurator and given the power of life and death by Caesar. In his administration lived a man named Judas who led many citizens to revolt. He said that they were cowards to pay taxes to Rome and to submit to mortal men rather than to God. Judas had formed his own peculiar sect and stood apart from other sect leaders at the time.

2. Now, there are three philosophical sects among the Jews. The first are called Pharisees, the second Sadducees, and the third, which claims to be the most severe, is called the Essenes. These Essenes are native Jews and have more love for one another than do the two other sects. The Essenes reject pleasure as evil, rather esteeming charity and suppression of the passions to be the highest virtues. They reject marriage, instead adopting the

children of others who are pliable and capable of being educated. They esteem them as their very own children and raise them up to be Essenes. They don't really oppose marriage, seeing that without it the human race could not continue, but they want to guard against the lustful behavior of women. They are persuaded that no woman keeps her vows to only one man.

3. The Essenes despise wealth and we admire their speaking against its accumulation. None has any more worldly goods than any other for they observe a law that anyone who joins their sect must put what he owns into the common pot. Because of this, none among them appear to be either impoverished or rich. Everyone's possessions belong to everyone else, so there is, as it were, one patrimony among the brothers. The Essenes also believe that oil defiles a person. If any of them is anointed without his consent the oil is wiped off of his body. They think perspiration to be a good thing, as well as to be clothed all in white. They also have stewards appointed to do menial tasks, not serving individuals, but serving them all.

4. The Essenes live in no particular town but may live in any town. If an Essene comes to live in another town they enjoy what their brothers have just as if it were their very own. They meet those they never met before as though they had known them for a long time. For this reason, when they travel, the Essenes carry nothing with them, fearing thieves. So in every city where the sect is present, one among them is appointed to provide clothes and other necessities for these travellers. They treat their bodies like children who are in fear of their parents. They don't allow one of their number to change clothes of shoes until they have worn out on their own over time. Neither do the Essenes buy anything nor sell anything to each other. If someone is in need the others simply supply what is required. Even though no recompense need be made each Essene may take what they want from whomever they please.

5. The Essene piety toward God is extraordinary. Before sunrise they never speak about worldly matters, but pray only certain prayers they learned from their forefathers, as if asking for the sun to rise. Then each Essene is set to work by their curator to exercise their skill in the arts, laboring with great diligence until the fifth hour. After that, they all meet together and

when they have dressed in white veils, they bathe in cold water. When this rite of purification ends they all go into a common building into which no other sect may enter. They then enter a dining room as unto a holy temple and quietly sit down. The baker sets loaves of bread in front of them and the cook brings each a serving of one type of food, setting it down before them. A priest then says grace for it is unlawful for anyone to eat before grace is said. This same priest, when finished eating, says grace again afterward. So, when they begin to eat and after they have finished, they praise God, who gives them food to eat. Then, rising, they set aside their white apparel and go back to work until evening. Then they return home for supper in the same way described. If any strangers come, they sit down with them. There is never any disturbance or loud speech that would pollute the house. Everyone speaks in turn. The silence of their houses is a real mystery to foreigners. It's due mainly to their continuing sobriety and the feeling that the food and drink allotted to them is more than sufficient for their needs.

6. Truly, the Essenes do all things according to the commands of their leaders. However, two things everyone does of his own free will: to come to the aid of anyone who needs it, and to show mercy. They are allowed free reign to come to the assistance of any deserving of it, and in need of it. Also, they give food to those who are hungry. However, they are not allowed to give anything to their own family without the approval of their leaders. Further, they justly restrain their anger and subdue their other passions. They are extremely faithful and ministers of peace. Whatever they say may be trusted more than if they added a vow. They esteem any shading of the truth to be worse than perjury in a court of law. They say that anyone who cannot be believed without swearing an oath to God already stands condemned. The Essenes are also very careful in their study of ancient writings, taking from them anything of advantage to both body and soul. They also seek such organic and mineral substances that may provide relief from illness.

7. But if anyone tries to join their sect, they don't admit him immediately. For a year an initiate must live as they live while remaining excluded from the rest. They give the initiate the previously mentioned belt and a white garment. When he has shown that during the first year he can observe their chastity and he draws closer to their way of life, the newcomer is

allowed to partake of their waters of purification, yet must still live apart from the rest. Following this one year demonstration of his fortitude the initiate's temper is tested for two more years and if appearing worthy at that time he is admitted to their common society. But before being allowed to eat with the rest he is obligated to take very serious oaths. First, he must vow to exercise piety toward God. Second, that he will treat all men justly. Third, that he will not harm anyone, either on his own initiative, or on the command of others. Fourth, that he will hate wicked men while giving the righteous his assistance. Fifth, that he will be faithful to all men, particularly to those in authority over him. For it is God who establishes government. Sixth, if he should find himself in authority he will at no time abuse such authority nor seek to raise himself above his subjects, either in the clothes he wears or in other fine ornamentation. Seventh, that he will always love the truth and admonish those who lie. Eighth, that he will keep his hands from stealing and his soul from illegal profit. Ninth, that he will hide nothing from his brother Essenes nor reveal any of their doctrines to those outside the sect, not even in peril of his own life. Moreover, he must swear that he will not communicate the Essene doctrines to anyone but those from whom he has received them. Tenth, that he will seek to preserve the Essene library and not take books from it nor divulge the means of the angel messengers. (The Essenes are believed to worship angels. See Paul's statement in Colossians 2:18.) These are the vows by which initiates were able to become members of the Essene sect.

8. If any member is found to have committed heinous sin, he is cast out of the sect. Those removed from the sect often die in a most grievous way. Since he is still bound by his vows and by the customs of the sect, he is also bound from all other food outside the sect. He is therefore forced to eat grass and to starve to death. For this reason, many who abandon the sect are received back again out of kindness in the throes of starvation. It is thought that the misery they have suffered, even unto the very brink of death, is sufficient punishment for their sins.

9. The Essenes are very just and fair in their judgment. Their courts are comprised of no less than 100 men. Once a judgment is rendered, it cannot be overturned. The name they honor most — save that of God himself—is the name of their founder, Moses. if anyone is found guilty of blaspheming Moses, he is executed. The Essenes also desire to obey

their elders and the majority. Accordingly, if ten are sitting together, one will not argue his case if the remaining nine do not agree with him. They also avoid spitting while around others either in the center of them or to the right-hand side. Moreover, they are more strict than the other Jewish sects in their obedience to the seventh day. They won't even start a fire on the Sabbath. Neither will they take a pot from its place and defecate in it. On other days they dig a small hole a foot deep with a small shovel that was given them when they joined the sect. Then they wrap their garment around themselves so as not to offend the divine rays of light. They then ease themselves down over the pit. Finishing, they scoop the dirt that was dug out back into the hole. They only do this in remote places which they choose for this purpose. Even though bodily functions are natural, they nevertheless have a rule to wash afterward, as if they had become defiled.

10. After an initiate's time of trial is over the new sect members are divided into four classifications. The junior members are considered far inferior to those who are seniors. In fact, if one of the seniors is even touched by a junior he must wash himself, as if he had been defiled by a foreigner's touch. The Essenes tend to live long lives some even more than 100 years because of their simple diet, and also by their way of life. They abhor life's miseries, overcoming bodily pain by the strength of their minds. They believe in life after death, that it will be glorious and much better than life in this world. The war with Rome gave much evidence as to how seriously the Essenes suffered. Although they were tortured and maimed, burnt alive and ripped to pieces, and underwent all sorts of instruments of torment to get them to blaspheme Moses, or to eat what was forbidden them, they could not be forced to do either of these things. Indeed, not once did they try to appease their tormentors nor did they shed tears. Instead, they smiled through the pain, laughing at and scorning those who inflicted the torments upon them. They offered their souls up to God with joy, expecting to receive them back again.

11. For the Essene doctrine is this: Bodies are corruptible and impermanent but the soul is immortal and lives eternally. The soul is ethereal and united to a bodily prison into which they are drawn by a natural enticement. But when is set free from its fleshly bonds, released from a lengthy bondage, it rejoices and flies away. This is close to the Grecian view that good souls live beyond the oceans in regions oppressed by nei-

ther rainstorms nor snowstorms nor intense heat. Such a place is refreshed by a gentle, westerly wind, perpetually blowing from the ocean. But evil souls live in dark, tempestuous caverns, full of never ending punishments. Indeed, it seems to me that Greeks have the same idea. They place their brave men on blessed islands—men whom they call heroes and demigods. But the souls of wicked, ungodly people go to Hades. Greek myths tell of people such as Sisyphus, Tantalus, Ixion, and Tityus are punished in such a place. They build their view on the presupposition that the soul is immortal. Therefore, exhortations toward virtue and from wickedness cause good men to conduct their lives even better, because of their hope in reward after death. On the other hand wickedness is restrained in evil men by their fear and the expectation that punishment of their evil deeds. Even though the evil may be hidden in this life it will be revealed and punished after their death. These are the divine doctrines of the Essenes regarding the soul, which are very appealing to those who have been made aware of them.

12. Among the Essenes were prophets who foretell the future. They read the holy books, use several types of purification, and are continually involved in the prophetic writings. Very rarely will they miss in their predictions.

13. Further, another order of the Essenes exists. This order agrees with their way of living, customs, and laws, but differ in their view of marriage. They believe that by not marrying, one cuts off an important part of human life—the prospect of children. They believe that if all men thought like the unmarried Essenes the entire human race would come to an end. However, they live with their spouses for three years. If they have three children during that time then they actually marry them. But they are careful not to have relations with their wives while the woman is pregnant. They do this to demonstrate that they didn't marry for pleasure, but for the sake of children. Women also bathe with their clothes on, as do the men with something around them. These are the Essene customs.

14. Now as to the other Jewish orders I mentioned: the Pharisees are those who are esteemed skillful in explaining Jewish law. They were first to be founded. These men believe in God's sovereignty over the affairs of men but attribute actions, whether good or bad, to man's power. These two

things, (sovereignty and free will), work together. The Pharisees believe that man's soul is incorruptible but that only the souls of good men who die are placed into new bodies. The souls of evil men are subject to eternal punishment.

The other order is that of the Sadducees. They don't believe in complete sovereignty, denying that God is even interested in men's deeds whether they are good or evil. They state that do good or evil is entirely a matter of human choice. Each man is free to act as he pleases. The Sadducees also deny the immortality of man's soul as well as any punishment or reward in Hades. Another difference is that the Pharisees are friendly to one another but the Sadducees are somewhat wild toward each other. Their relations within their own sect are so infelicitous, it's as if they are strangers. All this I mention regarding the three philosophic Jewish sects.

Chapter 9

Salome Dies. The Cities that Herod and Philip Built. Pontius Pilate Creates Disturbances. Tiberius Places Agrippa in Chains, but Caius Frees Him and Makes Him King. Herod Antipas Is Banished.

1. Now, Archelaus's ethnarchy had been reduced to a Roman province, and Herod's other sons—Philip and Antipas—administered their own tetrarchies. When Salome died, she bequeathed to (Caesar) Augustus's wife, Julia, her toparchy, Jamnia, and her palm tree plantation in Phasaelis. When the Roman government was transferred to Julia's son, Tiberius, following Augustus' death, both Herod Antipas and Philip continued to rule their tetrarchies. Augustus had reigned in Rome for fifty-seven years, six months, and two days. Philip built Caesarea at the origin of the Jordan River in the Paneas region. He also built Julias, a city in lower Gaulonitis. Herod Antipas constructed Tiberias in Galilee and another city named Julias in Perea, beyond the Jordan River.

2. Pontius Pilate, whom Tiberius sent into Judea, brought into Jerusalem by night images of Caesar called "Ensigns." These ensigns caused a great furor among the Jews the next day. Those Jews that passed near the ensigns were astonished at the sight of them, as further proof that Jewish law was being undermined. Their laws do not permit any images in Jerusalem. Not only was the Jerusalem citizenry in an uproar, but a vast number

from the countryside came running into the city. Displaying great zeal, they flew to Pilate in Caesarea and begged him to take the ensigns out of the holy city and uphold Jewish law as inviolable. When Pilate denied their request, they fell flat on their faces and didn't move for five days and nights.

3. The next day, Pilate sat at his tribunal in the marketplace and called the many Jews to him. He said that he wanted to answer their request but at the same time he called his armed soldiers to encircle the crowd. A cadre of soldiers with their swords surrounded the crowd three deep. The Jews, of course, were very concerned at seeing this unexpected development. Pilate then told them that unless they willingly accepted Caesar's images he would cut them all to pieces. He then signaled the soldiers to draw their swords. On hearing this order, the Jews fell down as one man, the multitude exposing their bare necks, and shouted that they would prefer death than have their sacred laws broken. Pilate was taken by surprise at this rigorous display of their superstition and ordered that the ensigns be removed from Jerusalem.

4. Following their removal, Pilate raised another disturbance by using the Jews' sacred treasure, called "Corban," to build aqueducts, through which he brought water into Caesarea from a distance of fifty miles. Again, the Jews were indignant at this action. So when Pilate came to Jerusalem, they gathered at his tribunal and raised an uproar. Warned in advance of the probable disturbance, Pilate ordered his troops, camouflaged in garments worn by private citizens, to mix in with the crowd of Jews. They were not to use their swords but instead to beat the rioters with clubs. Pilate then gave the signal for the surprise attack from his tribunal. The Jews were terribly beaten by the Roman soldiers. Many died by the beating, and many more were stomped to death. The carnage so amazed the rest of the crowd that they held their peace.

5. Meanwhile, Agrippa, the son of Aristobulus—who was murdered by his father, Herod the Great—came to Tiberius Caesar to make accusations against Herod Antipas. Before he brought formal charges, he stayed in Rome, making friends with several high-ranking men, but primarily becoming friends with Caius, the son of Germanicus, a private citizen. Agrippa held a feast for Caius as part of his friendship with him.

During that time, Agrippa stretched out his hand toward his friend Caius and exclaimed that he hoped Caius might be proclaimed emperor of the world. Tiberius later heard of Agrippa's outburst from one of the Agrippa's servants and became extremely angry. He placed Agrippa in chains and threw him in prison for six months where he was treated very severely. Tiberius died (from suffocation) after serving Rome for twenty-two years, six months, and three days (in 37 AD).

6. When Caius became Caesar, he released Agrippa and made him king of Philip's tetrarchy. Philip had died in the interim. When Antipas heard of this, it inflamed his lust for power. But his wife, Herodias, nagged and reproached her husband as slothful. She said it was because he was lazy that he had lost Philip's tetrarchy. For since Caesar had now taken a private person, Agrippa, and made him a king, how much more would Caesar have given the tetrarchy to Antipas, already a king! Her arguments began to make sense to Antipas, so he sailed to Rome to visit Caius. But Caius, through Agrippa's accusations, saw through Antipas's ambition and banished the tetrarch to Spain where he ultimately died with Herodias by his side. Meanwhile, Agrippa was given Antipas's tetrarchy, in addition to his own.

Chapter 10

Caius Commands that His Statue Be Set Up in the Temple, and What Petronius Did To Help the Jews.

1. Caius Caesar (aka Caligula) began to grossly abuse the office he had risen into. He believed himself to be a god and wanted everyone else to think of him as such. Caius threw many of the highest nobility out of Italy. Even the Jews felt his abuse. He sent Petronius to Jerusalem with an army to place statues of himself in the temple, ordering him to kill any Jews who opposed the statues and take the rest of the nation into captivity. God was not happy with Caius's orders. Nevertheless, Petronius marched from Antioch into Judea with three legions and many Syrian auxiliaries. Many Jews could not believe the rumors of a new war and those who did believe were in great distress. Terror shot through the nation as people sought to defend themselves, since Petronius's army had already arrived in Ptolemais.

2. Ptolemais is a Galilean maritime city built in the great plains and surrounded by mountains. The Galilean range extends about seven and one-half miles east of Ptolemais. The mountains to the south are about fifteen miles away in the Carmel region. Twelve and one-half miles north of Ptolemais are the highest mountains of them all, called "The Tyrianian Ladder" by the citizenry. A small river named the Belus runs about a quarter of a mile from Ptolemais. Near the river is a marvelous place called "Memmon's Monument." It is a round, hollow hole that collects sand out of which men make glass. Emptied by the many ships that fill their holds with its sand, the structure is filled again with common sand carried along from afar by the wind. In the hole, the common sand becomes sand for glass. But if removed from the cavity the sand becomes common sand again. This is a description of the area around Ptolemais.

3. But now large numbers of Jews assembled themselves in the plain surrounding Ptolemais with their wives and children. They pleaded with Petronius, first for protection of their sacred laws, then also for themselves. So great was their number, and so vociferous their requests, that Petronius left to meet other Jews in Galilee, leaving his army and the statues of Caius in Ptolemais. He then called the multitude and their leaders to Tiberias. There, he told them of Caesar's threats and of his military might to enforce those threats. He also called their requests unreasonable. After all, every other nation subjected to Rome had placed Caesar's images in their cities right alongside their own gods. The Jews alone opposed Caesar's statues and, therefore, it seemed to him like the action of revolutionaries risen up to do violence to Caesar.

4. But the Jews continued to insist that both their ancient laws and their customs did not permit them to make an image of either God or man. These images were not allowed anywhere, much less in the temple itself. To this Petronius replied, "Am I not also required to keep the laws of my own country? If I disobey Rome and spare you, isn't it right that I should die? Then, he who sent me here will declare war against you, even though I'm dead." Hearing this statement, the entire multitude cried out that they were ready to suffer for their law. Then Petronius, silencing the crowd, said, "Will you then declare war upon Caesar?" The Jews responded, "We offer sacrifices for Caesar twice a day and also for the citizens of Rome!" But, they continued, if Caesar would place images in their midst,

he would have to sacrifice the entire Jewish nation. They were ready to lie down and be slaughtered, even together with their wives and children. Petronius was amazed at this and pitied them all because of their intense religious fervor, and because they were willing to die for what they believed. He then dismissed the crowd having failed to sway it.

5. So in the days ahead, Petronius privately assembled the more prominent and powerful Jews. He also called the crowd to come together publicly. He threatened them again with Rome's power and Caius's anger, as well as reiterating the charge that Caesar had placed upon him and for which he was responsible. But since the multitude would not listen to reason, and because it was spring planting time [the crowd had lingered for fifty days without sowing their seed], Petronius called the Jews together once more. He told the Jews that he was willing to put himself at risk for them. He said, "Either God will help me to change Caesar's mind, and we both shall escape Caesar's wrath, or Caesar will continue to rage against you. But I am prepared to die for such a great multitude as you." Petronius then dismissed the throng, and the people prayed fervently for his success. Then Petronius marched his army out of Ptolemais and returned to Antioch.

From there, he sent a message to Caesar, telling him of the great commotion the images had created in Judea and of the people's pleas. Unless Caesar was prepared to lose both the country and all of its inhabitants, he must be prepared for the Jews to honor their law and countermand his previous order. Caius answered Petronius with great anger. He threatened Petronius with death for taking so long to implement Caesar's orders. But it so happened that the messengers who brought Caius's answer were caught in a storm at sea and were delayed for three months. Meanwhile, others enjoying safe passage from Rome brought news that Caius had died. Accordingly, the news of Caius's death arrived into Petronius' hands twenty-seven days before the message with the death threat.

Chapter 11

Concerning Claudius's Government and the Reign of Agrippa. Also Concerning the Deaths of Agrippa and Herod and the Children They Left Behind.

1. Caius (Caligula) had reigned three years and three months before he

was assassinated (in 41 AD). The Roman armies took Claudius and hurried him to Rome. But the Roman senate, led by consuls Sentius Saturninus and Pomponius Secundus, ordered the three regiments with them to preserve order in Rome. They then went to Rome with a great multitude to oppose Claudius by force. Because of Caius Caesar's onerous past treatment they wanted to set up a government of the aristocracy, as had once held power, or at least to vote for someone worthy of being emperor.

2. It so happened that Agrippa was staying in Rome at the time, so the senate asked him to consult with them. At the same time, Claudius also sent for Agrippa from his encampment, figuring Agrippa may be of some service should the need arise. So Agrippa, under the assumption that Claudius had already been made Caesar, went to meet with him. Claudius subsequently sent Agrippa to the senate to let them know of his intentions. He said that, firstly, he didn't order his soldiers to spirit him away to Rome. They had done it because of their zeal for him. He felt it right not to disappoint them, for if he did so, his own future might become uncertain. It was a dangerous thing to be called to serve as emperor. He further added that he intended to administer the government as a good prince and not as a tyrant. He would be satisfied only with the honor of the title— emperor—and coveted the senate's advice. Although he was not a moderate by nature, Claudius said that Caius's death served to demonstrate that once in the office, he should act with sober judgment.

3. Agrippa delivered the message, to which the senate responded, saying that since they also had an army and the wisest counselors as well, they would not volunteer to be slaves. When Claudius got their response he sent Agrippa to them again. Claudius stated that he could not bear the thought of betraying those who had vowed to be faithful to him, and so, though he did not want a war, if it came to that the battle should take place outside the city. It did not serve one's sense of piety to pollute Rome's temples with the blood of Romans, and particularly since it was due to the senate's unwise behavior. Agrippa delivered Claudius's message to the senate.

4. Meanwhile, one of the senate's soldiers drew his sword (while in their chambers) and cried out, "O, my fellow soldiers, why should we choose to kill our brothers—our own family—because they stand with Claudi-

us? Why should we not have as our emperor one who is without fault? He has many reasons to lay claim to the office, particularly among those of us who wage war!" This said, the soldier marched through the senate along with the rest of the soldiers there. The senate, judging that they had no way to turn to get out of the situation, hurried after the soldiers and proceeded to go to Claudius. But the men who had long backed Claudius met them at the city walls with drawn swords. There was certainly reason to fear the danger even though Claudius had not heard of his soldiers' actions. Unless he restrained the violent intentions of those arrayed against the senate he would lose the support of those over whom he wanted to reign. Instead of a populous country he would rule a sparse desert.

5. When Claudius heard Agrippa's report he ordered his soldiers to restrain from any violence, and then received the senators within his camp. He treated them well, and after some time went out with them to submit thank-offerings to God, which were proper with his being seated as the new emperor. Further, he then gave Agrippa the entire kingdom that his father had ruled and, in addition to the countries given Herod by Augustus, he added Trachonitis and Auranitis, and the kingdom of Lysanius. He decreed this all before the people, and ordered magistrates to have the order engraved on brass plaques and hung in the capitol. Claudius also bestowed on Agrippa's brother, Herod, the kingdom of Chalsis. Herod was Agrippa's son-in-law, as he was married to Agrippa's daughter, Bernice.

6. So now Agrippa became very wealthy by ruling such a large kingdom. He didn't waste his new found wealth on small projects but proceeded to encircle Jerusalem with such a large wall that when finished, it would be impervious to even a Roman siege. But Agrippa died in Caesarea before completing the wall. He had reigned three years, the same length of time he had ruled his other tetrarchies. Agrippa left behind him three daughters, born to him by Cypros, Bernice, and Miriamne, as well as a son named Agrippa, by Drusilla. Since the boy was very young, Claudius reduced the kingdom to a Roman province and sent Cuspius Fadus to be its procurator. Following Cuspius, Claudius sent Tiberius Alexander, who in honoring the Jews' ancient laws kept Judea peaceful. After this, Herod king of Chalsis also died, leaving two sons born to him by Agrippa's daughter, Bernice, named Bernie Janus and Hyrcanus. Herod had also sired Aristobulus by his former wife, Miriamne. A brother of Herod,

a private citizen also named Aristobulus, left behind a daughter named Jotape. These, as I have already said, were the grandchildren of Herod the Great's son Aristobulus, who with his brother, Alexander, were given birth by Miriamne and murdered by their father. Meanwhile, Alexander's progeny reigned in Armenia.

Chapter 12

More Trouble Under Cumanus that Was Begun By Quadratus. Felix Is Appointed Procurator of Judea. Agrippa Is Promoted from Chalsis to a Larger Kingdom.

1. After Herod king of Chalsis died, Claudius set Agrippa's son, Agrippa, on the throne of his uncle, Herod (the Great). Cumanus was named procurator of the rest of the Roman province. He succeeded Alexander under whom Cureanus had begun the troubles that were to bring ruination to the Jews. When the great crowds had come to Jerusalem for the feast of unleavened bread, a Roman cohort (480 soldiers) stood guard over the temple cloisters. [These men were always armed, keeping guard at such festivals to prevent any uprising the assembled crowds might foment.] One of their number proceeded to pull up his kilt and in an indecent way showed his (uncircumcised) private parts to the Jews, speaking in a way one might expect from such a posture. At this, the large crowd became inflamed, shouting to Cumanus to punish the man. The younger Jewish crowd, naturally given to rashness, began to throw stones at the soldiers who guarded the temple cloisters. This made Cumanus afraid that the entire crowd might attack him, so he sent for reinforcements, many of whom arrived on the scene shortly. The Jews then became greatly alarmed. They were beaten back out of the temple area, running pell-mell back into the city. Trying to get out of the enclosed space, they ran with such violence that many, squeezed together, were trampled under foot. Ten thousand Jews lost their lives. The festival, a cause for joy, became a source of mourning for the entire nation. Every family mourned for relatives who had suffered death.

2. Another catastrophe followed this one. It arose from a fracas caused by thieves on the public road at Beth-boron. A man named Stephen, one of Caesar's servants, was carrying some furniture when he was attacked by these thieves and taken prisoner. Hearing of the incident, Cureanus sent

some troops to go to the surrounding villages and bring back their inhabitants in chains. The charge against them was that they had not tried to go after the thieves, nor capture them. At this point, one of the Roman soldiers found a sacred book of Jewish law. He tore the book apart and proceeded to throw it, piece by piece, into the fire. Seeing this, the Jews went mad. It was as if the whole nation were set on fire. In their zeal for their religion they gathered a large assembly and ran as one man to Cumanus in Caesarea, demanding that the soldier had insulted both God and Jewish law, and that he be punished for his actions. Cumanus saw that the crowd would not be quieted unless he gave them what they wanted. So he ordered that the soldier be brought to him and dragged through the crowd of Jews who lusted for his punishment. Thus the man was executed. When this had taken place, the Jews left and went their way.

3. Soon thereafter, a fight broke out between the Galileans and the Samaritans at a place called Geman on the great plain in Samaria. While a great many Jews were traveling to Jerusalem for the Feast of Tabernacles, a certain Galilean was murdered. This act caused many Jews to run out of Galilee to fight the Samaritans. Even more Jews among them went to Cumanus and urged him to stop the battle before it became too enlarged. They insisted that he go to Galilee and punish the murderers. But Cumanus was busy with other matters so he sent them away without succeeding in their purpose.

4. When the murder became known in Jerusalem it enraged those at the festival, many of whom summarily left the feast. Without any generals to lead them, the angry crowd marched to Samaria. Shunning the magistrates set over them, they were led by a man named Eleazar ben Dineus, and by one Alexander. Both men were thieves and reactionaries. The crowd soon fell upon villages near the Acrabatene toparchy, killing sick people, as well as young and old Samaritans alike, setting fire to their homes.

5. Being informed of this, Cumanus led a troop of cavalry from Sebaste out of Caesarea to help those who had been attacked. He captured many of the men who followed Eleazar, executing many of them. Then many Jews ran out of Jerusalem clothed in sackcloth with ashes on their heads and tried to beg those who had gone off to fight the Samaritans to cease

and desist lest they provoke the Romans into attacking Jerusalem. They reasoned that rather than try to avenge themselves over the life of one Galilean, they should rather have compassion for their nation and their temple, their wives and their children, and not bring the danger of ultimate destruction upon them all. The Jews agreed with this reasoning and the angry crowd dispersed. Still, many continued to rob and plunder, hoping not to be punished. Rape and more disturbances took place over the entire country. The Samaritan leaders went to Tyre to see Syria's president, Ummidus Quadratus, asking that the Jews be disciplined. The leaders of the Jews, along with Jonathan—the son of Ananus, the high priest—also went to Tyre. They said that the Samaritans started the trouble in committing the murder. Cumanus had been told what had happened, but was unwilling to execute the murderers.

6. Quadratus put both parties off presently, saying that when he had the opportunity to travel to where the events took place, he would diligently inquire about the circumstances. Then, he went first to Caesarea with all of those whom Cumanus had captured alive. He then traveled to the city of Lydda to hear the Samaritan complaints, whereupon he sent for eighteen of the Jews who had attacked Samaria and had them beheaded. Two of their leaders, however, Jonathan and Ananias, the high priests, along with Ananias' son, Artanus, among other eminent Jews among them, he sent to Caesar. He did the same with the most superior of the Samaritans. Quadratus also ordered that Cureanus, the procurator, and Celer, the tribune, should also sail for Rome to tell Caesar what had happened. When the Syrian president finished all of these actions, he went up from Lydda to Jerusalem. Finding that the feast of unleavened bread was continuing without trouble, he returned to Antioch.

7. When Caesar in Rome had heard what Cumanus and the Samaritans said, he condemned the Samaritans, commanding that three of the most powerful among them be executed. He also banished Celer to Jerusalem in chains, to be tortured by the Jews then dragged around the city and beheaded. [It should be noted that Caesar Claudius's hearing took place with Agrippa in attendance. He zealously argued for the Jews, as did the other ranking men who stood by Cumanus.]

8. After this, Caesar sent Pallas's brother, Felix, to be procurator of Gal-

ilee, Samaria, and Perea. He promoted Agrippa from the kingdom of Chalsis to a greater office, giving him Philip's old tetrarchy, in which the cities Batanza, Trachonitis, and Gaulonitis stood. Further, Caesar added to Agrippa's rule the kingdom of Lysanius and the province of Abilene, which Varus had governed. But after reigning thirteen years, eight months and twenty days, Claudius died (of poisoning) and Nero became emperor (in 54 AD).Claudius had previously adopted Nero following his wife Agrippina's deceptions, even though he had a son of his own, Brittanicus, by his former wife Messalina. Claudius also sired two daughters—Octavia, whom he had wed to Nero, and Antonia, who he had by Petina.

Chapter 13

Nero Adds Four Cities to Agrippa's Kingdom; the Remaining Parts of Judea Were under Felix. The Disturbances Raised by the Sicarii, the Magicians, and an Egyptian False Prophet. The Jews and Syrians Fight at Caesarea.

1. Nero acted like a madman in many ways, much so because of the wealth he enjoyed had made him carefree. He used his good fortune to harm others. He murdered his brother and his wife, as well as his own mother. His barbarity also spread to other close relatives. Finally, being insanely motivated, he became a stage actor. I won't say any more about these things since others have reported them elsewhere. Rather, I shall turn only to those of Nero's deeds that concerned the Jews.

2. Nero gave the kingdom of Lesser Armenia to Herod's son, Aristobulos, and he added four cities to the toparchies belonging to Agrippa's kingdom. These were Abila, Julias in Perea, Taricha, and Tiberias of Galilee. He made Felix procurator over the remainder of Judea. A thief named Eleazar, who had ravaged the countryside with his henchmen for twenty years, was caught by Felix and sent to Rome with some of his men. Numerous others of Felix's followers were also caught by Felix and crucified.

3. When Judea had been purged of these robbers, another sort of thief sprang up in Jerusalem. They were called Sicarii, who murdered men in broad daylight within the city walls. They operated primarily at festivals, mingling with the crowds carrying concealed daggers. When they had stabbed one of their enemies and the man had fallen down dead, they

joined with the crowd in indignation at the murder. By this means they were able to avoid discovery. The first man murdered by the Sicarii was the high priest, Jonathan. Following that, many more were slain each day in the same way. The fear became very great in the city, even beyond the fear that the murders themselves should have caused. Men began to expect death at any instant, as do men in battle, so they were predisposed to look far ahead of themselves in the crowds, taking notice if any enemy might be seen in the distance. Many even began to distrust their friends if they noticed them nearby. But even in the midst of their fear, their suspicions and their caution, many felt the dagger. Such was the speed of the Sicarii and their cunning methods.

4. Another group of wicked men also came along. These men, although not as murderous in their actions, were nevertheless more wicked in their motives. They did more damage to the city's happiness than the murderers. These men deluded and deceived the people under the pretense of having Divine guidance. In reality, they had it in their minds to incite revolution. They caused the crowds in Jerusalem to act like madmen, leading huge numbers into the wilderness under the pretense that once there, God would give them signs leading to their liberation. Felix believed this to be the beginning of a general revolt, and therefore sent both armed infantry and cavalry to slay a great many of them.

5. Even greater was the damage done the Jews by an Egyptian false prophet. He was a cheat pretending to be God's prophet. He gathered around him 30,000 men he had deluded and led them to the Mount of Olives, where he prepared to force his way into Jerusalem. If he could conquer the Roman garrison and the people, he planned to place the city under his control using the men who had joined him. But Felix met the Egyptian with Roman power. When the battle began, the Egyptian and a few others snuck away, while most of his foolish followers were either killed or captured. The rest of his pseudo-army escaped and hid in their homes.

6. Now when all this commotion had quieted down, more difficulties arose. Like a diseased body, as one part is healed, another part becomes inflamed. A company of conmen and thieves assembled and persuaded the Jews to revolt, exhorting them to assert their freedom. They murdered many Jews who continued to obey Rome, saying that those who willingly

became slaves ought to be forcibly swayed from such inclinations. Their method of operation was to split up into different units and lay in hiding throughout the country. They then plundered and murdered the wealthy and set whole villages on fire, until all Judea felt the effects of their madness. The flame grew every day until it became a conflagration.

7. Another disturbance was to occur at Caesarea. The Jews who lived among the (Greek) Syrians there began to rise up against their neighbors. They moved under the pretense that they owned the city, since king Herod had built it. The Syrians agreed that the builder had been a Jew but they insisted that the city was Grecian because its statues and temples were not of Jewish design. This exchange increased contention between the Jews and the Syrians, which, as it escalated, resulted in a call to arms. The boldest among the sides marched out to fight. The older Jews were unable to stop those of their people who had become so violent. The Greeks thought it would be shameful to be beaten by Jews. The Jews were more wealthy and in better physical shape than their adversaries, but the Greeks had the advantage of being allied with the Roman garrison from Syria, which, being comprised of Syrian soldiers, stood ready to help them. On the other hand, the city fathers were concerned with keeping the peace. Wherever they found someone on either side pushing for a fight they flogged them and put them in irons. But this proved fruitless, for those who remained grew more and more exasperated and sought to deepen the conflict. Then Felix appeared in the marketplace in Caesarea commanding that the Jews go home, having been victorious over the Syrians. He threatened them if they disobeyed him, and disobey they did. So he ordered his soldiers to attack and a great many Jews were slaughtered. The Roman soldiers then stripped the dead of their possessions and the uprising continued. So Felix chose important men from each side of the struggle to be ambassadors to Nero, to present to the emperor their opposing viewpoints.

Chapter 14

Festus Succeeds Felix, Who Is Succeeded by Albinus then Florus. Florus's Barbarity Forces the Jews into the War.

1. Now it came to pass that Festus succeeded Felix as procurator. Festus made it his business to subdue the seditious men who had brought about

such trouble to the country. He captured most of the thieves and executed many of them. But Festus was soon succeeded by Albinus, a very wicked Roman who did not administer his office as did Festus. He used his high office to plunder the wealth of the people, burdening them with exorbitant taxation. Not only that, he permitted the relatives of the thieves in prison to bail them out. It made no difference whether they had been placed in prison by former procurators or mere city functionaries. No one remained in prison that had the ability to pay Albinus money. Accordingly, the Jews who fomented sedition against the Romans were many. The central source of the new troubles lay in those Albinus had released, as many like-minded people joined them. Every one of these wicked wretches was surrounded with a band of thieves. Albinus was like an arch-thief or a tyrant who abused his authority and plundered peace-loving citizens. The effect of all this was that the unfortunate souls who lost everything had to hold their tongues, when they had every reason to be angry and cry out. On the other hand, those who escaped theft had to remain quiet lest they be reduced to poverty like those who had suffered so greatly. So, every one remained silent and Albinus's tyranny was generally tolerated. But the seeds were being sown that would bring destruction upon Jerusalem.

2. Such was the wicked character of Albinus, who was succeeded by Gessius Florus. Florus made Albinus look like an angel in comparison. For Albinus did his evil works in private with lies and subterfuge, but Florus committed evil outwardly and with arrogance. It was as though he had been sent by Rome merely to execute people. Every sort of rape and injury was brought upon the Jews. In cases where pity would have been just, Florus was the most barbaric. He showed utter defiance in the face of unimaginable sin. He had no equal in disguising the truth. No one could have devised more subtle ways of deceit than did Florus. He considered it a petty offense to steal money from private citizens. Doing so, he ruined whole cities and entire populations all at once. In acting in such a way, Florus boasted throughout the country that anyone who wanted to become a thief could do so with impunity on the premise that they share their booty with him. Not surprisingly, his greed proved to be the ruination of entire toparchies that were made desolate during his reign. Many people left the country and fled to foreign lands.

3. While Cestius Gallus ruled in Syria as its president, no one went to

him to complain about Florus. But when Gallus came to Jerusalem as the Feast of Unleavened Bread approached, no fewer than three million people surrounded him and begged him to help stem the misery that had come upon their country. They shouted that Florus was a curse upon Judea. But Florus, who was present beside Cestius Gallus, laughed at the crowd. Finally, when Gallus had silenced the huge gathering, he promised the people that Florus would thereafter treat them more gently. He then returned to Antioch. Florus traveled with Cestius as far as Caesarea and lied to him. At that very moment, Florus was preparing to display his anger toward the nation by declaring war upon them. He did so with the purpose of hiding his extreme wickedness. He thought that if peace were to continue, the Jews might accuse him to Caesar. But if he could entice them to revolution he could hide his lesser crimes against them by bringing upon them so much greater misery. Therefore, he added more calamities upon the Jews daily, attempting to induce general rebellion.

4. Now it happened that the Grecians in Caesarea had been given the upper hand there, including the administration and judiciary of the city. They came down hard on the Jews living there. This was at the same time the war began, in Nero's twelfth year as Caesar, and Agrippa's seventeenth year, in the month of Artemisius [Jyar] (May 66 AD). The events that caused the war were by no means in proportion to the awful catastrophe the war brought down upon the Jews. Some Jews who lived in Caesarea worshipped at a synagogue owned by a Greek Caesarean. They had tried on many occasions to purchase the property, offering well above its market value. But the owner consistently dismissed their offers. He wanted to build other structures on the site as an affront to them. These workshops would leave the Jews only a narrow ingress into their synagogue. Access was therefore severely limited to their place of worship. Some of the younger, more passionate Jews went to the builders to try to get them to halt construction. The builders, of course, refused. The use of force was not an option, as Florus had forbade it. The Jewish leaders there were in great distress, so with Jonathan the publican as their spokesman, they went to Florus and persuaded him with a bribe of eight gold talents to stop the work. Florus, intent as he was on accumulating money, promised the Jews that he would do exactly what they asked of him. He then left Caesarea for Sebaste, leaving the matter to take its course. It was as if he had licensed the Jews to fight it out with the Greeks.

5. Now, on the following day, which was the seventh day of the week when the Jews were all crowding toward their synagogue, this incident occurred: A certain citizen of Caesarea, looking to cause trouble, took a clay pot and turning it upside down began to sacrifice birds. This was an affront to the Jews and their laws and an extreme provocation, as the synagogue was now polluted. The older and more moderate Jews thought it best to seek recourse through proper government channels. But the younger Jews were inflamed by the action and ready to fight. Also ready to make trouble were some Gentile troublemakers in Caesarea who had covertly sent the man to offer the sacrifices. So the fight began. Jucundus, cavalry leader, who had been ordered to prevent the violence, went to the synagogue and took the clay pot away, attempting to do as he was ordered. But he became overwhelmed by the Caesareans. The Jews then grabbed their law books and fled to Narbata, about seven miles from Caesarea, to a place they owned. John the publican and twelve of the older Jews returned to Florus and complained bitterly about the situation. They again asked the procurator to help them. They reminded Florus of the eight talents they had previously given him. (The eight Roman talents weighed approximately 568 pounds or just over 9,000 ounces.) But Florus turned on the Jews and had them arrested and put in prison. He accused them of taking the books of their law out of Caesarea.

6. Meanwhile, the citizens of Jerusalem became more distressed about these events but restrained themselves from any action. Florus took the opportunity to blow the flame of war into a conflagration. He sent for 17 talents (1,207 lbs) to be taken from the Jews' sacred treasure, saying that Caesar wanted them. Hearing news of this, the Jews were stunned and ran as one man to the temple, crying wildly as they ran. They called upon Caesar to free them from Florus's tyranny. Some of the more rebellious cried out against Florus, reproaching him vociferously. Some men took a basket and went begging for alms for Florus as if he was destitute and miserably poor. But Florus, far from being ashamed of his lust for money, became more enraged and was provoked to take even more. Instead of going to Caesarea to extinguish the uproar there, Florus marched quickly to Jerusalem, bringing the might of Rome in order to get what he wanted. He believed that by his threats and by terror he could subjugate the city.

7. But the people hoped Florus would be ashamed of his threats. They planned to meet his soldiers with applause and cheers, in order to appear to receive Florus submissively. But prior to their ruse, Florus sent the centurion, Capito, with fifty soldiers to drive them back. This made the crowd unable to show themselves as welcoming to the man they had so vehemently denounced earlier. Florus said that the Jews, if they were truly courageous and wanted to speak openly with him—to laugh at him to his face and show their love of liberty—they shouldn't just use words but speak with their swords and spears as well. The people were amazed at this statement. Then when Capito's cavalry unit burst into their midst they had to disperse before meeting Florus and their plan to show their submissiveness to him. Each Jew went to his own house and spent the night in fear and confusion.

8. So Florus then went to his quarters in the palace. The next day he took a seat in his tribunal there. The high priests and the more powerful and prominent men of the city came before him. He advised them that if the criminals who had begun the disturbance were not brought to him, they too would suffer his vengeance. But the high-ranking Jews argued that the people were peaceably disposed. They begged the procurator to forgive those who had spoken wrongly. It was impossible in such a large multitude to distinguish them from the rest. Everyone was sorry for what had been done, and denied any part in it for fear of what might follow. They asked Florus to bring peace to the nation and to take such advice that would preserve Jerusalem for the Romans. Rather, for the sake of many innocent people the guilty should be forgiven, and the majority not be harmed for the actions of only a few.

9. Florus became even angrier when he heard this. He called on his soldiers to wreak havoc in the Upper Market-place, killing all who were there. So the soldiers, agreeing that this order would also help them in their search for treasure, went beyond the place where Florus had ordered them to go, and began forcing themselves into houses, murdering the occupants. The citizens fled along the city's narrow lanes; slaughtered in their tracks if caught by the soldiers. Every method of pillage and murder were used. Many innocent citizens were brought before Florus, whom he first abused with the whip and then crucified. The total number slain that day, including wives and children, [they didn't even spare infants],

was 3,600. What made the catastrophe even worse was a new device of Roman brutality. Florus did what no Roman governor had ever done. He took men of the equestrian order, born as Jews but of Roman dignity, and had them nailed to crosses in front of his tribunal.

Chapter 15

Bernice Pleads in Vain with Florus to Spare the Jews. How the Flames of War Were Quenched and then Rekindled by Florus.

1. About this time, Agrippa went to Alexandria to congratulate Alexander on obtaining the government of Egypt from Nero. Agrippa's sister, Bernice (Bern-EE-cay), came to Jerusalem and witnessed first hand the wicked behavior of the Roman soldiers. Deeply offended by it, she sent the captains of her cavalry units and her guards to Florus, begging him to stop the slaughter. But Florus would not agree with her request, nor had he any respect for her high office. Florus continued to value only the advantages that his plundering had gotten him. As the Roman soldiers' madness broke out again in dreadful violence, Bernice was caught in the midst of it. They tortured and slew those caught before her very eyes. Indeed, they would have murdered Bernice also, had she not found refuge in the palace, spending the night surrounded by her guards for fear of the soldiers' wrath. She stayed in Jerusalem to fulfill a vow she had made to God. Now, it is normal for those who have been afflicted with illness or other injuries to make vows. For thirty days they are to make sacrifices, abstain from wine, and shave their heads. These things Bernice did. She now stood barefoot before Florus's tribunal and begged him to spare the Jews. He paid no attention. In fact, she believed that she too was in danger of being executed.

2. This all took place on the 16th of Artemisius (June, 66 AD). The next day, the agonized Jewish crowds ran together to the Upper Market-place and cried out at the top of their lungs for those who had perished. The loudest of their lamentations were directed toward Florus. The more prominent Jews, including the high priests, were frightened at the crowd's outburst. They fell at the throng's feet and begged them to quit lest they provoke Florus to some new unrelenting bloodbath beyond that which they had already suffered. The crowd complied immediately in reverence of the men who had asked them, hoping that Florus would not come after

them again.

3. Florus was disappointed when the disturbance stopped. So he tried to rekindle the flame sending for the high priests along with others of influence. He told them that the only way he would believe that the people would not make further disturbances such as these was if they were to go out and meet the two cohorts of troops who were ascending into Jerusalem from Caesarea. While the distinguished Jews were encouraging the crowds to do so, Florus sent a message to the cohorts' centurions that none of their men were to return the Jews' greetings. If the Jews subsequently responded in a way that seemed aggressive, the soldiers should make use of their weapons upon them. Now the high priests then gathered the multitude in the temple and asked them to go out to the Roman cohorts saluting them in a civil manner before there was no other way out of the situation. But the more rebellious within the crowd would not agree to do this. On the contrary, remembering those who had been slaughtered earlier led them to act more contentiously.

4. At this time, all of the priests and servants of God brought out the holy vessels. They were wearing the ornamental garments in which they administered their sacred duties. The harpists also came out bearing their musical instruments. They, together with the hymn singers, fell on their faces before the multitude, begging them to preserve their holy ornamentation and not to provoke the Romans into stealing the sacred treasures. The high priests had sprinkled copious quantities of dust on their heads. Their garments had been torn to shreds. They implored by name all of the powerful men among them, and the throng surrounding them, not to betray their country by even one small offense, allowing it to be laid waste by those who wanted to do exactly that. They said, "What good will it do to for the soldiers to receive greetings from the Jews? What change will it bring if you don't go out to meet them? If you salute the cohorts in a civil way, won't that prevent Florus from starting a war? Won't you save your country and be free from all future suffering? Besides, isn't being swayed by a few warlike men among you a sign of your feeble leadership abilities? It would be much better if you forced these to act more sedately."

5. By the force of these words, which were said both to the crowd and to the seditious men among them, some were restrained by the arguments

and some out of the respect that had been paid them. So, composed and silent, the crowd went out to meet the soldiers. When they had come upon the cohorts, the crowd saluted them. But when no corresponding greetings came from the soldiers, the more seditious men among the Jews shouted out against Florus. This outburst was the centurions' signal to start the attack. The soldiers surrounded the throng, striking them with their clubs. Those trying to flee were trampled by the cavalry horses. A great many fell dead by Roman swords, and many more died crushed under the feet of their own people. At the gates to the city, the Jews crushed one another as they tried to cram into the small opening. Those who lost their feet were suffocated or stomped to pieces by those Jews still standing. The carnage was such that many who died could not be recognized by their relatives for proper burial. The soldiers also beat those they caught with clubs, showing no mercy. The Romans pushed the crowd through the rising ground surrounding the pool at Bethesda and farther on to the temple. Their object was to seize the Antonia tower, as Florus also wanted to take possession of these places. He brought his guards out of the king's palace and ordered them to go as far as the Antonia citadel. But the Jewish mass turned back upon him and stopped his advance. Many Jews stood on housetops throwing darts at the Romans, who took many casualties from the missiles raining down upon them. Finally, because the Romans could not get through the crowd that had sealed the narrow streets, they retired to their encampment at the palace.

6. But the rebellious Jews feared that Florus would make another attempt to take possession of the temple through Antonia. So they climbed up to the temple cloisters that joined Antonia and cut the cloisters down. This move cooled Florus's lust for a time. He was eager to get at God's treasury in the temple by going through the Antonia tower, but when the cloisters fell, he stopped his attempt. He then sent for the high priests and the Sanhedrin, telling them that he was withdrawing from Jerusalem but that he would leave them as large a garrison as they desired. They promised not to begin any more trouble as long as he would leave them one garrison, but not comprised of any men who had just fought against the Jews. The crowd still bore enmity against that unit because of what they had suffered. So Florus gave the Jews the garrison they had requested and with the rest of the Roman troops returned to Caesarea.

Chapter 16

Cestius Sends the Tribune Neopolitanus to Jerusalem to Observe. Agrippa Speaks to the Jews, Attempting to Divert their Intentions of Making War Upon Rome.

1. Meanwhile, Florus contrived another plot to obligate the Jews to begin a war. He sent Cestius a false accusation that the Jews had revolted against the Roman government. Florus complained that they had begun hostilities, lying that the Jews had been the culprits when, in reality, they were the only people to suffer. But the prominent Jerusalem Jews did not keep silent. They too wrote to Cestius, as did Bernice, protesting about the unlawful practices against the city of which Florus had been guilty. Upon reading both messages, Cestius consulted with his subordinates about a course of action. Some believed that Cestius should march to Jerusalem with his army. There, he could either punish the uprising if it were real. On the other hand, if Florus was wrong, Cestius could put Roman affairs there on a more sure foundation if the Jews would only continue peacefully. But Cestius thought it best to send one of his close associates to analyze the state of affairs in the city and return with an honest account of Jewish intentions. So the president sent one of his tribunes, Neopolitanus. The tribune met with king Agrippa at Jamnia as he was returning from Alexandria, introducing himself and describing his mission in Jerusalem.

2. The high priests in Jerusalem, along with the Sanhedrin and the powerful men of the city, came to congratulate Agrippa on his safe return. After they had paid their respects and told him of their own troubles, they complained of Florus's barbaric treatment. Agrippa was indignant at what he heard but then subtly transferred his anger toward the Jews. He pitied them but wanted to moderate the exalted thoughts they had of themselves. He wanted to believe that they had not been treated as badly as they said, so as to restrain them from taking vengeance. These great men of the Jewish nation, with more wisdom and wealth than the rest of their countrymen, understood that Agrippa intended his rebuke for their welfare. Meanwhile, many people came from as far as seven and a half miles from around Jerusalem to congratulate both Agrippa and Neopolitanus. But before these arrived, the widows of the men slain in the latest crisis came mourning and lamenting their husbands' deaths.

Hearing their weeping, the people joined in and begged Agrippa to help them. They also cried out to the tribune and complained of the many miseries suffered under Florus. They showed the king and the tribune the desolate city marketplace and the many houses plundered by the Romans. They persuaded Neopolitanus, through Agrippa, to walk around the city accompanied only by a servant—to wander from place to place to see for himself the manner in which the Jews submitted to the rest of the Romans. It was only Florus that they hated because of his excessive barbarity against them. So Neopolitanus walked around and observed the people's good disposition. Then he went up to the temple where he gathered the multitude around him and commended them highly for their faithfulness to Rome, earnestly appealing to them to continue to keep the peace. Then, having performed such aspects of Divine worship permitted him, Neopolitanus returned to Cestius in Antioch.

3. The throng of Jews then addressed Agrippa and the high priests, asking them to send ambassadors to Nero to speak against Florus. They were concerned that their silence might lead Rome to believe that they were guilty of the slaughter and wanted war. If they didn't show Nero whose fault the conflict was, he might think they started it. The Jews said that if Nero denied them ambassadors, they would not be silent, but speak openly about the matter. But Agrippa thought it would be dangerous to select men to go to Rome to accuse Florus. On the other hand, he didn't want to deny their request, as the assembled Jews were obviously of a disposition to fight. Therefore, he called the multitude into a large gallery. He placed his sister, Bernice, in the house of the Asamoneans in view of the crowd. (She probably stood on the balcony.) The house overlooked the gallery at the passage to the upper city where a bridge joined the temple to the gallery. The balcony was visible from the gallery below it.

4. Then Agrippa began to speak: "Had I thought that you all zealously wanted to go to war against Rome and that the more pure and sincere people among you wanted the same, I would not have come out to you to speak. Nor would I have been so bold as to give you counsel. For all speeches that try to persuade men to any course of action are worthless when those who hear are predisposed to do the contrary. But some of you want war because you are young and inexperienced in war's misery. Others among you are for a fight because of the unreasonable expectation

of regaining your liberty. Still others are earnestly bent on war because in the confusion that warfare affords they might plunder those who are too weak to resist them. Therefore, I thought it proper to assemble all of you and say to you what I think would accrue to your advantage.

"Perhaps those who desire war will gain wisdom and change their minds. Hopefully, the best men among you will not be injured by the evil conduct of some others. Let not any among you condemn me in case what I say is unpleasant for you. But my words will in no case be remembered, even by those who agree with me, unless you will all remain silent. I am well aware that many have made tragic lamentations about the harm coming from your procurator's hand and about the glorious advantage of regaining your liberty. Before I inquire among you as to who among you want war, and speak to you about who it is that you must fight, I must first correct some misconceptions that some have harbored.

"If you insist on avenging yourselves on those who have harmed you, why do you still pretend that this war is to be for the cause of liberty? But if you believe that all servitude is intolerable, what purpose would your complaint against your particular governor serve? For if they treated you moderately, servitude would still be an unworthy thing. If you consider the several reasons we might use to justify fighting, there is precious little to justify war. Your first reason is the accusations you have made against your procurator. Well, you ought to be submissive to those in authority over you and not provoke them in any way. But when you vociferously accuse men of small offenses you excite the men you accuse to rise against you. Their aggravation may make them stop hurting you privately and to a lesser degree, yet lay waste to you all publicly. Nothing reduces the force of lashes as to bear them with patience. When those harmed bear their affliction in silence, it keeps those who inflict the harm from further doing so.

"But let us agree that the Roman ministers have severely and unjustly harmed you. Yet not all Romans have injured you, not even Caesar against whom you would declare war. Caesar does not intend to send wicked men to govern you. They in the west cannot see for themselves those in the east. Nor is it easy for them to even hear what's going on in these parts of the world.

"Now it is absurd to make war against a great many people for the sake of only one people and particularly to do so against a strong people for such a little cause. And it is even more absurd when the other side doesn't even understand your complaints against them. Further, the crimes you speak of may be corrected soon. The same procurator will not rule forever. His successor may arrive with more moderate inclinations. But as for war, once it has begun it is not easily stopped nor waged without its usual disasters. However, as to your desire of recovering your liberty, you are unreasonably late. Your forebears ought to have worked hard in former times so as not to have lost it. The first time Judea was enslaved was the most difficult to endure and the struggle against slavery was just. But the slave who has once been brought under subjection and then runs away is more resistant to correction than one who loves liberty. For the time was appropriate for fighting when your people rose against Pompey when he first entered Judea. You might never have allowed the Romans to enter Jerusalem. Back then, both our ancestors and their kings were in better circumstances than we. They were richer and stronger and had more courageous hearts and yet even they could not defeat the Roman army. Now you have grown accustomed to obedience down through the generations. Your circumstances are also much inferior to those who were first enslaved. Will you now venture to oppose the entire Roman empire?

"The Athenians once set fire to their city to preserve their liberty. They pursued proud prince Xerxes when he sailed over the sea, which could not contain him as he led such large land forces as to conquer Europe. He made his enemies run away like fugitives in a single ship, striking a great part of Asia—the lesser Salamis. And yet, now these same Athenians are slaves of Rome. Roman orders have become law in this principal capital of Greece. Those Lacedemonians are the same. They achieved great victories at Thermopylae and Plataea. With King Agesilaus, they owned every corner of Asia. Yet they are now content to be under the Roman thumb. The Macedonians still delight in what great men their Philip and Alexander were, seeing that the latter promised them a worldwide empire. But these same people are now greatly changed. They now obey those who rule the world in their place. Additionally, there are 2,000 nations who have had more reason than we to claim complete freedom, and yet they also submit to Rome.

"You are the only people who believe it a disgrace to serve those to whom the rest of the world has submitted. On what sort of army do you rely? Where are the weapons you depend on? Where lies your fleet of ships whereby you might control the Roman seas? And where are the funds that will allow you to overcome them? I ask you: Do you suppose to make war against the Egyptians? The Arabians? Will you not carefully reflect on the Roman empire? Will you estimate your own weaknesses? Haven't you been often beaten by neighboring armies? How can you then fight against the Romans who have proved invincible throughout the inhabited world? The Romans even want to lay claim to land beyond that! The Euphrates isn't a sufficient boundary to the east, nor the Danube in the north. As to their southern limits, they have walked all over Libya, an uninhabited country, as is Cadiz, their westerly limit. In fact, they've looked for another uninhabited land beyond the ocean and have taken their weapons as far as the British Islands that were never before known.

"What, therefore, do you pretend to do? Are you more wealthy than the Gauls, stronger than the Germans, wiser than the Greeks, and more in number than all the rest of men upon the earth? What confidence is it that elevates you to oppose Rome? Perhaps you will say, 'It is hard to endure slavery.' I agree. But how much harder is it than for the Greeks, who were once thought of as the noblest people under the sun? The Greeks, even though living in a large country, are in subjection to six bundles of Roman rods. The same subjection faces the Macedonians, who have a more just claim to freedom than do you. What about the five hundred cities in Asia? Do they not submit to a single governor and to the same number of rods? Why do I need to remind you of the Heniochi and the Tauri nations, those that live in the Bosphorus? And the nations around Pontus and Meotis who previously had no king but are now subject to three thousand armed men? There, forty large Roman ships keep the sea peaceful, which before was very stormy and not navigable. How strong a plea would the people of Bithynia, Cappadocia, Pamphylia, the Lycians and Ciliceans make for their freedom? But these are made to pay Roman tribute even without a Roman garrison. What circumstances are the Thracians under, whose country is so wide it takes five days to cross it and seven days to travel its length? It is a very unpleasant place, more defensible than yours, and so cold as to prevent foreign attack. Don't they also submit to the two thousand Roman troops garrisoned there? What about the Illyrians, who live in the country adjoining Thrace as far as Dalmatia and the Danube? Are they not governed by barely two legions, which also halted an invasion by the Dacians? What about the Dalmatians, who have frequently revolted trying to regain their liberty? They were never able to be thoroughly subdued, as time and again they assembled their army together and rebelled. Yet now they remain silent under one Roman legion.

"If any people enjoyed advantages that could provide for a successful revolt, it would be the Gauls. They are completely walled around by natural barriers. On the east are the Alps. On the north is the Rhine River. On the south are the Pyrenees and on the west is the ocean. The Gauls have such obstacles against possible invasion and include some three hundred and five nations in their midst. One might say that they have fountains of domestic happiness flowing from within, sending out streams of gladness over most of the world. But even the Gauls pay tribute to Rome, which draws its prosperity from them. They put up with this not because their minds are weak or effeminate or because they come from poor breeding. After all, they fought for eighty years to preserve their liberty. But because they have great regard for Rome's might and even more for Rome's good fortune, which in their eyes is more powerful than Rome's weapons. These Gauls are, therefore, enslaved by twelve hundred soldiers, even less than the number of their cities. Nor has gold mined in Spain been enough to support the war to keep them free. Nor were they protected by a vast distance from Rome by either land or sea. Nor could the warlike tribes of the Lusitanians and Spaniards escape. The Romans were as inexorable as the ocean tides, which the ancients thought terrible.

"No, the Romans have exported their power beyond the Pillars of Hercules and have walked among the clouds in the Pyrenean mountains subduing those nations. And one legion is sufficient to guard these people although they were difficult to conquer, being so far from Rome. Who among you has not heard of the multitude of Germans? To be sure, you have seen them frequently, strong and tall, since the Romans have sent them everywhere in the empire. Yet these Germans, who live in a large country, whose minds are stronger than their bodies, whose souls despise death, and whose rage is fiercer than wild beasts, whose border is secured by the Rhine River, are tamed by eight Roman legions. Those taken captive became Roman slaves. The rest of their people had to save themselves by running away. Do you, who also depend on Jerusalem's walls, think about the wall the Britons' had? The Romans sailed away and conquered them even though they were surrounded by an ocean and live in a land as large as a continent. Now, four legions are sufficient to guard this huge island. But why waste more breath on this matter? The Parthians, the most warlike of nations and masters of many more nations by their mighty forces, send hostages to the Romans. Whereby you might see if you will, that even in Italy, the finest nation in the east, in the hope of peace submits to serving Rome.

"Now when almost every other person under the sun submits to Roman might, will you be the only people to make war against them? And do so without regarding the Carthaginians' fate? They, while boasting of their noble Phoenician ancestor, the great Hannibal, fell by Scipio's hand. Nor have you thought of the Cyrenians who are of the Lacedemonians. Nor of the Marmaridite, the people spread throughout the arid deserts. Nor of the Syrtes, a land even when described, sounds horrible. Then there are the Nasamons and the Moors and the immense multitude of the Numidians. None has been able to stop Roman advances. And as to the (African) continent, fully one-third of the habitable world, whose nations are almost without number, and which is bounded by the Atlantic Ocean and the Pillars of Hercules, which feeds an innumerable number of Ethiopians; all these have been totally conquered. Besides its annual crop of fruits and vegetables which support the many Roman citizens for eight months of the year, this holding pays all sorts of tribute and raises revenues to fill all the Roman government's needs. None of these, unlike you, think these injunctions a disgrace even though only one Roman legion lives among them.

"Indeed, why should I show you the power of Rome over these remote countries when you already have a good example in Egypt, right next door. This country extends all the way to Ethiopia and Arabia the Happy, bordering on India. Egypt has seven and a half million men in addition to the people of Alexandria. We know this from their poll tax revenues. Yet Egypt is not ashamed to submit to Roman governance. Alexandrians have a great temptation to revolt. The city is full of people and wealth and is very large. It is almost four miles in length and one and a quarter miles wide. Alexandria pays more tribute to Rome in one month than you do in one year. And besides what it pays in coin it also enough to send corn to Rome annually to support it for four months. It too is walled on all sides, either by impassable deserts, seas having no harbors, or rivers and lakes. Yet none of these barriers has proven too strong for the Roman good fortune. However, the two legions encamped in that city provide the reins which control the remote areas of Egypt, as well as those parts where the noble Macedonians live.

"Where then are the people you must hire as auxiliaries? Must you get them from the uninhabited places of the earth? No. All parts of the habitable earth are under Roman control. Unless your hopes extend far beyond the Euphrates and think that those Jews who live in Adiabene will come to your aid. But surely they will not embarrass themselves by joining an unjustifiable war. If they follow such bad advice, will the Parthians allow them to take part? For the Parthians are more concerned about maintaining the truce that exists between themselves and Rome. If the Parthians allow any of those countries they rule to march against the Romans, it will break the treaties between the two countries.

"What remains, therefore, is this: You do have recourse to God's assistance. But isn't He also on the Romans' side? How could they establish an empire so vast without the help of God's providence? Reflect on its impossibility! If you want to preserve your zealous religious oblations, and they have proven difficult to protect even when you fight against a weaker nation, how can you hope to have God's help when you transgress His law? You make Him turn His face away from you! And if you do continue to observe the Sabbath days and rest from all labor on that day, won't you be easily overcome as your forefathers were by Pompey? He pressed his siege of Jerusalem more strenuously on those days when the city rested. But if in wartime you disobey Jewish law, I cannot imagine any reason that will encourage you to make war. For you only have one concern: that you do nothing against your forefathers. And if you voluntarily violate laws given by God how can you then call upon Him to help you? Now all men who go to war do so depending upon either Divine or human assistance. But if you go to war without either of these aids, you are obviously choosing destruction. Why don't you just murder your children and wives with your own hands? Or go ahead and burn your beloved city? Because if you do these things in your madness, at least you will escape the disgrace of being defeated.

"But it is far better, my friends while the ship is still in port, to look ahead to the waiting turbulence and not sail out of port into the teeth of a hurricane. For we may justly pity those who are trapped by unforeseen circumstances. But for him who charges into obvious ruination, he gets no pity, only disgrace. And, of course, it's unimaginable that you might enter a war agreeing to terms of surrender at its beginning. Can you imagine that when the Romans have overpowered you that they will treat you with more moderation or as an example to other nations and not burn your holy city and utterly destroy your entire nation? Those of you who may survive the war will have no place of refuge, since the Romans are already masters of every nation or are threatening to become so. Nor indeed is the danger restricted to you Jews who live here, but also to Jews everywhere. For no nation exists on earth that does not have some of you among them whom your enemies will execute in the event you go to war. Every city that contains Jews will be filled with slaughter for the sake of you few men, and their executioners will be pardoned. But if this slaughter does not take place, consider how wicked it would be to take up arms against those who have been so kind to the people of your race. Have pity, then, if not on your wives and children, but on your sacred city and its sacred walls. Spare the temple. Preserve for yourselves this holy house with its holy furnishings. For if the Romans overpower you, they will no longer withhold their destruction, particularly since you repay their restraint from doing so in the past has been repaid with such ingratitude. Look at your sanctuary and the holy cherubim of God. Look at this our common country. I have withheld nothing that would accrue to your preservation. If you follow my advice, we will all enjoy peace. But if you indulge your passions, you will run into these dangers of which I personally shall be free."

5. When king Agrippa had finished speaking, both he and his sister, Bernice, wept. Seeing their tears calmed some of the restless crowds but many still cried out. Their fight was not against the Romans, they said, but only against Florus because of what they had suffered at his hand. Agrippa replied that they had already made what looked like war upon Caesar by not paying him his tribute and for cutting off the temple cloisters from the Antonia tower. He advised them that by rejoining the cloisters to Antonia they would avoid the appearance of a revolt. They should also pay the tribute due Caesar, for neither the citadel nor the tribute belongs to Florus but to Caesar.

Chapter 17

The Jewish War Against the Romans Begins. The Rebel Manahem Leads for a Time, then Is Killed.

1. The people listened to Agrippa's advice, and many went up into the temple with the king and Bernice to begin to rebuild the cloisters. The rulers and senators also split up and went into the villages collecting

tribute money. They soon collected the 40 talents they were short. So Agrippa's words put a temporary stop to the threatened revolt. Moreover, he tried to convince the multitude to obey Florus until Caesar sent his successor. But at this the crowd became more unruly, shouting insults at the king. They had Agrippa thrown out of the city. Some of the more warlike had the audacity to throw stones at him. When Agrippa saw that those who wanted war continued in their violence, he became very angry at the indignities he had suffered. So he sent the Jewish leaders, as well as their more powerful men, to Florus in Caesarea. He thought Florus might appoint tribute collectors for the nation while he went back to his own kingdom.

2. At this same time, some of the Jews who had been central in exciting the people to go to war made an assault on the fortress named Masada. Using treacherous means, they captured the fortress killing all the Roman soldiers stationed there, and placing their own men in the fortress to hold it. Meanwhile, Eleazar, the son of Ananias the high priest, a very bold young man who at the time was governor of the temple, persuaded those who administered God's service there to reject any gifts or sacrifices from foreigners. This move precipitated the actual war with Rome, because the Jews had rejected sacrifices for Caesar. When many high priests and officials begged them to reopen the sacrifices foreigners brought for their leaders, the administration would not listen. Eleazar and his administration relied very heavily on the people and had the majority of those who sought war behind them. Above all, they had the highest regard for Eleazar, the temple governor.

3. Regarding this situation, the powerful men gathered and met with the high priests and the head of the sect of Pharisees. They worried that the steps leading to war with Rome were becoming irreversible. So they met together to determine what if anything they could do. Accordingly, they decided to see if words would help to calm the more warlike factions. So they assembled all the people at the bronze gate of the temple's inner court that faced the rising sun. They began by showing the crowd their most intense disapproval over the attempt at revolution and for their desire to bring the conflict upon the nation. They said that the reason for cutting off foreign gifts and sacrifices was unjustifiable. They told the crowd that their forefathers had primarily decorated the temple using do-

nations from foreigners. They had always accepted these gifts from Gentile nations and had never rejected anyone's sacrifice. That would have been of utmost impiety. These foreign gifts could be seen easily around the temple, which they had placed there, and which had been there for a long time. Cutting off foreign gifts and sacrifices would aggravate Rome to make war upon the Jews. The temple government had forgotten God's laws and was preparing a new form of worship. They were determined to run the risk of having the city destroyed for ungodliness if they only allowed Jews and not Gentiles to worship there. If such a law were to be imposed only against a solitary person, he would be rightly indignant seeing its inhumanity to him. The leaders further said that while the people may not hold the Romans or Caesar in high regard, to forbid the reception of their oblations might find the Jews in a situation that they, too, had no place to offer their own sacrifices. The city will lose its stature unless the insurgents quickly get wisdom, restore the former sacrifices, and apologize to those who have been turned away before the matter is reported to those whom it will injure in the future.

4. As the conversation wore on, the priests who were knowledgeable in Jewish customs testified that all of their forefathers had received sacrifices from foreign nations. But not one of the revolutionaries would listen to them. In fact, the ministers who forbade foreign gifts and sacrifices would not go about any of their religious duties. Rather, they prepared to start a war. So the more powerful Jews recognized that the rebellion had become too difficult to stop. They saw danger coming from the Romans, who might be first to witness the seditious actions, and tried to save themselves. They sent out emissaries. Some went to Florus in Caesarea, the central figure being Simon, the son of Ananias. Some also went to Agrippa. Chief among them were Saul, Antipas, and Costobarus, all kinsmen of the king. They asked both Florus and Agrippa to march with their respective armies into Jerusalem to stop the sedition before it got any further out of hand. Of course, their message was good news to Florus's ears. Because he wanted to kindle the rebellion, he gave the emissaries no answer at all. Agrippa also remained neutral. Agrippa was on both the Jewish and the Roman sides. He wanted to preserve the Jewish people for Rome and the city and temple for the Jews. But he also knew that the conflict was not to his advantage. So he sent 3,000 horsemen from the people in Auranitis, Batanea, and Trachonitis under Darius, his cavalry

commander, and Philip, the son of Jacimus, the general of his army.

5. Hearing of Agrippa's forthcoming help, the powerful Jews and the high priests—as well as that portion of the citizenry that wanted peace—took courage and seized the upper city of Mount Zion. The rebels held the temple and the lower city. Stones from slings rained continually between the two sides. Occasionally, Agrippa's troops fought hand-to-hand with the rebels. The rebels proved bolder in their attacks, but the king's soldiers fought with greater skill. The soldiers attempted to capture the temple and drive out the men who had profaned it. At the same time, Eleazar led the rebels in trying to gain the upper city. For seven days, the slaughter raged. Neither side would give ground.

6. The next day was the festival of Xylophory. On that day, it was the custom of the Jews to bring firewood to the altar so that the fire there might not be extinguished. Those holding the temple area excluded their opponents from the religious custom. When the people came together, the Sicarii crowded in among the weaker people and began to do their evil work. (The reader will recall that the Sicarii were those thieves and murderers who hid daggers or swords called Sicae under their garments and stabbed the unwary in crowds.) Now the Sicarii grew bolder. The soldiers were soon overpowered by their number and bravery and were driven out of the upper city by force. Other rebels set fire to the high priest Ananias's house and to Agrippa and Bernice's palace. Then they set fire to the city archives making quick work of their own debt records, thereby releasing themselves from any further obligation to repay them. The fire was also set to gain favor with the vast number who owed money, so that these poorer men might be persuaded to join in the fight against their more wealthy creditors. So the record-keepers fled and the records burned. When they had burned the nerve center of the city, the arsonists turned to the more powerful Jews and the high priests, some of whom hid in subterranean vaults. Others fled with Agrippa's soldiers to the upper palace, shutting its gates after them. Among these were the high priest, Ananias, and the emissaries who had gone to Agrippa. Meanwhile, the rebels became content with their victory and the buildings they had incinerated and went no further.

7. But the next day, which was the fifteenth of the month, Lous (roughly

August), the rebels attacked the Antonia tower. They laid siege to the Roman garrison there for two days before capturing the tower and executing the garrison. They then set the citadel ablaze and marched on the palace where the king's troops had fled. They then divided into four companies and attacked the walls. Those inside had neither the courage nor the ability to counterattack because of the overwhelming superiority of the rebels. But they fanned out into the breastworks and turrets and shot at those laying siege. Many rebels fell at the wall's foot. The battle raged night and day. The rebels figured that those inside the palace would fail due to hunger while those inside thought the rebels would soon grow exhausted from the strenuous siege.

8. In the meantime, a man named Manahem ben Judas, called "The Galilean," was a very clever logistician. [He had formerly insulted the Jews under Cyrenius saying that they were subject to God first, then to the Romans.) Manahem took some of his strongest men and went to Masada. There, he broke open King Herod's armory and supplied not only his own men, but other robbers as well. These became his personal guards, and he returned to Jerusalem looking like a king. He became the leader of the rebellion and gave orders to continue the siege. But the rebels lacked the proper siege apparatus. It was also impractical to undermine the walls because of the many darts that rained down upon them. Instead, they began a tunnel from a great distance away, ending underneath one of the palace towers, making it unstable. Having accomplished that, they set fire to the wooden substructure and left it burning. When its foundations were burned thoroughly the tower suddenly fell. Running into the collapsed tower, the rebels were met by another wall that was hastily constructed by the defenders. The defenders had known that the tower might fail in the undermining, so they raised up more fortifications. The rebels thought they had been successful, but when they saw the unexpected barrier they were dismayed. However, the defenders then sent a message to Manahem and to the other rebel leaders asking for terms of surrender in which they might come out safely. Surrender terms were granted only to the king's soldiers and to the rebels' own countrymen who subsequently left their positions. But the Romans were left alone. They lost heart, as the rebel numbers were vastly superior to their own. The Romans also considered that to give the rebels the right hand of security would be an insult to themselves. Even if they did give it they couldn't trust the rebels. So the

Romans deserted their positions and retreated to the royal towers, those called Hippicus, Phasaelus, and Miriamne. But Manahem and his followers entered the area from which the Romans fled and killed as many of them as they could catch before they got up to the towers. They took what the Romans left behind and set fire to their bivouac. This battle ensued on the sixth of the month Gorpieus [Elul] (roughly August-September).

9. The next day, the high priest, Ananias, was found hidden in an aqueduct. The rebels executed him together with his brother, Hezekiah. Then the rebels besieged the towers, sealing them off lest any of the Roman soldiers should escape. The victory over the strongholds and the death of the high priest encouraged Manahem to become dreadfully cruel. He believed he had no opponent to challenge his leadership, but he was no more than a tyrant without the support of his subordinates. When Eleazar and the rebels conferred together, the men said it wasn't right to revolt against the Romans to secure their freedom and then to deny the same freedom to their own people. It was unreasonable to support a leader more malevolent than they, even though Manahem was not personally guilty of innocent bloodshed. They said if they had to choose a leader it would be better to grant that position to anyone but Manahem. So the men attacked Manahem as he went up to worship in the temple. He was dressed in royal regalia and surrounded by his armed followers, and supported by the rest of the people. They threw stones at him, thinking that if they were to kill Manahem, the entire rebellion would fail. Manahem and his men resisted for a while but when they realized how large the crowd was that was attacking them they fled, every man for himself. Those captured were executed on the spot, while those that hid themselves became the target of search parties. A few of them were able to escape to Masada, including Eleazar ben Jairus, Manahem's relative. Meanwhile, Manahem ran to a place called Ophla alone, where he cowered in fear. But they took him alive and brought him to the front of the multitude. They tortured him in various ways then executed him. They did the same to some of his lieutenants, including Manahem's main instrument of tyranny, a man named Apsalom.

10. As I said previously, the people assisted in the manhunt, hoping that Manahem's capture might slow down the rebellion. But still others wanted war but hoped to find a way to pursue it with less danger, now that

they had killed Manahem. Truly, when the people wanted to stop besieging the Roman soldiers, the rebels wanted to press the siege all the more. That is, until the Roman general, Metilius, sent a message to Eleazar asking the rebel leader to spare their lives and give the Romans safe passage. In return, the Romans would give up their arms and the rest of their possessions. The rebels agreed to the request and sent Gorion ben Nicodemus, Ananias ben Sadduk, and Judas ben Jonathan to them to pledge them security by their right hands under oath. So, when Metilius brought his troops out, the rebels did not touch them. Nor was there any sign of impending treachery. But as soon as the soldiers had all laid down their shields and swords according to the terms of surrender, and were obviously leaving with no apparent intent to fight, Eleazar's men violently attacked them, surrounded the soldiers, and killed them all. The Romans didn't try to defend themselves, nor did they cry out for mercy. They only screamed about the breach of the surrender terms and the Jews' oaths of surety. And so all the Romans were brutally murdered except Metilius. The general begged for mercy, promising that he would convert to Judaism and be circumcised. They let him live but no one else. This loss to Rome was only a minor one, as just a few were killed out of their enormous army. But it did portend the Jews' own destruction. The Jews now publicly mourned that all-out war was inevitable now that Jerusalem was polluted with such abominable crimes for which they could certainly expect retaliation. The city was filled with sadness, and every one of the most moderate men among them was under great stress, knowing that they would all be punished for the rebels' wickedness. It so happened that the murder of the Roman soldiers took place on the Sabbath, a day when all Jews rest from their labors and worship God.

Chapter 18

The Disasters and Slaughters that Came Upon the Jews

1. On the very same day and hour of the soldiers' massacre in Jerusalem, the people of Caesarea rose up and slaughtered the Jews among them. One would think that it had all happened under Divine providence. In the space of one hour, 20,000 Jews were killed. Caesarea became devoid of Jews. Florus caught those who tried to run, sending them to the galleys in chains. The entire nation was enraged when they heard of the carnage in Caesarea. The Jews divided up into several bands and wasted Syrian

villages and their neighboring cities of Philadelphia, Sebonitis, Garasa, Pella, and Scythopolis. After that, Gadara, Hippos, and some cities in Gaulonitis were destroyed, while fire demolished others. From there, the Jews went to Kedasa of the Tyrians, to Ptolemais, Gaba, and Caesarea. Neither were Sebaste in Samaria nor Askelon able to withstand the violent attacks upon them. When those cities were burnt to the ground, the Jews completely demolished Anthedon and Gaza. The Jews pillaged many villages surrounding these cities and an enormous slaughter was made of the men trapped within them.

2. The Syrians killed as many Jews as they lost. The Syrians executed many other Jews caught in their cities. The barbarism took place not just because of Syria's longstanding hatred of the Jews, but also from fear for their own lives. The ferment in Syria was dreadful. Every city was a war zone, with each divided into two camps. If one was spared the other was destroyed. So the days went by in bloodshed and the nights in fear–that time was worse than the daytime. When the Syrians thought they had defeated the Jews, they then suspected the Judaizers as opponents. Neither side wanted to kill those whom they only suspected to be combatants, but they feared that when they mingled among them they might attack. Further, greed had become a reason to kill any opponent, even those who seemed harmless. For, without fearing punishment, many looted the personal effects of the slain and stole the property of those killed in their own homes just as if their home was a formal battlefield. Honor came to the ones stealing the most, as having overcome the most enemies. Cities filled with unburied dead bodies became a common sight. The bodies of old men lay beside those of infants, all scattered on the ground. Dead women also lay amongst them, uncovered in their nakedness. The entire province was filled with unspeakable atrocities while the dread of even more heinous barbarity hung over every head.

3. So far, the conflict had raged between Jews and foreigners, but when the rebels made excursions into Scythopolis they found Jews standing with the men of that city, arrayed in battle formation as their enemy brothers. They preferred safety from the Romans before their relationships to other Jews, and fought against their own countrymen. The Jewish eagerness to fight the rebels was so great that the men of Scythopolis suspected them of treachery. They were afraid that should the rebels assault their city in the

dark, many of the Jews who stood with Scythopolis might make amends to their brothers and defect. So the men of Scythopolis demanded that the Jews with them demonstrate their loyalty. The Jews were ordered to take their families out of the city to a nearby grove of trees. The Jews did this without suspecting anything. The people of Scythopolis then waited for two days in order for the Jews to feel secure. But on the third day the citizens seized the opportunity to cut every Jews' throat. Some died surprised by the attack, and some died as they slept. The number of Jews killed exceeded 13,000. After the slaughter, the Scythopolitans pillaged the entire camp.

4. The story of what happened to Simon, the son of Saul, a Jew who had been of some repute, also deserves telling. This man's strength and his audacity distinguished him from his peers. But he abused them both in the evil he did to his own countrymen. Every day Simon would sneak into Scythopolis and kill a great many of the Jews of the city, causing others to flee. His actions turned the tide of battle. But he got what he deserved for the murders he committed. When the men of Scythopolis surrounded the grove of trees and were throwing their darts at the Jews there, seeing that their number was too great, Simon drew his sword but did not attack them. Instead, he cried aloud with great emotion, saying, "O you people of Scythopolis, I deserve to suffer for what I have done to you. After giving my word that I would be faithful to you, and then to have killed so many of you, my own relatives. We now deserve the treachery of aliens for we have acted wickedly against our own nation. I will therefore die by my own hand, polluted wretch that I am. For it is not fit that I die by the hands of our mutual enemies. Let my suicide be both punishment for my great crimes, and a testimony commending my courage. And, so that none of my enemies may boast of killing me and that no one may mock me as I fall." When Simon had said this, he looked at his family standing around him with eyes full of both pity and rage. His family consisted of his wife and children and his aged parents. He first caught his father by his grey beard and ran his sword through him. Simon then did the same to his mother who willingly died. After that, he did the same to his wife and children, all almost offering themselves up to his sword, so desirous were they of escaping death from enemy hands. When he had finished, Simon stood upon the bodies of his family so that the Scythopolitans could see him. Stretching out his right hand so as to be seen by

the throng, he rammed his sword into his own bowels. This young man, Simon, was to be pitied, both because of his strength and for his courage. But because he had sworn of his faithfulness to foreigners against his own people, he got what he deserved.

5. Besides this murder in Scythopolis, other cities also arose against their resident Jews. The men of Askelon killed 2,500 Jews. In Ptolemais, another 2,000 were slain and many put in chains. The people of Tyre also put a great many Jews to death and threw even more into prison. The people of Hippos and those of Gadara executed the bolder Jews and arrested others whom they feared. The rest of the Syrian cities did the same, both out of fear and hatred. Only the people of Antioch, Sidon, and Apamia spared those Jews living among them, not allowing them to be enslaved or murdered. Perhaps the Jews in these cities were spared because their numbers were of such great size as to prevent any attempt to harm them. But I think the main reason for their safety was the compassion felt for the more moderate Jews. The Gerasans also did not harm the Jews living there. The Jews who wanted to leave the city were escorted safely as far as the city limits.

6. A plot also developed among the Jews in Agrippa's kingdom. Agrippa had gone to visit Cestius Gallus in Antioch, but had left one of his friends, Noarus, in charge of public affairs. Noarus was a relative of king Sohemus (probably an Arabian). Seventy men came out of Batanea, who were known as prudent Jewish gentlemen and elders in their families. They requested they be left with a cadre of soldiers so that if anyone rose against them they would have sufficient forces to protect themselves. But Noarus sent some of Agrippa's guards out by night and killed all seventy. He did this without Agrippa's approval. Noarus was a greedy man, choosing in his wickedness to abuse his own countrymen even if it meant the ruination of Syria. His cruelties were also illegal. Agrippa got wind of what Noarus had done and would have had him assassinated, but for his high regard for king Sohemus. But Noarus was immediately discharged from his temporary office.

But as to the rebels, they took the fortress named Cypros above Jericho, cutting the throats of its garrison and utterly demolishing the fortifications. About this same time, a throng of Jews at Machorus persuaded the Roman garrison there to evacuate the place and surrender it to them.

The Romans agreed to leave on certain conditions, being fearful that the place could be easily taken by force. When they had negotiated their safe passage, they gave up the stronghold. The people of Machorus then emplaced a garrison for their own safety and held it in their own power.

7. In Alexandria, its citizens' insurrection against the Jews continued unabated. This had been going on since the days of Alexander the Great, who had found the Jews ready to help him defeat the Egyptians. As a reward for their assistance, Alexander gave the Jews the same privileges in the city as the Greeks. This honorarium continued through Alexander's successors, who set apart a special part of the city where the Jews could live without being polluted by Gentiles. They were then not to be intermixed with foreigners, as had previously been the case. The Jews also had the further privilege to be called Macedonians. When the Romans took possession of Egypt, neither the first Caesar nor none following him ever thought of diminishing the honors that Alexander had bestowed upon the Jews. But conflicts perpetually arose between the Jews and the Greeks. The city fathers continually punished those taking part, yet the Greeks' mistreatment of the Jews grew even worse. At this time especially, since there was trouble brewing in other places, the problems in Alexandria reached a boiling point. At one juncture, the Alexandrians held a public meeting to discuss emissaries they were preparing to send to Emperor Nero in Rome.

When a crowd of Jews came flocking to the theater to join the meeting, the Greeks spotted them outside and immediately caused an uproar, saying the Jews were enemies come to spy on the proceedings. On hearing this, the Greeks in the theater rushed out to attack the Jews. The Greeks killed some Jews as they tried to escape. Three Jews were captured and hauled away, to be burned alive. Then all the Jews rose up and came as one man to rescue these three. They threw rocks at the Greeks but then took their lamps and rushed violently into the theater, threatening to burn alive all who were there. They would have done it, too, had not Tiberius Alexander, the city's mayor, got the Jews to calm down. He didn't have to use force, but sent some of the city fathers into the theater, who begged the Jews to quiet down lest they provoke the Roman army to rise against them. But the more rebellious among the Jews laughed at Tiberius's requests and threw insults at him.

8. So, when Tiberius saw that these rebellious Jews would not be pacified until they met with some tragedy, he sent two Roman legions based in the city against them. An additional 5,000 soldiers who had just arrived from Libya went with them. The Jews were to be devastated. Not only were the soldiers ordered to wantonly kill them, but to take their possessions and set fire to their homes as well. These troops rushed madly into the section of the city called the "Delta" where the Jewish people lived. The soldiers did as ordered, yet not without suffering bloodshed on their own side as well. The Jews massed in defense, putting the best armed men in the front line. They were able to resist the Roman surge for a long time, but once they fell back, they were destroyed unmercifully. The destruction was complete. Some were caught in the open field while others were forced to take cover in their homes—that were then plundered and set on fire by the Romans. No mercy was shown to little babies, nor was any given to the aged. The Romans proceeded to slaughter Jews of every age until the Delta overflowed with blood and 50,000 Jews lay dead in heaps. None would have remained alive had they not gone to Tiberius Alexander to beg for mercy. Alexander had pity on them and ordered the Romans to withdraw. Those soldiers of discipline stopped killing immediately. But the Alexandrian populace bore such hatred for the Jews that they had difficulty making them discontinue. It was a hard thing to order them to leave their fallen comrades.

9. This was the awful catastrophe that fell upon the Jews in Alexandria. Meanwhile, Cestius Gallus thought he would no longer lie still since the Jews were up in arms everywhere. So he took the entire twelfth legion out of Antioch. Out of those remaining, he selected 2,000: six cohorts of infantry and four cavalry troops, in addition to the auxiliaries sent by the local kings. Antiochus had sent 2,000 horsemen and 3,000 soldiers on foot, and as many archers. King Agrippa sent the same number of infantry and 1,000 cavalrymen. Sohemus also followed suit with 4,000 men of arms, one-third of which was cavalry but the majority archers, and marched to Ptolemais. There was also a vast number of auxiliary troops gathered together in the free cities, who though lacking in some martial arts skills, made up for the lack in their eagerness and their hatred of the Jews. Agrippa himself joined the growing army both as a guide in the march over the country and to direct his army's movements. Cestius took a portion of his forces and marched to Zabulon, a fortified city in

Galilee called the "City of Men." The region divides the country of Ptolemais from our nation. He found that the men of the city had deserted it, having fled in scores to the mountains. The city had received all sorts of good things, so Cestius allowed his soldiers to pillage the buildings and then set them on fire. Zabulon had been a place of admirable beauty. Its houses were like those in Tyre and Sidon and Berytus. After this, Cestius Gallus ran over the entire countryside seizing whatever he happened upon and setting fire to the villages in his path. He then returned to Ptolemais. While the Syrians—mainly from Berytus—were busy plundering, the Jews took courage again knowing that Cestius had retired from the field. They made a surprise attack on those left behind and slaughtered about 2,000 of them.

10. Cestius then marched from Ptolemais and went to Caesarea. But he sent part of his army ahead to Joppa. Cestius ordered that if the latter could take Joppa by surprise they should hold it. But if the citizens got wind of a pending attack they should wait for him and the rest of the army. So a portion of the army marched quickly along the seashore to Joppa, while the rest went overland. They came upon Joppa in a pincers movement and easily took the city. The inhabitants had made no preparations to defend Joppa, so the soldiers fell upon them and killed every man, woman, and child. The number slain was 8,400. The troops then pillaged and burned the city. In the same way, Cestius sent a cavalry unit to the Narbartene toparchy located next to Caesarea. He devastated the countryside and killed a great number of its people, plundering and burning villages as he went.

11. Then Cestius sent Gallus, the commanding officer of the twelfth legion, into Galilee, giving him forces sufficient to conquer that region. Sepphoris, the strongest city in Galilee, received Gallus joyously. The thieves and rebels there fled to Mt. Asamon, near Sepphoris in the middle of the country. So Gallus sent his men up against them. From defensive positions above the Roman assault, the rebels easily threw their darts down upon them, killing about two hundred. But circling around the mountain the Romans got behind their enemy and took the advantage. Then Cestius soon defeated the rebels and thieves. Wearing only light armor, the Jews could not hold out against the more heavily armored Romans. When defeat was inevitable, the Roman cavalry made escape

impossible. A few rebels and thieves hid in the rocks but over 2,000 of their number were killed.

Chapter 19

What Cestius Did Against the Jews and how upon Entering Jerusalem, He Made an Incredible, Unexpected Retreat

1. Gallus saw nothing else in Galilee that looked like insurrection, so he returned with his army to Caesarea. Meanwhile, he marched his entire army from Caesarea to Antipatris where he had heard that a sizable Jewish force held a certain tower called Aphek. He sent a portion of his army on to attack them, but the frightened Jews there dispersed before they clashed. So the Romans, finding the Jewish camp deserted, burned not only the camp but also the surrounding villages. Cestius then marched from Antipatris to Lydda but found the city devoid of its men. The entire population had gone up to Jerusalem for the Feast of Tabernacles. So he killed the fifty Jews who showed themselves and then burned the city. Marching onward, Cestius ascended near Bethoron, pitching his camp at a place called Gabao about six miles from Jerusalem.

2. When the Jews in Jerusalem realized that the war was coming to their city, they left the festivities and took up arms. Encouraged by their great number, the Jews charged boisterously and haphazardly into battle. They had given no consideration to the fact that it was the Sabbath day, a day the Jews hold in high regard. But in their rage they forgot their sacred day and charged violently into the oncoming Roman forces. The Jews broke through the Roman ranks, slaughtering men right and left as they charged. The Jews might have been annihilated Cestius's entire force if the Roman cavalry and reserve infantry had not made a flanking maneuver and come to his aid. As it was, five hundred fifteen Romans died. Of that number, four hundred were on foot and the rest on horseback. The Jews lost only twenty-two. Of these, the most valiant were related to King Monobazus of Adiabene. Their names were Monobazus and Kenedeus. Next to them were Niger of Perea and Silas of Babylon, the latter a deserter from Agrippa's army. When the Jews could make no further progress, they retired to Jerusalem. But one Simon ben Giora led a force that

attacked the Romans again as they were ascending to Bethoron and put their rear guard in a panic. Simon captured many beasts of burden carrying weapons and led them into the city. Cestius delayed further action for three days. Meanwhile, the Jews had seized Jerusalem's high places and set watches at the city gates. They resolved not to rest as long as Cestius remained close to the city.

3. Agrippa soon realized that the Romans were likely to be in danger. A huge multitude of Jews had seized the mountains surrounding Jerusalem, so he planned to use diplomacy. He thought that if he couldn't persuade them all to stop fighting, perhaps he might be able to separate the moderate rebels from the more inflamed. So he sent Borceus and Phebus, those of his men best known to the Jews, to promise them that Cestius would give his oath that should they throw down their arms and come over to the other side, they would each receive a full pardon. But the more violent rebels, fearing that the rest of the multitude might go over to Agrippa out of fear for their lives, determined to attack and murder the emissaries immediately. Accordingly, they killed Phebus before he could utter a word. Borceus received a minor wound and was able to save himself by running away. The rest of the Jews were very angry with the rebels who had done this and beat them with stones and clubs, driving them into Jerusalem.

4. But now Cestius thought that these new disturbances among the Jews afforded him a good opportunity to attack. Moving with all of his troops he put the Jews in full retreat, pursuing them to the city gates. He then made camp on a hilltop called Scopus, or Watchtower, not quite a mile from the city. He wasted three days hoping that the Jews inside the city would yield to some degree. Meanwhile, Cestius sent much of his army to neighboring villages to seize corn. On the fourth day, which was the 30th day of Hyperbereteus, (the Hebrew Tishri, or roughly October-November 66 AD), Cestius deployed his troops in full battle array and marched them into the city. The rebels kept the people silent, but were themselves frightened by the good order of the Roman troops. The rebels then left the outer portions of the city and retreated into Jerusalem's inner city, and into the temple itself. When Cestius arrived into the city, he set fire to the area called Bethesda, also called Cenopolis [or the new city]. He then set up camp against the royal palace. If Cestius had attempted to force himself into the walls at this time, he would have been successful, and the

war would have ended at once. But Florus had bribed Tyrannius Priseus, the army's quartermaster, along with several cavalry officers. They kept Cestius from the proceeding with the assault. This was the reason that the war lasted so long and from which the Jews suffered ultimate disaster.

5. In the interim, the high priest Ananus, Jonathan's son, persuaded many of the ranking men in Jerusalem to invite Cestius into the inner city. They were ready to open the gates for him. But Cestius dismissed the offer partly out of his anger at the Jews and partly because he didn't believe Ananus's party to be sincere. He delayed his decision to such an extent that the rebels sensed treachery and threw Ananus and his followers from the wall. The rebels then pelted them with rocks, driving them into their homes. The insurgents then spaced themselves out along the walls and cast their darts at the Romans who tried to climb. The Roman troops made their assault upon the walls for five days but to no avail. But the very next day, Cestius chose a number of his best men, along with bowmen, and attempted to break into the northern part of the temple. The Jews beat them off of the cloisters and threw them back several times before the Romans got near the wall. Finally, the sheer volume of Roman darts made the defenders retire. Meanwhile, the Romans' first rank had braced their shields above them and against the wall, as did all coming up from behind. Those bringing up the rear did likewise, forming a testudo [tortoise shell], which protected them from the rebel darts that slid off the testudo and did no harm. So the Roman soldiers, now able to undermine the wall while protected, began preparations for setting fire to the temple gate.

6. With this battle going on, terrible fear seized the rebels to the degree that many of them hurriedly left the city as if total defeat was pending. On their retreat, the remaining Jews gained courage. They went into that part of the city the rebels had left vacant in order to open the gates for Cestius to come to their aid. Cestius could have taken the city easily by continuing his siege, but did not. It was, I suppose, because of God's anger at what was taking place in the city and its temple that kept Cestius from putting an end to the war that very day.

7. It also never occurred to Cestius how low morale was among the besieged, nor how encouraged the moderate Jews were because of his pres-

ence. So losing any assurance of taking Jerusalem without disgrace, Cestius recalled his soldiers from the city for no apparent reason. The thieves were encouraged when they realized that Cestius had ordered this unexpected retreat. As Cestius moved away from the city, the rebels chased his rear guard and killed a considerable number of his cavalry and infantry troops. Cestius then spent the night at the Scopus camp, then retreated farther the following day. By his retreat, Cestius gave an open invitation for his enemies to continue harassing his rear guard, which they did, destroying many men. The rebels also attacked the Roman retreat on both flanks, hurling their javelins obliquely. The rear guard never moved around to fight the onslaught from behind, thinking them too numerous to overcome. Nor did the main body of Romans counterattack the threats on their flanks. They were overburdened with weaponry and were afraid of breaking their tight ranks. Meanwhile, the Jews were not so encumbered and continued their thrusts against the Romans. These Jewish tactics were why the Romans suffered so much without being able to avenge themselves. So they were harassed as they marched, and their ranks became disorderly. Those falling out of ranks were slain immediately, among whom was Priscus, commanding the sixth legion, Longinus the tribune, and Emilius Secundus, the commander of a cavalry troop. So it was with great difficulty that the Romans got to Gabao, their former camp, and with great losses to their equipment. Cestius stayed there for two days and was greatly concerned about his next move. On the third day at Gabao, he saw that the Jews had been greatly reinforced. Everywhere he looked he saw more Jews. Cestius then understood that any further delay would prove fatal. The longer he stayed in Gabao; the more enemies would be upon him.

8. Therefore, in order to hasten his retreat, Cestius ordered to leave behind anything that might slow down the march. So they slaughtered the mules and other creatures, except those carrying their weapons and other war machines that would be of use. They were also afraid that the Jews might seize them as they marched. Cestius then force-marched his army as far as Bethoron. The Jews eased their attacks when the Romans passed through wide-open spaces, but when the column got penned up descending through narrow passages some of the Jews took positions in front of the column that made passage difficult. Others pushed the rear guard units down the hillsides. The main Jewish force spread out along the nar-

rowest part of the passage and covered the Roman army with their darts. The foot soldiers did not know how to defend themselves under such circumstances, and the cavalrymen were even more confused. The missiles rained down in such profusion that to march in ranks along the roadway was impossible. Not only that, but the slopes were so steep that they prevented the Roman cavalry from going on the attack. The precipices on each side fell off into deep valleys into which Romans frequently tumbled. There was no place to run, nor there was there any way for the Romans to defend themselves. Finally, so great was their distress that they were in the utmost despair, mournfully shouting their laments. Meanwhile, the Jewish shouts were joyful, encouraging one another as the sounds echoed back from the cliffs and valleys. It was a noise that intermixed joy and rage. The situation was such that the Jews almost captured Cestius's entire army, had not night intervened. The Romans then fled to Bethoron where the Jews surrounded them and waited for morning.

9. The next morning Cestius realized that he did not have room to march his entire army out of Bethoron. So he devised a plan of escape. He selected four hundred of his bravest soldiers and placed them at the strongest point of his fortifications. He then gave the order that when they went up at the morning change of guard they were to raise their ensigns to make the Jews think the entire army was still there. Meanwhile, Cestius took the rest of his forces and quietly marched almost four miles. When the Jews realized the trick and that the camp was empty, they charged the four hundred who had tricked them, hurled their darts, and killed them all. They then took off after Cestius. But he was already far ahead, having begun his march well before dawn. The Romans increased their speed as the sun rose, and in fear and trepidation left their siege engines behind, the catapults for throwing boulders, as well as the majority of their war machinery. So the Jews kept chasing the Romans as far as Antipatris, but seeing Cestius was too far ahead, went back and commandeered the war machinery and robbed the dead bodies. When they had gathered up all the booty the Romans had left behind, they ran singing back to Jerusalem. The Jews had only suffered a few casualties, but had killed 5,300 Roman infantrymen and three hundred eighty cavalrymen. The defeat occurred on the eighth of Dius [the Hebrew Chesvan] (roughly October-November), in the twelfth year of Nero's reign (66 AD).

Chapter 20

Cestius Sends Ambassadors to Nero. The People of Damascus Kill those Jews Living among them. After the Jews stop Pursuing Cestius, they Return to Jerusalem and Make Ready its Defense, Assigning many Generals to their Armies, and Particularly Josephus, the Author of these Books.

1. After Cestius's catastrophe, many Jews got out of the city like rats leaving a sinking ship. The brothers Costobarus and Saul, along with Philip ben Jacimus—commander of King Agrippa's army—left Jerusalem and went to Cestius. [Later, we shall tell how Antipas, who had been besieged by the Jews in the king's palace but had refused to flee, was slain by the rebels.] At their request Cestius then sent Saul and his friends to Nero. They were to inform Caesar of the great trouble they were in and to place blame for the combustion of the war on Florus. Hoping to ameliorate the danger he was in, Cestius attempted to provoke Nero's anger against the Roman procurator.

2. Meanwhile, the people of Damascus proceeded to slaughter the Jews among them when they heard of the Roman massacre. They had gathered the Jews of the city in the public square out of suspicion of a revolt. The citizens of Damascus saw no downside to the mass slaughter, except one. No one trusted his own wife, most of whom had become adherents to the Jewish faith. Their greatest concern was how to hide the executions from their wives. So they fell upon the Jews and silently slit their throats. No one interrupted the slaughter as 10,000 unarmed Jews died in the space of one hour.

3. Following their return to Jerusalem, the Jews who had chased Cestius began to harass others who favored Rome. They also persuaded other Jews to join ranks with them. A great many gathered together in the temple and appointed leaders for the coming conflict. Joseph ben Gorion and the high priest, Ananus, were chosen as city governors with the primary responsibility of repairing the city walls. They chose not to elect Eleazar ben Simon to that office, even though he possessed all the booty taken from the Romans and the money seized from Cestius. Eleazar also held a substantial portion of the public treasury. The people saw Eleazar as rebellious—a man whose followers behaved as his personal guards. However,

the people needed Eleazar's money. So, in part due to their need but also to Eleazar's subterfuge, the vote was circumvented, and Jerusalem submitted to Eleazar in all public policy matters.

4. The people also chose army generals for Idumean operations. The picked Jesus, son of one of the high priests, Sapphias, and Eleazar, son of the high priest Ananias. They also sent orders to Niger, who was the then governor of Idumea. His family was from Perea beyond the Jordan River, and therefore was known as "The Peraite." His orders were to obey the aforementioned commanders. Nor did the Jews neglect other areas of the country. Joseph ben Simon was sent to Jericho as a general officer, as was Manasseh to Perea, and John the Essene to the Thamna toparchy. Lydda, Joppa, and Emmaus were added to his territory. They elevated John ben Matthias to governor of the Gophnitica and Acrabatene toparchies, and Joseph ben Matthias of the two Galilees. (Joseph ben Matthias is none other than the author, Josephus, who begins speaking of himself in the third person.) His rule would include Gamala, the strongest city in that northern area.

5. Each of the commanders and governors exercised his authority willingly and judiciously. But one stood out above the others. When Josephus arrived in Galilee, his first act was to obtain the good will of its people. He thought it would accrue to his overall success even if he might fail in other endeavors. He was also cognizant that should he distribute some of his power to the leading men of the area he would make them close allies. Also, he reckoned that if those well-known citizens of the country executed his orders he would gain friends among all the people. So he chose seventy of the most prudent older men and designated them to be rulers in Galilee. Josephus also chose seven judges in each city to hear minor cases. He and the seventy elders would hear the most important cases involving life and death.

6. When Josephus had finished establishing the legal processes for the people's dealings with one another, he began making provision for their protection from foreign invasion. Knowing that the Romans would invade Galilee, he built walls around Jotapata (Yo-TAP-uh-tuh), Bersabee, and Salamis. Besides these cities, he fortified Caphareccho, Japha, and Sigo, as well as Mount Tabor, Tarichee, and Tiberias. He also constructed

walls around Lake Gennesar (the Sea of Galilee), in southern Galilee. He also fortified places in northern Galilee, including the place called "The Rock of the Achabari," as well as to the towns Seph, Jennith, and Meroth. In Gaulonitis, Josephus fortified Seleucia, Sogane, and Gamala. He gave the people of Sepphoris the exclusive right to build their own walls because he realized that they were not only wealthy, but they were ready for war and didn't need his supervision to prepare them. Gischala presented a similar case. The city already had a wall built by John ben Levi with Josephus's consent. But Josephus gave personal attention all the other fortifications, working together with the various builders and giving the necessary instructions. He also assembled a Galilean army of more than 100,000 young men, which he armed with old weapons that he had gathered and prepared for them.

7. Josephus became convinced that the source of Roman invincibility lay chiefly in the soldiers' obedience to orders and through consistent drilling in the use of their weapons. He thought that before he taught his army how to use their weapons, which they would gain with experience, he would insert many officers into their ranks. He had observed that Roman obedience in the ranks was engendered by the multitude of their officers. Therefore, Josephus, in dividing his army into ranks after the Roman pattern, appointed many lower-ranking officers. He also distributed the army into various sized units – in tens, hundreds, thousands, and even larger units. He also taught them to communicate with one another. Trumpets called and recalled the soldiers. He taught his men how to spread the army's flanks, to execute wheeling maneuvers, to support one another when one unit was losing ground, and to run to help those who most threatened. Josephus also instructed the troops in moral courage and bodily strength. Above all, he trained them for war against the Romans. They were going to fight men who were very disciplined. The Romans were men who by their courage and strength had conquered the entire known world. He told the Galileans to commit themselves to a hard discipline even before they faced actual battle. They should cease to indulge themselves in any former mischief, such as theft, robbery, pillaging, and fraud against their own countrymen. They shouldn't think that just because their families were in harm's way that this would play to their advantage. Wars are best fought by men who have clear consciences. Men who have harmed others in private life will fear retribution in battle to say

nothing of God, who will also be their enemy.

8. And so Josephus continued to admonish his troops. He then chose a number of men ready for battle: 60,000 on foot and two hundred fifty on horseback. Besides these men in whom most of his trust resided, Josephus hired 5,500 mercenaries. He also used six hundred men as personal bodyguards. The army, except for the mercenaries, was well provisioned by the cities as each city sent only half of their fighting men to do battle while the other half remained at home. While one part went to war the other farmed. The corn received by the men under arms was paid for by the security the army provided the city.

Chapter 21

Concerning John of Gischala. Josephus Uses Strategies against John's Plots against Him, and Recovers certain Cities that had Revolted.

1. So, as Josephus was engaged in administering affairs in Galilee, a treacherous man arose from Gischala. His name was John ben Levi. John was a cunning scoundrel, far exceeding what one would ordinarily expect of other important men. John had no equal in wickedness. He began life in poverty, which bred in him his wicked plans as he grew older. He told lies easily and was very convincing at making them believable. John considered it a virtue to deceive others and practiced deceit even upon those he loved. He was a hypocrite who only pretended to be human. Where he could smell a profit, John didn't hesitate to shed blood. His greed and treachery carried him to the heights of depravity. He was particularly adept at theft. Over time, John's henchmen grew in number. At first, only a few joined his gang but as he continued in his evil ways, they became more and more numerous. John took special efforts to ensure that members of his band were never caught in their mischief. He chose four hundred of the physically strongest and most courageous among them who were also highly skilled in the martial arts. He formed this band out of men formerly from the vicinity of Tyre, vagrants who had run away from their villages. By this gang of four hundred, John laid waste to all Galilee, angering a great many who were busy preparing for war against the Romans.

2. John's lust for wealth took precedence over his other desire to command troops or otherwise advance himself. But when he discerned that Josephus was highly pleased with his enterprising skills he persuaded the Galilean general to trust him to repair the walls of his native city, Gischala. In that work, John received much money from the city's rich citizens. Then he contrived a very shrewd plan. He spread the lie that the Jews in Syria were required to import their olive oil exclusively from other nations. He then asked Josephus to allow him to have the sole rights to export oil to the Syrian border. So John bought four amphorae of olive oil for four Tyrian (from Tyre) drachmas. That drachma has the same value as the Attic (Athenian) drachma. (Note: The Grecian amphorae were of different sizes but rarely held more than about thirteen gallons each. So John had bought about fifty gallons.) Then he sold a half-amphora in Syria for the same four drachmas (an 800% profit!). Now, olive oil was very plentiful in Galilee and particularly so at the time. So, having the sole right from Josephus to export the oil and in exporting great quantities, he gathered an immense fortune. John then used this money to Josephus's disadvantage, the very man who had given him the exclusive export license. John figured that if he could topple Josephus he could himself become governor of Galilee. So John ordered his thieving band of highwaymen to step up operations. He reasoned that by increasing trouble in the land he might be able to catch Josephus in one of his traps. As the governor came to his country's assistance, John could assassinate him. Or if Josephus chose to overlook John's raids, John could accuse Josephus of neglect of duty in protecting the people. John also spread a rumor that circulated widely. He said that Josephus was turning all the affairs of state over to the Romans. The villain devised many such traps for Josephus in order to bring the governor to ruin.

3. Meanwhile, a few young men of the village called Dabaritta, who kept guard in the Great Plain, laid a trap for Agrippa and Bernice's steward, Ptolemy. They stole everything Ptolemy had with him, among which were many expensive garments, a few silver cups, and six hundred pieces of gold. Unable as they were to hide all of this loot, they took it to Taricheae and gave it to Josephus, expecting a reward. But Josephus was not pleased. He blamed them for violating the king and queen and gave the loot to Eneas for safekeeping, the most powerful man in Taricheae. His intent was to return it all back to its rightful owners when the time

was right. This action brought Josephus into grave danger, for the young thieves were angry with the governor for having received no reward. They thought Josephus would now send the stolen goods that had caused them great effort back to Agrippa and Bernice. In revenge, the young men ran by night back to their villages and shouted for all to hear that Josephus had betrayed them. They also caused unrest in all the cities around them, to the degree that 100,000 armed men crowded together in the hippodrome in Taricheae and raised a vociferous outcry against Josephus. Some cried out that the governor should be removed from office while others wanted him burned alive. John also joined in the commotion, as did Jesus ben Sapphias, the mayor of Tiberias. Not only did Josephus's friends become terrified at the violent intentions of the crowd, but also his personal bodyguards fled—save for only four who stayed with him. They woke Josephus just as the multitude was to set fire to his house.

The four bodyguards tried to warn Josephus to run, but Josephus wasn't surprised at being deserted, nor at the crowd that had gathered against him. He leapt out to stand in front of the multitude with his clothes torn and ashes on his head. His arms were behind him and his sword was hanging on his neck. Seeing him like this, Josephus's friends, particularly those of Taricheae, pitied him. But the men from outside the country and those from neighboring cities, to whom Josephus government seemed a burden, reproached him and demanded that he immediately produce the money he owed them. They also called for him to confess to betraying them. They imagined from his pitiable appearance that he was ready to confess everything to them. To the crowd it was obvious that Josephus put himself in this sad condition to beg for their pardon. But Josephus's humble appearance was only a part of his strategy. Josephus planned to set the different factions in the crowd against one another. However, he first promised to confess everything, so the governor was permitted to speak. He began:

"I neither intend to send this money from the young guards back to Agrippa nor keep it for myself. I have never had any of your enemies as my friend; neither did I ever consider doing anything that would be disadvantageous to you to be an advantage to myself. But you, O people of Taricheae, I knew that your city stood in more need of fortification than other cities, and I wanted money for the construction of walls. I was also fearful that the people of Tiberias and other cities might hatch a plot to seize the money. Therefore, I kept the money privately that I might use it to surround the city with a wall. But if this doesn't please you, I will produce what was brought to me and leave it to you all to take it. But if what I have done pleases you, you may do as you want to the one who had only your best interests at heart."

4. On hearing this, the people of Taricheae loudly commended Josephus. But those from Tiberias and the rest of the crowd called him insulting names and threatened his life. So both sides stopped quarreling with Josephus and took up against one another. Josephus was encouraged by the 40,000 citizens of Taricheae who stood with him, and so he spoke openly to the whole crowd. He reprimanded them for their rashness, telling them that he intended to use the money to build walls around Taricheae, but would increase security in other cities as well. For no city should be lacking in funds if they would only agree as to how they should be spent, and not allow themselves to be angry at the man who got them the money.

5. Hearing this, those in the crowd who had been fooled left angrily. So Josephus went back into his house. But 2,000 of the men then armed themselves, stood outside and verbally and assaulted Josephus. So Josephus arrived at yet another strategy. He climbed up to the roof of his house and with his right hand motioned the gathering to be quiet. He said to them, "I don't understand what you want. Nor can I hear you for all the confusion of your voices." But Josephus then said that he would comply with their demands if they would only send a few of their number in to talk about it. When the leaders of the 2,000 heard this, they entered the house. Josephus then invited the men into the rear of the house and had them beaten until their clothing fell off. In the meantime, the crowd outside shuffled around, supposing that the negotiations were simply taking a long time. Then suddenly Josephus had the doors swung open and the bloodied men thrown into the street. The crowd was so terrified at this sight they threw down their weapons and ran away.

6. But John's envy of Josephus grew even stronger when he heard of his escape, whereupon he devised a new plot against the governor. He pretended to be sick and wrote a letter to Josephus requesting leave to use the hot baths in Tiberias, that he might try to improve his health. So Josephus, not suspecting another of John's plots against him, wrote to the mayor of Tiberias requesting lodging for John and other necessary things. After two

days in Tiberias, John used these favors to his advantage. Using fraudulent means and bribes, he persuaded several of the people in Tiberias to revolt against Josephus. Silas, who had been appointed mayor of Tiberias by Josephus, immediately wrote to the governor informing him of the plot. When Josephus received the letter he marched all night, arriving in Tiberias the next morning, where a large crowd met him. But John, who suspected that Josephus was coming for him, sent one of his friends to say that he was sick in bed and could not come to pay his respects. So Josephus gathered the people of Tiberias into the stadium. Speaking from an elevated platform, the general told them about the letter he had received from Silas. Meanwhile, John secretly sent some armed men with orders to assassinate Josephus. When the people saw the armed men about to draw their swords, they shouted warnings to Josephus. Seeing the weapons almost at his throat, Josephus stopped speaking and fled with his men to the lakeshore. He commandeered a ship in the harbor, leapt onto it with two of his bodyguards, and sailed away into the Sea of Galilee.

7. Now the soldiers who had come with Josephus to Tiberias took arms and prepared to attack John's men. But Josephus feared that a civil war might bring the city to ruination merely for the sake of a few unruly men. So he sent a few of those with him to tell his soldiers only to provide for their own defense and not to kill anyone or accuse anyone of starting the disturbance. The men obeyed his orders and lay silent. But the people of the surrounding country, when they heard of the plot against Josephus and of its instigator, they arose in great numbers to attack John. But John had fled to Gischala, his native city. Meanwhile, the Galileans, now many tens of thousands of them, ran to Josephus from their respective towns. They cried out that they had come against John, who had plotted against their common interests. They were prepared to burn not only John, but also the city that had welcomed him. Josephus told the crowd that while he appreciated their good will, he intended to subdue his enemies by more peaceful means than slaughter. Josephus then told the multitude that any city that didn't reject John within five days time would be burned to the ground, homes and families alike. He made an exception by name of those cities that were represented in the crowd. Hearing this, 3,000 of John's supporters left him immediately. They came to Josephus and threw down their arms at his feet.

Meanwhile, surrounded by 2,000 Syrian renegades, John decided to use more subtle methods to achieve his ends. He secretly sent messengers to Jerusalem accusing Josephus of a power grab. He wrote that the governor would soon attack their city unless means to prevent it were forthcoming. Most of the people in Jerusalem knew the letter was a hoax. But several of the leaders and rulers who were envious of Josephus secretly sent money to John for him to raise an army of mercenaries to fight the general. They also issued a decree recalling Josephus as governor. But thinking the decree might not work, they sent 2,500 armed men with it. They also sent four high-ranking persons: Joazar ben Nomicus, Ananias ben Sadduk, along with Simon and Judas, both sons of Jonathan. These men were all adept at public speaking. The Jews hoped that they could turn the people of Galilee against Josephus. Their orders were that should Josephus surrender voluntarily, they were to allow him to come to Jerusalem to give an account of himself. But if Josephus was obstinate and insisted on remaining governor, they should treat him as an enemy. Now, Josephus's friends had forewarned him that this army was coming against him, but they did not give him the reason for its coming, as it had been kept secret. Because of this, four cities rose up against Josephus: Sepphoris, Gamala, Gischala, and Tiberias. But Josephus quelled their revolt without bloodshed. He met and routed the four Jerusalem emissaries and captured the best of their army. He then sent the four back to Jerusalem. But the Galileans were angry and zealously anxious to slaughter both the four and their army, along with those who had sent them. But they all escaped their hand.

8. Meanwhile, John hid behind the walls of Gischala in fear of Josephus. Within a few days, the people of Tiberias rose up once more and invited king Agrippa to be restored to authority there. But when Agrippa did not arrive at the appointed time, and when a few Roman cavalrymen appeared, the people ordered Josephus out of the city. The revolt in Tiberias soon became known in Taricheae, where Josephus had brought all his troops to gather corn. Josephus was then undecided whether to march to Tiberias against the revolutionaries or to stay put. He was afraid that Agrippa's soldiers might prevent him from entering Tiberias if he lingered, as they might arrive in the city at any time. Additionally, Josephus didn't want to march on the next day, which was the Sabbath. So he made plans to outwit the rebels. First, he ordered the gates of Taricheae closed so that no one would be able to leave the city to inform the Tiberians

of his secret plans. He gathered two hundred thirty boats on the Sea of Galilee and placed no more than four sailors in each boat. Josephus then sailed quickly to Tiberias, but kept the fleet far enough from the city that no one there could easily see the vessels. He ordered the boats to tack back and forth in that place while he and only seven unarmed men sailed close enough to the city walls so that they could be clearly seen. When his enemies—who continued to insult him from the city walls—saw him, they were astonished. They thought that all of the boats were full of armed men and threw down their weapons. They sent out signals of surrender and begged Josephus to spare the city.

9. Upon seeing their fear, Josephus further terrified the rebels with threats and insults. Instead of taking arms against the Roman enemy, they had insisted upon investing their energy on internal dissensions. They were doing exactly what the Romans wanted them to do! They had endeavored to capture the one man who had desired only their safety! They were not ashamed to close the city gates against the very man who had built their walls! Josephus then pledged to speak with any emissaries who might give just cause for their behavior and, therefore, save the city from destruction. Hearing this, ten of Tiberias's most prominent men came to the shore and boarded one of Josephus's boats. He then ordered his men to take the ten a great distance from the city. Next, Josephus ordered that fifty senators, men of the greatest influence, should also come to him and give him some security on behalf of the city. After this had taken place, and using other pretenses, Josephus ordered all of his boats to sail closer to the city. Filling each of them with Tiberians, he ordered that each vessel sail immediately for Taricheae and to place the captives into the prison there. Finally, the whole fleet set sail to Taricheae with the entire senate of Tiberias, some six hundred men.

10. Then the citizens of Tiberias began to shout that only one man, Clitus, had been the instigator of the revolt. They begged Josephus to spend his anger on him alone. So Josephus, who didn't want to kill anyone, commanded one of his bodyguards, Levius, to leave his boat and cut off both of Clitus's hands. But Levius feared the large enemy crowd and refused to go. Meanwhile, Clitus saw that Josephus was enraged and ready to leap out of his boat in order to execute the punishment himself. So, from the shore Clitus begged that he be allowed to keep one of his hands.

Josephus agreed as long as Clitus himself would perform the amputation. So in great fear of Josephus, Clitus drew his sword and cut off his left hand. So did Josephus take the people of Tiberias prisoner and recover the city with empty boats and seven guards. A few days later he also took Gischala, which had revolted along with the people of Sepphoris, and allowed his soldiers to plunder the city. But then he gathered all of the loot together and returned it to the Gischalans. He did the same with Sepphoris and Tiberias. After subduing these cities, Josephus thought it wise to first plunder the cities to teach them all a lesson, and then to give their property and money back to them in order to gain their good will.

Chapter 22

The Jews Get ready for War while Simon ben Gioras Plunders the Countryside.

1. And so the uprisings in Galilee were brought to an end. Ceasing their internal squabbles, the people began to prepare for war against Rome. In Jerusalem, the high priest, Artanus, and many of the powerful men who were against Roman occupation, repaired its walls. The citizens also made a great many instruments of war. They filled the entire city with darts and all sorts of armor forged on the anvils. Many of the young men were engaged in military exercises, and everywhere were signs of the conflict to come. Yet the more realistic citizens were very sad. Many who saw the destruction ahead wailed loud lamentations. Many of those who craved peace saw omens of the evil to come. Those that wanted war saw these same omens as suited their own desires. The very state of Jerusalem, even before the Romans laid siege, was that of doom and destruction. Meanwhile, Ananus tried to get the rebel faction to lay aside their preparations for a spell and focus rather on their own best interests. He tried to restrain those who were the most zealous but their desire for violence was too much for him. We shall speak later about the end that Ananus was to come to.

2. Meanwhile, in the Acrabbene toparchy, Simon ben Gorius assembled a great many of those who backed rebellion and began to ravage the countryside. He harassed both the rich men's homes as well as their bodies, practicing blatant and open tyranny against his own government. When Artanus and other leaders in Jerusalem sent an army against Simon, he

retreated into the thieves' den at Masada and holed up there. Then Simon and the highwaymen with him plundered the Idumean country until Ananus and his other enemies were slain. Ultimately, Judean rulers were so overwhelmed with the number of dead bodies and the continual ravaging of the countryside that they raised an army and placed garrisons in each village to protect them from Simon and his henchmen. This was the state of affairs in Judea at the time.

The Jewish Wars

Book III

Containing the Interval of about One Year from Vespasian's Coming to Conquer the Jews to the Taking of Gamala

Chapter 1

Nero Sends Vespasian into Syria to Make War upon the Jews

1. When Nero learned of Rome's troubles in Judea he was privately terrified and dismayed, as one might expect. But publically, the emperor appeared strongly confident and extremely angry. He blamed the Roman defeats on command negligence, while belittling any heroism that might be attributed to the Jews. As the man who bore responsibility for the entire Roman empire, Nero saw fit to shrug off such misfortunes. He pretended to have inner strength that superseded such adverse circumstances. Yet the state of Nero's soul was clearly indicated by his speed in attempting to rectify the situation in the east.

2. As Nero was deliberating as to who would command the troubled eastern theater, he saw no general but Vespasian equal to the task. Vespasian would be best suited to not only punish the Jews for their rebellion, but to keep the rebellion from spreading to other nations in the region as

well. Only Vespasian could sustain the burden of such a major campaign. He was growing old in his current encampment, and had spent his life in warfare of one sort or another. Vespasian was the one who had brought peace to the western empire, bringing it all in subjection to Rome following the German uprising. He had also regained Britain by force of arms; an achievement little recognized by Vespasian's father, emperor Claudius. The latter had been given credit for that achievement even though he had nothing to do with it.

3. So Nero reckoned that Vespasian's age, experience, and great skill to be good omens. Not only that, but the general's sons were personally devoted to him, and they were young enough to learn and serve under their father's care. Perhaps Providence was also intervening, paving the way for Vespasian's later rise to become emperor. So, Nero ordered Vespasian to command the Syrian armies, sending him eastward with the great honors and compliments befitting a man of his high rank. Thus, Vespasian sent his son, Titus, from Achaia, (a region in western Greece), where he had been with Nero, to Alexandria with orders to bring back the fifth and tenth legions. Meanwhile, Vespasian passed over the Hellespont (a narrow strait in northwest Turkey), going overland to Syria. There he assembled the Roman forces, along with a large auxiliary force gathered from the area's kings.

Chapter 2

The Jews Suffer a Great Slaughter near Ascalon, Vespasian Arrives in Ptolemais

1. The Jews were so elated and encouraged by Cestius's unexpected defeat they could not contain their zeal. But like any people puffed up by their good fortune, they extended their campaign to more remote areas of the region. They assembled a multitude of their best troops and marched on Ascalon, an ancient city sixty-five miles from Jerusalem. Ascalon had been an ancient Jewish enemy, and for this reason the Jews were determined to attack it first and with great speed. Three men led the excursion: Niger, called "The Peraite," Silas of Babylon, and John the Essene. Ascalon was a strongly walled and fortified city, but had no allies nearby. The Roman garrison consisted of one cohort (about five hundred men), consisting of infantry and cavalry troops commanded by Antonius.

2. In their rising anger, the Jews marched faster than normal. It was as if they had covered only a short distance when the Jews approached the city and came within its shadow But Antonius was aware that a pending Jewish attack was imminent. Therefore, fearing neither the Jews' number nor boldness, he took his cavalry out of Ascalon and received their first assaults fearlessly. When the Jews had crowded up to the very walls of the city, Antonius beat them back. The Jews were unskillful at warfare, but not so the city's defenders. The Jewish force consisted of foot troops trying to fight men on horseback. They had no particular discipline, while attacking Romans fought as a cohesive unit. The Jews were also poorly equipped, attempting to do battle with men who were fully armed. They fought more out of their rage than by sober counsel, exposing themselves to soldiers who were precisely obedient, responding to every order without question.

So, the Jews were easily defeated. As soon as their advance line was in disarray, they fled before Antonius's cavalry. The reserve Jewish troops, arriving behind the first wave and closing in on the city walls, fell to the swords of their own comrades. Jews became enemies of their own brothers. They kept this up until they were all forced to withdraw before the onslaught of Antonius's horsemen. The Jews were disbursed over a very wide plain, becoming easy targets for the Roman cavalry. A vast number of the Jews were slaughtered. If they ran away, they were overrun. If they turned back and driven together again, the Romans slew them in bunches. The Romans surrounded the Jews, driving them every which way, firing arrows point blank and killing scores.

In their distress, the Jews viewed themselves as small in number, while the Romans, though also few in number, seemed to them like a great multitude. The Jews were afraid of the shame of running away yet they also hoped for a change of fortune, and fought zealously. The Romans drew strength from their success. The fight lasted until sundown, when 10,000 Jews lay dead. Two of their generals, John and Silas, were wounded, along with the majority of those who were left alive. Niger, the remaining commander, fled to Sallis, a town in Idumea. Also wounded in the battle were several Romans.

3. The Jewish spirit was unbroken by the horrible calamity. On the contrary, the losses they sustained at Ascalon heightened their resolve for

future encounters. Looking at all the dead bodies under foot, they were enticed by their earlier glorious victories to return to Ascalon and destroy it. So, when they had partially convalesced from their wounds, they massed their forces and attacked the city again with greater fury and in much larger numbers. But their former bad luck followed them, as did their lack of skill and preparation for battle. As they marched, Antonius laid ambushes for them in the mountain passes. Unexpected snares found them surrounded by cavalry before they could form for battle. The Jewish losses numbered 8,000. The rest of their army ran away. Niger was among them, but still fought bravely in his retreat. The Romans maintained contact and pressed the Jews brutally, driving them to a strong tower in a village named Bezedel. Not wanting to waste time laying siege to the fortress, Antonius and the Romans set fire to the city walls, unwilling to allow the courageous Jewish commander to escape. While the tower was ablaze, the Romans left rejoicing, believing Niger to be dead. But Niger had leapt out of the tower into the darkness of a cave beneath it and was saved. In three days time, Niger called out to those searching for his body to give him a decent burial. When he appeared alive, the Jews were joyously surprised as if God's providence had saved Niger to be their commander in future battles.

4. Vespasian's army left Antioch for Syria, where he found King Agrippa awaiting his arrival. [Antioch is the largest city in Syria and without doubt is the third most prominent city in the Roman Empire in terms of its antiquity and prosperity.] Vespasian marched to Ptolemais and was met by the citizens of Sepphoris of Galilee, who were at peace with Rome. These people had previously met with Cestius Gallus and pledged their loyalty to him, being sensitive about their own safety and understanding Roman power. Cestius had given the city the security of his right hand and the protection of a Roman garrison. So, the citizens of Sepphoris graciously received General Vespasian and promised to help him in the fight against their own countrymen. The general gave them as many infantry and cavalry troops as he thought sufficient to oppose any possible Jewish disturbance there. Indeed, now that a war was beginning, the threat of losing Sepphoris—the largest city in Galilee—was not a minor one. It had been built as a natural fortress and its fidelity to Rome might well provide for the security of the whole nation.

Chapter 3

A Description of Galilee, Samaria, and Judea

1. Phoenicia and Syria surround both Upper and Lower Galilee. They are bounded on the west by the city limits of Ptolemais and Mount Carmel. That mountain once belonged to Galilee but now belongs to Tyre. Adjoining Carmel is Gaba, also known as "The City of Horsemen," because the horsemen discharged by King Herod lived there. Samaria and Scythopolis bound the two Galilees on the south, as far as the Jordan River. To the east lie Hippene and Gadaris, along with Gaulonitis and King Agrippa's territory. Lower Galilee extends in length from Tiberias to Zebulon on the east to Ptolemais on the western coast. Lower Galilee runs from a village named Xaloth in the great plain, as far as Bersabee. Upper Galilee is as deep, going as far as the village of Baca, where the Tyrian land begins. Its length runs from Meloth to Thella, a village near the River Jordan.

2. These two Galilees comprise a very sizable land. Many alien nations surround them yet the Galileans have been able to offer strong resistance against any enemy attack. The Galileans are trained for warfare from mere infancy and its population has always been numerous. Nor has the land ever lacked courageous men. The soil is rich and productive and contains many fruit tree groves. In fact, its fruitfulness compels even the laziest among the people to cultivate the soil. And so no part of the land lies fallow. Every inch is farmed by the populace. Additionally, because of the rich soil, the cities and many villages of Galilee are full of people. The very smallest among them contains no fewer than 15,000 inhabitants.

3. In short, if anyone thinks that Galilee is smaller in physical size than Perea, he must acknowledge that Galilee is superior in substance. For the entire country is capable of being cultivated and is blooming everywhere. Perea, though larger geographically, is mostly desert and rough terrain. It is vastly inferior to Galilee in the production of row crops. And yet, a moist soil does exist in some parts there that produces fruit. Planted in its plains are a variety of trees, chief among which are the olive tree, the date palm, and the grapevine. Perea has sufficient water from seasonal streams that flow in arroyos out of the mountains. Perea also has springs that never run dry, even when the seasonal torrents fail them in the dog

days of summer. Perea's length runs from Macherus to Pella, and its width from Philadelphia to the Jordan. Its northern border is at Pella, as we've said, and its western border is by the Jordan. Moab sits on Perea's southern frontier and its eastern limits reach to Arabia, Sebonitis, as well as Philadelphene and Gerasa.

4. As to Samaria, it lies between Judea and Galilee. It begins with Ginea, a village in the great plain. Samaria ends at the Acrabbene toparchy and is similar to Judea in geography. Both countries are comprised of hills and valleys, are wet enough for agriculture, and are very fruitful. Trees, both wild and cultivated, are in abundance and are full of fruit in the fall. The trees are not watered naturally by many streams, but receive their moisture from a regularly abundant rainfall. The rivers and streams that flow through Samaria and Judea contain very sweet water. Their grasses are excellent and feed cattle that yield more milk than in other places. But the best sign of excellence of the two countries is that they are both full of people.

5. Within Samaria and Judea lies the village called Anuath, also known as Borceus. Anuath marks the northern boundary of Judea. If one looks at the southern length of Judea, it is bounded near Arabia by a village the Jews call Jordan. However, Judea's breadth runs from the River Jordan to Joppa on the coast. Jerusalem is in the center of the country. Some here have called the city "Navel" of Judea, which it truly is. Neither is Judea exempt from seaside delights. Its shore extends northward as far as Ptolemais.

The country was once divided into eleven regions, of which Jerusalem was the uppermost. It presided over the surrounding country as the head presides over the rest of the body. Of the other Judean cities that ruled over several toparchies, Gophna was the next in line, followed by Acrabatta, Thamnia, Lydda, Emmaus, Pella, Idumea, Engaddi, Herodium, and Jericho. After these presiding cities came Jamnia and Joppa. The regions of Gamala, Gaulonitis, Batanea, and Trachonitis come next, which are all part of Agrippa's kingdom. Trachonitis begins at Mount Libanus and the headwaters of the Jordan River, and reaches east to west to the Lake of Tiberias (the Sea of Galilee). Its length runs from the village called Arpha as far as Julias. A mixture of both Jews and Syrians live there. And so, I have with utmost brevity, described the country of Judea and those countries around it.

Chapter 4

Josephus Tries to take Sepphoris but Is Repelled; Titus Arrives in Ptolemais With a Great Army

1. Auxiliary cavalrymen numbering 1,000 along with 6,000 auxiliary infantrymen were sent to the people of Sepphoris under the Roman tribune, Placidus. They split into two units and made camp. The infantry was placed in the city to guard it while the cavalry camped in the plain. The cavalry continually marched back and forth through the countryside, bringing serious problems to Josephus and his men. They plundered places surrounding the city and captured anyone who dared to reveal themselves. Because of this, Josephus attacked Sepphoris before its citizens could turn against the rest of Galilee. Ironically, he hoped to take the very place he had recently fortified with a strong wall. He failed in his hopes. Josephus not only had too weak a force to overrun the city, but he also failed to prevail upon Sepphoris's citizens to surrender the city to him. By his actions, Josephus angered the Romans to such a degree that they declared war on all Galilee. By both day and night, Roman troops torched villages and towns in the plains, persistently killing anyone who even appeared to be capable of fighting, and enslaving the weaker people. Fire and blood consumed all Galilee. Nothing in the country was exempt from misery and calamity. The people's only refuge was to flee to the other cities that Josephus had fortified.

2. Meanwhile, Titus had sailed from Achaia to Alexandria earlier than the winter season would normally permit. There he took command his most eminent fifth and tenth legions. Marching quickly northward, Titus came swiftly to Ptolemais, where he found his father, Vespasian. He joined his two legions with his father's fifteenth legion. Five cohorts came up from Caesarea with one troop of Roman cavalry and five other cavalry troops from Syria. These ten cohorts each had 1,000 infantry, but the other cohorts had no more than six hundred infantrymen and one hundred twenty horsemen. A substantial number of auxiliary soldiers also joined the Romans, supplied by the kings Antiochus, Agrippa, and Sohemus. Each king had contributed one thousand troops on foot, the majority of whom were archers. So, the entire army, including the kings' auxiliaries

totaled 60,000—both infantry and cavalry combined, not counting the servants. These servants followed the army in vast numbers. Trained for war like the regular troops, the servants were undistinguishable from the regular fighting men. They served their masters in peacetime and underwent common dangers in times of war. They were inferior neither in skill nor strength. They were merely subject to their masters.

Chapter 5

A Description of the Roman Armies and Roman Camps; and Special Commendations of the Roman Armies.

1. One cannot but admire the Roman precaution in providing household servants for themselves. They serve not only in common life, but are also of great advantage in wartime. If one only sees the other areas of Roman military discipline, he will be forced to confess that their acquisition of such a large empire is due not to mere luck to their courage. They don't just pick up their weapons when war threatens, nor do they stop training in peacetime. Rather, they always have their weapons close at hand, never resting from war exercises, and never waiting until war causes them to take up arms. No difference exists between their exercises and actual warfare. Every day and with great diligence, each soldier trains as if in a real battle. This training is the reason they don't tire easily. This consistency is so ingrained that neither disorder nor confusion can steal from it, nor fear drive them out of it, nor hard work exhaust them. This solid discipline always conquers enemies who do not practice it. One might call their military exercises "un-bloody battles," and their battles "bloody exercises." Neither can an enemy surprise the Romans with a sudden attack. As soon as they have marched into enemy land, the Roman soldiers do not begin to fight until they have fortified their camp. The wall they build is substantial and complete. Some troops remain outside the encampment while those inside arrange themselves systematically. If the fortified ground is uneven, it is first made level. The camp is always square, and many carpenters are required to erect a wall and interior buildings.

2. The Roman camp is designed to contain tents for the men. The outer circumference resembles a wall arranged with towers at equal distances from each other. Located between the towers are engines for throwing darts and javelins. These engines sling stones as well as any other missile

that can annoy the enemy. All stand ready for their unique operation. The Romans also construct four gates into the camp. Each side has one that is large enough for their horses to use, and wide enough to counterattack should the need arise. Very wide and useful streets divide the interior of the camp. The officers' tents are in the center and in the midst of these is the general's tent, resembling a temple. The camp looks like a quickly constructed city with its marketplace and handicraft tradesmen. There are seats for the officers, where, if conflicts arise among the men, they can be heard and judged. The camp, then, and all in it is totally encompassed by a wall very quickly built by their many skilled workmen. If need be, a trench is excavated around it all at a breadth and depth of six feet.

3. When the Roman camp has been made secure, they live together by companies in quiet and uniformity. The Romans manage their affairs in confidence and good order. Each company has its own wood, corn, and water brought in as needed. They all eat together and never alone. Trumpets sound the times for sleeping, rising, and watches. In fact, nothing is done without these signals. In the morning, each soldier reports to his centurion, who in turn reports to his tribune to salute him. In the same way, the superior officers report to the army general who gives them their password and orders, which they will carry back to all under their command. The same is true when they are in a battle. They can maneuver very quickly, whether in modes of attack or orderly retreat.

4. Now when the time comes to abandon the camp, the trumpet sounds, and everyone gets busy. At the first trumpet, they all take down their tents and get ready to evacuate the camp. Then the trumpets sound again to get ready to march. At this signal, the soldiers place their baggage on their mules and other beasts of burden and stand ready to march. They also set fire to the camp at this time to keep their enemies from using it. It will be easy to build another camp somewhere down the road. Then the trumpets sound a third time, and they leave the camp. This trumpet also warns those that may find themselves late and out of ranks when the army marches. When all is ready, the crier stands at the general's right hand and asks the men three times in their own language whether or not they are ready to go out to do battle. To which the men reply in loud and cheerful voices, "We are ready!" Even before the question is asked, the men almost shout their response. They yell as if filled with a kind of warlike fury. At

the same time, they lift their right hands toward the sky.

5. When all this is accomplished, and the camp is left behind, the army marches silently and in good order. Every man keeps in step as if they were going to war. Breastplates and helmets protect the foot troops. Every man carries two swords on his hips. The sword on the left is much longer than the other, for that on the right is no longer than a span (about nine inches). Some infantrymen are chosen out of the rest to protect the general, each armed with a lance and a circular shield. The rest of the foot soldiers carry spears and an oblong shield, plus a saw and a basket, a pick and an axe, a leather strap and hooks, and three days' rations. A footman carries almost as much as a mule. The horsemen have a long sword on their right side and carry a long javelin in their hand. A shield lays across the horse's side next to a quiver, containing not less than three darts with wide points, each the size of spears. They also wear helmets and breast-pieces, as do the infantry. Those cavalrymen who are picked to guard the general have exactly the same armor as the other cavalrymen. The general always leads his legion forward surrounded by these men who are chosen by lot to be his guards.

6. So, this is how the Roman troops march and rest, and also how they carry the several weapons they employ. But when they are ready to fight they leave nothing to chance or to be executed without prior planning. In every case, they first meet and agree on any action, and they execute their plan. For this reason, they rarely commit errors. If a mistake is made it is easily corrected. They believe that any errors they make following their prior planning is better than a rash victory owing to luck alone. Running rashly into battle without a plan might blind them to any errors made. But creating a battle plan, even though the plan might fail, allows them to see their mistakes in planning or execution and to learn from those mistakes. For, if by mere chance victories may come, they do not have their origin in the victor but mere luck. But in planning well—if sad, unexpected events occur during the battle, the Romans have comfort in knowing that they did their best to prevent them.

7. Roman military training in weaponry grows soldiers with strong bodies and courageous souls. Fear of punishment also steels their minds for warfare. For their regulations make not only desertion a capital offense,

but also even the slightest slothfulness or idleness can bring a punishment of death. And the Roman generals are even harsher than their regulations. They deflect any criticism of cruelty toward those condemned by the great rewards they bestow upon the most valiant soldiers. The soldiers' readiness to obey is exceedingly strong even in peacetime, and not merely ornamental or simulated.

In a battle, the whole army moves as one body. So tightly coupled are their ranks, so sudden are their maneuvers, so sharp is their obedience to orders, so quick is their response to the ensigns, and so nimble are their hands at all their work, that what they are about to suffer is borne with great patience. We do not know of any examples where the Roman army has lost a battle or even when come close to losing, either by the size of the force facing them, enemy strategies, the difficulty of the terrain, nor by sheer luck. Their successes have been more certain than if fortune alone had granted them. Therefore, in a situation where planning goes before action, and where after receiving the best advice that advice is followed by so solid an army, is it any wonder that the limits of Rome's empire extend to the Euphrates on the east, the (Atlantic) ocean on the west, the fertile regions of Libya on the south, and the Danube and Rhine rivers on the north? One might say that Rome's possessions are not inferior to the Romans themselves.

8. I have given the reader this description of the Roman army not so much to commend them as to comfort those who have been conquered by them. I also want to deter others from attempting to rise against their government. This discussion of Roman military affairs may also be of use to the curious who have a desire to learn. I now return from my digression.

Chapter 6

General Placidus Attempts to Take Jotapata (Yo-TAP-it-tuh), but is Beaten Back. Vespasian Marches into Galilee.

1. Vespasian and his son, Titus, had delayed for a time at Ptolemais in order to get their army ready. Meanwhile, Placidus had overrun Galilee and executed many of the Jews he captured. He realized that in trying to avoid capture, many Jewish warriors ran behind the walls of the fortified cities built by Josephus. So, Placidus marched furiously against Jotapata,

the strongest city of them all. He figured that if he could take Jotapata by a sudden, surprise attack, he would gain great honor for himself among the other generals, and give Rome the advantage in future campaigns. If the Romans could take the strongest of Galilean cities, the rest would be so afraid they would surrender. But Placidus was mistaken. The men of Jotapata knew of Placidus's plan to attack them, and sallied out of the city to surprise his forces. They fought the Romans vigorously on the ground where they were least expected. Being more numerous, prepared for battle, and extremely motivated, the Jews knew that both their country and their wives and children were in danger. They easily put the Romans to flight. Many Romans were wounded, seven mortally. But their retreat was orderly and their wounds only superficial due to the armor they wore. Also, the Jews threw their weapons from a distance and did not get close enough to engage in hand-to-hand combat. They were only lightly equipped with weaponry while the Romans were more heavily armed. Three Jews died in the fight, and a few more were wounded. So Placidus, finding his force too weak to assault Jotapata, turned and marched away.

2. Vespasian was ready to attack Galilee, so he marched out of Ptolemais in the order of marching normally used by the Romans. He ordered the lightly armored auxiliaries to march first along with the archers, in order to prevent any sudden parries by the enemy. These men could also scout the forests for anything that looked suspicious, and were capable of repelling ambushes. Following them marched the fully armed Roman troops, both infantry and cavalry. The first of these was a detachment of ten men out of every hundred who carried, besides their arms, everything necessary to survey the evening's campsite. After these marched the engineers, men who were to make the roadways even and straight, to plane down any rough spots and to fell any trees that might hinder the army's progress, so that the men would not tire out. Behind these men, Vespasian set his carriages and those of the other field commanders, along with a good number of cavalry guards. Following these men came a select body of infantrymen, cavalrymen, and pike men. Then marched the unique cavalry of his own legion. Each legion of these privileged outfits had one hundred twenty horsemen. Then came the mules that pulled the siege engines and other war machines of that nature. After these came the cohorts and tribunes, surrounded by soldiers chosen out of the ranks. Then marched the ensigns and flags carrying the image of an eagle, which is at the head

of every legion. The eagle is the king and strongest of all birds. It was for Rome a symbol of dominion and an omen that they would conquer all against whom they marched. The trumpeters follow these sacred ensigns. Then comes the main body of troops in their squadrons and battalions, six men abreast. A centurion walked with them to keep the men in step. The servants of the legion followed the infantry and led the mules and other beasts of burden that carried the soldiers' baggage. Finally, behind this long line of marching groups came a multitude of mercenaries. Then, to bring up the rear and provide security for the march came a rear guard of both armored infantry and a large body of cavalry.

3. And so Vespasian marched with his army to the Galilean border. He pitched his camp and had to restrain his soldiers who were eager to go to war. He also displayed his army before the enemy in order to frighten them and afford them time to repent and change their minds before the battles began. At the same time, Vespasian made preparations to attack the fortified cities of Galilee. Indeed, the very sight of Vespasian's army caused many to recant their notions of revolt. Everyone was in awe of his force. The men in Josephus's camp were in a city called Garis, not far from Sepphoris. When they heard that the war had come close to them and that the Romans would attack in hand-to-hand combat, they dispersed and fled. Their retreat occurred even before the battle had begun, and even before the Roman forces came into view. Only Josephus and a few others were left behind in the Jewish camp. Realizing that he did not now have a large enough army to engage the enemy and that the Jews' morale was at an ebb, and that the majority of Jews would, if possible, rather seek terms than fight, Josephus considered the entire war effort lost. He then determined to get as far from danger as possible. So he took those who remained with him and went to Tiberias.

Chapter 7

Vespasian Marches to Jotapata Following His Victory at Gadara. After a Long Siege, Vespasian Takes the City.

1. So, Vespasian marched on Gadara and took the city in his first attempt, finding Gadara lacking many men fit for battle. He entered the city and killed all of the young men. The Romans had no mercy whatsoever on any age because of their hatred for the nation and because of the Jews'

treatment of Cestius in his retreat from Jerusalem. Vespasian then set fire to Gadara, and to all the villages and towns nearby, some of which were unoccupied. Those that did have inhabitants, he made slaves of its citizens.

2. When Josephus retreated to the city he chose as most secure, the people of Tiberias panicked. They could not imagine that he would run anywhere unless he thought defeat by the Romans to be inevitable. Indeed, at that point they were correct in understanding his thoughts. He saw where the affair was headed, and the only way of escape for the Jews was repentance. But even though he believed the Romans would forgive him, Josephus would have rather died many deaths than to betray his country or to dishonor the supreme command of the army to which he had been entrusted. Neither could he see himself living happily with those against whom he was sent to wage war. He determined, therefore, in a letter to the prominent Jews in Jerusalem, to give an exact account of affairs in Galilee. He did not want to overemphasize the power of the enemy and make them appear unbeatable. Nor did Josephus want to make Rome's power sound less than it was. That might make the Jews want to fight when they were predisposed to seek peace. He also sent a message that they should send word immediately if they considered coming to terms with the Romans. Otherwise, if they were resolved to wage war they must send him an army of sufficient strength to fight the Romans. Accordingly, Josephus wrote these things and sent messengers hurrying away to Jerusalem.

3. Vespasian seriously wanted to demolish the city of Jotapata. He had received intelligence that the majority of the Jewish army had retreated there and that it was a very secure hiding place for them. So he sent both infantry and cavalry units to level the mountainous and rocky road. The road was accessible by foot but impossible by horseback. But the road builders accomplished the task in four days, opening up a wide path for the army. On the fifth day, which was the 25th day of Artemisius (The Jewish month Jyar: roughly April to May), Josephus travelled from Tiberias to Jotapata. His arrival raised the low morale of the Jews in the city and changed Vespasian's plans somewhat. A Jewish deserter told the Roman commander of Josephus's arrival in Jotapata. Hearing this, Vespasian sped up his march, supposing that if he was able to take Jotapata he could

largely end the war, particularly if he could get his hands on Josephus. So Vespasian took the news to be of great advantage to Rome. He believed the circumstance to have been brought about by God's providence; that by His own accord the finest Jewish commander was holed up in a place where his capture was possible. Accordingly, Vespasian sent Placidus with 1,000 horsemen, along with Ebutius, an eminent decurion in both strategy and action, to surround Jotapata in order that Josephus might not slip away.

4. On the next day, Vespasian and his entire army followed the advance party, arriving outside Jotapata late in the evening. He brought his troops north of the city and pitched camp on a small hill less than a mile from its walls. The Roman commander made every effort to worry the enemy. This was indeed the case. Not one Jew dared go beyond the city walls. The Romans, who had marched all day, did not attack immediately. In order to block every exit, they encircled Jotapata with a double line of infantry and with a third line of cavalry behind them. This caused the Jews to give up any idea of escaping, which prompted more boldness. Nothing makes a man fight more desperately in battle than to have his back against the wall.

5. The Romans began their assault on the following day. Meanwhile, many of the Jewish defenders had left the city and had formed a line of resistance in front of the walls to face the Romans there. So, Vespasian sent his archers and stone slingers to work against them, their missiles coming from a great distance away. Jotapata looked vulnerable, so Vespasian and his main body of foot soldiers prepared to launch their assault. But Josephus, in great fear for Jotapata, counterattacked out of the city in great numbers, driving the Romans from the walls. The Jews performed boldly and heroically, yet suffered as many casualties as the Romans. While the Jews were toughened by their waning hopes of survival, the Romans were pressed on by a sense of shame. They possessed better skill and strength, but the Jews had fear to arm themselves and lead them to fight furiously. The battle lasted all day, and only ended when night came. The Jews had wounded a great many Romans and killed thirteen. Of the Jews, seventeen were killed, and six hundred wounded.

6. On the following day, the Jews charged out of the city and attacked

the Romans again. They fought even more desperately than before, encouraged as they were by their good showing the previous day. But they found the Romans also ready to fight more desperately, for their sense of shame fueled their fury, viewing their failure to secure a quick victory as tantamount to defeat. The fighting continued for five days, as the Jews of Jotapata sallied out from the walls to hit their enemy, then fought fiercely behind their fortifications. They weren't afraid of the Roman strength, nor were the Romans discouraged by the difficulties they faced in taking the city.

7. Almost all of Jotapata lay next to a steep escarpment, and was almost completely surrounded by sheer cliffs and deep valleys. Looking down from the walls, one can barely see the valley bottoms. Only on the north side where most of the city lies is there level ground. Josephus had previously fortified the city with a wall so high that an enemy could not scale it. Jotapata is also surrounded by mountains. The city can hardly be seen until one approaches it. Jotapata held a formidable defensive position.

8. Therefore, in Vespasian's attempt to overcome the city's natural defenses as well as the Jews' bold defense, he resolved to press the siege vigorously. He called the officers under him to a war council and asked them the best way to prosecute the siege to Roman advantage. They thought it best to build an embankment against that part of the wall where it was practical. So, Vespasian sent the entire army out to gather the necessary materials. The men cut down all the trees on the surrounding mountainsides and assembled a huge mountain of stones. Some brought hurdles or coverings to spread over the work to protect the builders from the effects of Jewish spears, darts, and arrows that rained down from the wall. Others dug up soil from neighboring hillsides and brought it to the builders. Every Roman was busy in one way or another. No one was idle. But the Jews threw large stones and all sorts of missiles from the city walls upon the hurdles. The noise of the falling flotsam was so violent it impeded the work.

9. Vespasian then set up one hundred sixty engines for releasing stones and darts to cover the city. Each one was then set to work, attempting to dislodge those Jews who manned the city walls. The darts made an unholy racket as they hit. Stones weighing as much as sixty pounds were thrown from machines built for that purpose, along with flaming firebrands, and a

multitude of arrows. The walls became so dangerous that none of the Jews could safely come onto them. Neither could they venture into the other parts of Jotapata that were being bombarded by the engines' missiles. A great many Arabian archers began their work alongside the engines, fighting with the Romans who threw their spears and slung their stones. But the Jews did not lie still. When they could not cast their weapons from high places on the walls, they sallied out of the city in groups like bandits to pull away the shelters that covered the Roman workers, killing those left unprotected. When the workmen retreated, the Jews pulled away the dirt that comprised the siege embankment, burning the wooden parts and as well as the shelters. At length, Vespasian realized that the spaces between the embankments were problematic. So he united the protective shelters as one piece and joined together the armies in front of the walls as one unit. This prevented the Jewish isolated sallies.

10. When the Roman embankments rose close to the wall's battlements, Josephus thought he had to oppose the Roman embankment to provide more security for the city. So he gathered his workmen and ordered that they build the walls higher. The workmen complained that it would be impossible with so many missiles being thrown at them. So the commander invented a sort of cover for them. He ordered that piles be driven and that the hides of freshly killed oxen be strung between the piles. These hides would yield and flex as the stones hit them. The spears and darts would slide off harmlessly, and the firebrands hurled would be quenched by the skins' moisture. He brought the skins up to the workmen and the work to elevate the wall continued in safety. Construction continued day and night until the wall was over 40 feet high. Josephus also built several towers along the wall and fitted the wall with strong battlements. This immensely discouraged the Romans who, in their opinion, should have already been in the city. They were amazed at Josephus's contrivances and the courage of the people in the city.

11. Vespasian became visibly irritated by the subtlety of Josephus's strategies and the boldness of Jotapata's citizens. Being encouraged by the elevated wall, the Jews renewed their raids against the Romans. Every day they sent out small squads using the same methods that thieves use. They stole whatever they could lay their hands on and set fire to the Roman works, until at last, Vespasian ordered his army to cease fire. He resolved

to stop fighting and rather to starve the city into surrendering. He believed that lacking provisions, the city would be forced to ask him for mercy. Or, if they had the courage to hold out until the very end, they would die of starvation. Additionally, as the people grew weak from famine, he would be able to defeat them more easily. The Roman general continued to order that no one leave the city.

12. The besieged city had plenty of corn and other necessities, but they lacked water. No spring existed within the city walls. The citizens were normally more than satisfied with rain water, but it was summer and rain rarely fell in that season. So even during the earlier siege they were very concerned about satisfying their thirst. Soon all water supplies were depleted, and the citizens began to lose heart. Josephus had seen to it that the city was well supplied with other necessities and that the men were of good courage. Wanting to extend the siege longer than the Romans expected he ordered water rationing. But the people considered the scanty water ration even more difficult for them than no water at all. Not being able to drink until their thirsts were slaked made the people want water all the more. They were more discouraged than if they had no water at all. The Romans knew of the water situation in Jotapata. For when they stood above the city, beyond the wall, the Romans could see people gathering to get their rationed water. Whenever the people gathered like this, Roman javelins rained down upon them, killing many.

13. Vespasian hoped that the city's water supply would soon run dry, and its people forced to surrender. But Josephus was determined to break that hope. He ordered that the people wet many of their clothes and hang them out on the battlements to dry. Soon, the entire wall was covered with water running down its side. When the Romans saw that the people of Jotapata apparently had so much water they could afford to throw it away, they became discouraged once more. They had thought the city cisterns were about to become empty. This made the Roman general think again about starving the city into capitulation, so he ordered the siege to begin again to compel the city's surrender by force of arms. This is exactly what the Jews wanted. They had no hope of escape and preferred death in battle over death by hunger or thirst.

14. But Josephus had devised another plan to supply all the city's needs.

The Jewish Wars 201

A certain very rough, uneven cliff existed beside the city, so steep it could hardly be climbed. Because of its inaccessibility, the Romans didn't bother to guard it. So Josephus sent some men down into the valley that lay west of the city with letters to the Jews he knew in the surrounding countryside. He received word back that the Jews would bring whatever the city needed, and lots of it. So he advised them to creep along by the night watch when they came to the city with their backs covered with sheepskin with its wool outward. If anyone saw the Jews in the darkness, they might be thought of as dogs. The city received this assistance until the Roman pickets perceived the ruse and began to guard the steep cliff.

15. Josephus concluded that Jotapata could not hold out for long and that his own life would be at risk if he continued to resist the Romans. So he gathered the city's prominent men and discussed how they might escape. When the rest of the citizens caught wind of the meeting, they gathered around Josephus and begged him not to leave and desert them. They argued that they depended entirely upon his leadership and said they still had hope for the city's deliverance. If he would stay, everyone agreed to undertake any effort with cheerfulness because of him. His staying would also provide some comfort for them even should they be overwhelmed. They also argued that fleeing his enemies and deserting his friends did not become Josephus. He would be abandoning the city as a captain from a sinking ship in a hurricane. By leaving, he would cause the city to drown. When he arrived, it had been safe and quiet. Once Josephus left, all resistance to the Romans would cease. Only in him was their trust placed.

16. Josephus argued that he must leave the city for their sake, not to provide for his own safety. If he stayed, there was little he could do while they were relatively safe. Once the Romans overran them he would die for no purpose. But if he was free from the siege he could bring them relief. For he would immediately gather a great number of Galileans from all over the country and draw the Romans away from Jotapata by another war. He didn't see any advantage to them by his staying in the city. In fact, his presence would only serve to provoke the Romans into a harder siege as they primarily wanted to capture him. Once the Romans heard that he was gone from the city the Romans' eagerness to take the city would wane. So after hearing this, the children and older men and carrying their infants came mourning to Josephus. They fell down before him, grabbing

his feet and holding him tight. They begged him with loud weeping that he would share in their future. I think they did this not out of envy that he might be spared but out of the hope that they, too, would be delivered. The people thought that a misfortune could not overcome them as long as Josephus stayed with them.

17. Josephus thought that if he made the decision to stay it would be due to their plaintive requests. But if he decided to leave they would arrest him. His pity for his friends in their distress also drained him of his eagerness to leave them. So he resolved to stay in Jotapata. Arming himself with their common despair, he said to them, "Now is the time we must begin to fight in earnest, when no hope of deliverance remains. It is a brave thing to prefer glory before life, and to set about some such noble undertaking that will be remembered for years to come." Having said this, Josephus went to work immediately. He led an assault through the city gates against the enemy pickets, then ran as far as the Roman camp itself and ripped their tents to shreds, setting fire to their works. In this way Josephus never quit fighting, neither the next day nor the day after that. He went on with this strategy for numerous days and nights.

18. When Vespasian saw that the Romans were being shamed by these sallies, he ordered his troops not to engage such desperate men as nothing provokes courage like desperation. [The Romans were ashamed to be made to flee in the face of Jewish attacks. But if they made the Jews withdraw, the Roman heavy body armor slowed their pursuit. So the Jews, when they had finished a mission and before the Romans could catch them, ran back into the city.] But the Jews' violence would be quenched when it became clear that they weren't making any progress, much like a fire is quenched when all the wood has burned. The Romans thought it right to gain victory as cheaply as possible, since they are not forced to fight, but merely to expand or maintain their power. So Vespasian repelled the Jewish sallies by his Arabian archers and his Syrians with slings and other men throwing stones. Nor did the Roman general stop his offensive engines which caused the Jews much suffering. When the engines threw their stones or javelins the Jews within range were unable to escape. But the Jews continued to press hard against the Romans fighting desperately without regard to either body or soul. When one Jew tired another took his place.

19. When Vespasian realized that his own camp was being besieged by the Jewish assaults and when his lines were not far from the city's walls, he brought forth his battering ram. A battering ram is a huge wooden beam, not unlike a ship's mast. The ram is suspended by ropes passing under its middle hung from another beam above it, like the balance in a set of scales. It is braced by strong beams on either side, like a cross. When the ram is forced backward by a great many men pulling as one, and then pushed forward by those same men, it batters the walls with it prominent iron head, making a violent noise. No strong tower nor wide walls can resist any more than its first few blows. Any construction will ultimately yield to the battering ram. Vespasian wanted to use the ram at the beginning of his siege when he was so eager to take the city. The ram lay unused in the field as the Jews gave them no rest. So the Romans brought even more war engines forward near the walls in order to clear the walls of the Jewish defenders. Other Romans tried to empty the walls by hurling stones and javelins at the Jews. The Arabian archers and Syrian slingers also drew closer to the walls. This brought matters to such a head that none of the Jews could so much as mount the walls. So the Romans brought the battering ram forward. It was completely encased in protective covering, and the upper part thoroughly shielded with animal skins. Both the ram and those operating it were secure. At the very first ram strike, the wall shook, and the people of Jotapata screamed as if they were already overwhelmed.

20. When Josephus saw the ram continually battering the same place on the wall, he realized that the wall would soon crumble. He, therefore, designed a plan to reduce the engine's force. He gave orders to fill sacks with chaff and to hang them down in front of the place where the ram was hitting the wall so that the chaff might lessen the blow by its cushioning effect. This plan greatly delayed the Roman attempts to batter down the wall. If they moved their engine to another part of the wall, the Jews simply moved the sacks and placed them in front of the ram's strokes with the result that the ram was diverted, and the wall remained unharmed. But then the Romans fashioned an apparatus of long poles with hooks on its end with which they cut in two the lines holding the sacks of chaff. When the ram regained its momentum, the newly built wall began to give way. Josephus and the others now had only fire with which to de-

fend themselves. They gathered whatever dry materials they could find, sallying out of the gates in three groups, setting fire to the machines and to the Roman banks as well. The other Romans were so surprised at the Jews' boldness they were overcome with apprehension and were at a loss as to how to offer assistance. The fires set by the Jews also prevented help from coming from the rear. The materials used were dry and mixed with bitumen, pitch, and brimstone, to the effect that the flames exploded immediately. What had cost the Romans much exertion was consumed in the space of one hour.

21. It was here a certain Jew appeared I find worthy of mention and commendation. He was called Eleazar ben Sameas, from Saab in Galilee. This man picked up a huge boulder and threw it down from the wall onto the battering ram. It hit with such force that the metal head broke off of the ram. Then Eleazar leapt down into the midst of the Romans, picked up the head and nonchalantly carried it back up to the top of the wall. He did all this as an easy target for the missiles of the enemy. Not surprisingly, his naked body was wounded by five darts. But he showed no signs of fatigue when he stood on top of the wall in a moment of great defiance. Then he crumpled up in his wounds and fell down with the ram's head. Then two brothers, Netir and Philip, both Galileans from a village called Ruma, then displayed their courage. They leapt down upon the 10th legion with such noise and violence as to throw their ranks into confusion and put all they met to flight.

22. After these courageous deeds, Josephus and the rest of the defenders took torches and burned the Roman war machines with their protective coverings, together with other materials of the 5th and 10th legions which had fled. The other legions quickly buried their equipment and any combustible materials. However, in the evening the Romans erected the battering ram again and put it against that part of the wall they had damaged earlier. At that point, a dart from a certain Jew defending the city hit Vespasian in the foot. The wound was only minor as the dart had traveled a great distance from the wall to Vespasian. However, when the troops near him saw the blood, the wound caused great disorder among them. Fearing the worst, the word of the general's wound spread quickly through the Roman army. Many soldiers left the siege and came running to Vespasian in shock and fear for him. Before them all ran Titus out of

concern for his father. The Romans were all in mass confusion out of their love for Vespasian and for his son Titus's agony. But the father soon put an end to his son's fears and calmed his troops. He showed that he wasn't bothered by the pain and that his wound was only slight. Vespasian encouraged his legions to fight the Jews more violently, for every man was now willing to expose himself to danger immediately to avenge his general. The men shouted encouragements to one another and charged back to the walls.

23. Neither Josephus nor those standing with him deserted the walls even though many fell dead upon one another as the missiles cast by the Roman engines rained down upon them. On the contrary, they assaulted those manning the battering ram beneath them with swords, stones, and fire. But they fought with little effect, dying one after the other in the dark. The Jews could be seen by the Romans while the Romans remained unseen by the Jews. The light of the fires lit by the Jews illuminated only themselves, as if in broad daylight, making them easy targets for enemy darts. The Jews were blind to the engines toward the rear, so that the missiles coming at those on the wall were invisible and impossible to avoid. The force with which these engines threw their numerous rocks and darts wounded several Jews at the same time. And the racket was thundering as the stones hit the wall carrying away the battlements and ripping the corners off the towers. No army could have withstood the onslaught of the huge boulders thrown against the Jews.

    A few incidents of that night can explain the force of the Roman barrage. One of the men standing by Josephus was decapitated by an incoming stone. His head was flung to a place over a mile from where he was hit. A pregnant Jewish woman coming out of her house was hit in the abdomen by a stone thrown by one of the engines. Her unborn child was thrown over 300 yards by the force of the projectile. The noise generated by the Roman engines was as awesome as the stones and missiles they threw. Then there was also the thud of the dead bodies when they hit the ground below the wall. Even more dreadful were the screams of the women inside Jotapata, as they mixed with the moans of the wounded men on the wall. The surrounding mountains echoed the noise back into the city. A great many of those who fought bravely for Jotapata died just as bravely. Many more were wounded. The ground where they fought was awash in blood. So many corpses littered the wall's base that one could have ascended to the top on their backs. When morning came the wall still stood even though it had been battered without intermission. Those inside the walls covered their bodies with armor and began to rebuild its collapsed portions before the Romans could bring up their siege scaffolding.

24. The next morning, after giving his men a brief rest, Vespasian gathered his army to storm Jotapata. He wanted to draw the Jews off of the places where the wall had been damaged. So he ordered the bravest of his cavalry troops to dismount and placed them three-deep against the ruined wall. They were armored from head to toe with lances in hand. The men were to begin their ascent as soon as the ladders were in place. Behind them, Vespasian placed the best of his infantry, while the remaining cavalry were spread out around the walls and into the surrounding hills in order to prevent anyone's escape once the city was in Romans hands. Behind the cavalry, he placed his archers, commanding them to be prepared to fire their arrows. He gave the same command to those armed with slings and to the machine operators. Then Vespasian ordered the men to take up the ladders and have them ready to set up against the unscathed parts of the wall. The strategy was to have the Jews inside the walls busy trying to stop the ascent by these ladders while the dismounted cavalry climbed into the city unopposed. The rest of the defenders were also inundated with darts, and in this way his troops were to be given access into Jotapata.

25. But Josephus saw through Vespasian's strategy. He set old men and men whose energy was spent along the undamaged parts of the walls, expecting no problems there. He then set the fittest of his men at the places where the walls had given way. In front of them all, he placed six men drawn by lot with himself at the most dangerous position. Josephus also ordered that when his men heard the legions' shouts they were to stop up their ears to avoid being frightened. Also, to avoid the enemy darts, the men should get down on their knees covering themselves with their shields, and perhaps retreating somewhat until the archers ran out of arrows. When the Romans begin to ascend the walls, all the men should leap out and meet them suddenly with their weapons poised. Everyone should strive to do his best, not just to defend the city, because it was now impossible to preserve, but to avenge Jotapata when the Romans had finally destroyed it. They should now visualize how their old men are to

be slaughtered, and how their wives and children are to be instantly murdered by the Romans. Before all of this happens the men should spend every ounce of their fury, knowing that these catastrophes are coming, and pour their fury out upon their enemies.

26. So Josephus positioned both of these two groups. The other helpless citizens—women and children—had seen Jotapata surrounded by this huge Roman army. They also had seen the city walls broken down, the enemy with their swords shining in the hills around them, and the Arabian archers with their arrows. These helpless ones mourned in a final death-cry of distress, as if destruction had already come upon them. But Josephus ordered the women to close their doors lest they should render their fighting men impotent by their pity. He commanded that they be silent and threatened them if they were not. Meanwhile, he stood in the breach where he had stationed himself. He took no notice of the Romans bringing ladders to other places, but waited bravely for the shower of Arabian arrows to come.

27. The trumpets of the several Roman legions blared as one, and the army shouted ferociously. The Arabian arrows flew thick into the city. But Josephus's men remembered his orders, stopping up their ears and covering their bodies against the arrows. The siege engines had not yet begun to work when the Jews suddenly ran out to take them before their operators could put them in action. As the Roman soldiers began ascending the walls, the great battle commenced. The Jews were in extreme danger but fought courageously, not discontinuing the struggle until they fell dead or had killed their opponent. But the Jews grew tired from the constant physical combat. They had no reserves in support. The Romans, though, continually replaced exhausted men with fresh reserves. When one man was thrust back another quickly came to take his place. They fought side-by-side with shields joined, encouraging and protecting one another. In close formation, they became an impregnable force, and climbing the wall pushed the weary Jews backward.

28. In utmost distress and need, Josephus nevertheless used the situation to help him think. [Necessity is the mother of invention especially when aggravated by despair.] He gave orders to pour boiling oil on the Roman dismounted cavalry. As soon as the oil was prepared the men brought up

great quantities of it and poured it on the heads of the Romans climbing the wall. They threw their pots down still hissing from the fire. The oil burned the Romans so severely they fled from their positions on the battered wall in horrible pain. The oil ran down their bodies from head to toe and under their armor, eating away at their flesh. It burned like fire itself as it clung to their skin and slowly cooled. The Roman troops could not free themselves from the burning oil, cooped up as they were in their helmets and breast armor. They could only thrash about in their torment as they fell back from the bridges they had set up. So the Romans crouching beside the battered wall were beaten back through the troops pressing forward. The retreating troops became easily wounded by the lances of those advancing.

29. The Romans lost no courage in this brutal setback. nor did the Jews lose their resourcefulness. When the Romans saw their friends thrashing about in torment, their desire to incur revenge on those who had poured the oil grew vehemently. Every Roman then cursed the man in front of him as a coward if he was hindered in any way from reaching the wall. Meanwhile, the Jews arrived at a new strategy to prevent the Roman ascent. They poured boiling fenugreek (from a local plant) on the scaffolds and planks to make the Romans slip and fall back. Neither those climbing up nor those going down could keep their footing. Some fell backwards and were crushed under the feet of those ascending. Many fell down upon the scaffolding they had raised, and when prone, the Jews killed them. When the Romans saw they could not keep from slipping and falling, the Jews were freed from hand-to-hand combat and shot their arrows. So in the evening, Vespasian recalled the soldiers who had suffered so severely. Many died while even more lay wounded. But no more than six Jews of Jotapata died although more than 300 suffered wounds. The fight took place on the 20th of Desius [Sivan]. (June-July, 67 A.D.)

30. Vespasian comforted his angry men. He found them sullen and wanting to return to action. So he ordered that the embankments be raised even higher. They were to erect three towers, each fifty feet high and covered with iron plates on each side. The iron would make the towers more stable because of their great weight, and would be incapable of catching fire. He set the towers on the embankments and put on them lighter war engines capable of casting stones and missiles at the enemy, along with

the dart throwers and archers. He also placed his strongest and bravest men, with their slings, behind the protective iron battlements. They were to cast their stones at the Jews on the wall who were easily seen from the Romans' high position. So, the Jews became unable to escape the missiles raining down upon them from above. They were also unable to return missiles to men they could not see. The towers were so far above them that a hand-thrown javelin could barely reach it. The towers' iron plates made them impossible to damage using fire. The Jews were forced to run out of the city walls and fall upon the Romans, who shot at them. In this way the people of Jotapata resisted the Roman siege, even though a great many of them died each day. They could not retaliate in any meaningful way nor keep the Romans out of the city without placing themselves in a death trap.

31. About this same time, Vespasian sent Trajan to attack the city called Japha near Jotapata. The citizens of Japha had sensed they also could oppose the Romans, encouraged as they were by the length of time Jotapata had been able to thwart the Roman siege. Trajan then commanded the tenth legion. Vespasian gave him 1,000 horsemen and 2,000 troops on foot. Arriving at the city, Trajan found it well fortified, to be taken only with difficulty. Besides its strong natural positioning, Japha was encompassed by a double wall. But then he saw the people of Japha running out of the city, ready to fight him. He advanced and joined the battle, chasing the Jews inside their first wall. The Romans followed close behind and entered the gates with them. When the Jews tried to escape through the second wall, those inside shut them out, being afraid that the Romans would then force themselves into the city along with their own men. Clearly, it was God who brought the Romans into Galilee as punishment, exposing every one of Japha's citizens to annihilation.

The crowd of citizens trapped between the walls hammered at the inner gates earnestly, calling the names of those inside the city, begging them to let them in the gate. Yet their throats were cut as they screamed. For the Romans had shut the gates to the outer wall, and their own friends had closed the inner gates, so they became trapped between the walls and were slaughtered in great numbers. Many Jews were run through by their friends' swords, and many fell on their own swords, in addition to the vast number killed by the Romans. The courage of those trapped failed, for, besides the enemy's onslaught, their own kinsmen had betrayed them.

This completely broke their spirits. They died, not cursing Rome, but cursing their own fellow citizens. Soon, all 12,000 were annihilated. Trajan figured that no one else in the city was capable of resistance. Although there might be a few left, he thought they would be too timid to fight, so he backed off and reserved the honor of taking the city to Vespasian. He sent messages to the general and asked him to send his son, Titus, to complete the victory over Japha.

Imagining that some fighting still might be necessary, Vespasian sent five hundred horsemen and 1,000 foot troops with Titus, who went quickly to Japha. He put his troops in line, setting Trajan on his left flank, leading the Romans in their siege. The soldiers brought ladders up against the city walls on every side. The Galileans inside Japha opposed the Romans on their walls for a season, but soon retired into the inner city. On their retreat, Titus's troops leapt into the city, and it looked as though Japha was theirs. But the men remaining in Japha gathered together and fought the Romans in the narrow streets. The women threw whatever they could find on Roman heads. The fight wore on for six hours until all of Japha's fighting men lay dead. The rest of the men had their throats cut, some in the open air and some in their own homes, young and old alike. No male remained in Japha except infants, who, along with the women, were carried away as slaves. The number of those killed, both in the city and the former fight, was 15,000. A total of 2,130 became captives. The Galilean calamity at Japha took place on the 25th day of Desius [Sivan] (roughly May-June, 67 AD).

32. The Samaritans didn't escape their share of misfortunes during those months. They gathered at Mt. Gerizim, which they regarded as a sacred mountain, and camped there. This collection of courageous men obviously threatened war. But their wisdom had not apparently been increased by the miseries that had fallen upon their neighboring cities. Not heeding the Roman victories, the Samaritans' march was pure insanity, as if they depended upon their weaknesses to prepare them for any battle that might flare up. So, Vespasian thought it best to prevent their movements and to cool their eagerness. Although Samaria had garrisons stationed throughout the country, many of them had traveled to Mt. Gerizim. Their assembly gave the Romans grounds to believe they meant war. Vespasian, therefore, sent Cerealis, the commanding officer of the fifth legion, to Mt. Gerizim with six hundred cavalry and 3,000 infantry. Since the Sa-

maritans held the high ground, Cerealis hesitated to climb the mountain to engage the enemy. So, he surrounded the base of Mt. Gerizim with Roman troops and kept watch all that day. Now, the Samaritans were without a water supply and the summer sun beat down upon them mercilessly. [They had forgotten to even provide themselves with life's basic needs.] In fact, some Samaritans died they very day of thirst while others who preferred slavery to death deserted to the Romans. Cerealis knew from their reports that those remaining on the mountain were seriously demoralized by their misfortunes. So the Roman commander went up the mountain, after he surrounded it with his troops, and invited the Samaritans to come to terms with him, giving them the security of his right hand. Cerealis assured them that if they laid down their arms they would save themselves, as he would then prevent any harm from coming to them. But when not one Samaritan was forthcoming to surrender, Cerealis attacked and killed all 11,600 of them, to the last man. This occurred on the 27th of Desius [Sivan]. These were the catastrophes that fell upon the Samaritans that day.

33. The people of Jotapata still held on bravely, bearing up under the miseries that destroyed all hope. On the forty-seventh day of the siege, the platforms built by the Romans were becoming higher than Jotapata's wall. On that last day, a deserter was taken to Vespasian. He told the general how few Jews remained in the city and how weak they were. The people had grown so worn out with lack of sleep and the perpetual fighting that they would be helpless against any force that might come against them. They would also be vulnerable to a secret assault. For, on the night's last watch, when sleep used to come for the weary men on duty, the watch was accustomed to falling asleep. So, the deserter advised Vespasian to attack at that time. But Vespasian was suspicious. He knew how faithful the Jews were to one another, and how they hated the punishments that might be inflicted upon their brothers, particularly since the people of Jotapata had already suffered such torments. So the Romans tortured the deserter in a fiery trial, yet he would not change his story. As the Jew was crucified, he smiled down at his tormentors. However, Vespasian soon thought that at least some of the deserter's testimony might be true. He also thought that it wouldn't hurt to try to take the city on the last watch, even if the report turned out to be false. So, he commanded his men to take the deserter down from his cross and hold him in custody. Vespasian

then prepared his army to attack.

34. The army marched silently on Jotapata on the night's last watch. Titus himself was the first to enter the city, along with one of his tribunes, Domitius Sabinus, accompanied by a few men from his fifteenth legion. They cut the throats of the sleeping watch and entered Jotapata very quietly. Following Titus's party came the tribunes Cerealis and Placidus, leading those under them. With the fortress overrun and the Romans in the city center in broad daylight, their presence was still unknown to the Jews, most of whom were fast asleep. Just then, a deep fog chanced to fall upon the city, keeping those who were awake from seeing the situation until the whole Roman army was upon them. It was then that the citizens realized the catastrophe. As they were being slaughtered, the people understood too late that the city had been overrun. The Romans, remembering the suffering they had endured during the long siege, neither pitied nor spared anyone. They drove the people down over the precipice, slaughtering them as they went forward. The narrow streets prevented those Jews who could still fight from defending themselves. They lost their footing on the edge of the cliffs, overpowered by the mass of Roman soldiers pressing forward from the fortress. Many Jews, even some of Josephus's chosen men, were provoked to commit suicide, resolving not to be slain by the Romans. Unable to kill any of their enemies, many Jews met on the edge of the city and killed themselves.

35. However, some Jews who had observed the Romans sneaking into the city ran from them as fast as they could run. They got into one of the towers on the north side of Jotapata and held out for a short time there. But surrounded by a host of Romans, and too weak to kill themselves, they willingly offered their necks to those who fought them. The Romans might have boasted that this last thrust against Jotapata was without any casualties. But a certain centurion named Antonius was killed by deceitful means. Here's how it happened. One of the great numbers of Jews who had fled into the caverns below the city asked that Antonius give him the right hand of security and preserve his life by helping him out of the cave. Whereby Antonius reached out his hand, which the Jew knocked aside, stabbing the centurion in the stomach with a spear, killing him.

36. The Romans slaughtered every one of the citizens who appeared open-

ly that day. On subsequent days, the Romans searched out hiding places and fell upon those in the caverns. They massacred people of every age except infants and women. They took 1,200 captive. A total of 40,000 Jews were slaughtered in the city's capture, in addition to those killed during the preceding siege. Then Vespasian gave the order to demolish Jotapata. All of its fortifications were burned to the ground. And so Jotapata was taken in the thirteenth year of Nero's reign, on the first day of the month Panemus [Tamuz] (roughly July, 67 AD).

Chapter 8

Josephus Is Discovered by a Woman and Is Willing to Surrender Himself to the Romans; the Discourse He Had with His own Men when they Tried to Prevent it; what Josephus Said to Vespasian when Brought Before the General; and how Vespasian Treated Him Afterward.

1. The Romans searched for Josephus out of their hate for him and also because Vespasian wanted him taken alive. He figured that if Josephus were in Roman hands the better part of the war would be over. They searched for him among the dead and looked into the city's most concealed recesses. But when the city was first taken, Josephus had been assisted by supernatural providence. Finding himself in the midst of the Roman forces, he withdrew from them and leapt into a deep pit. Adjacent to the pit's bottom on one side was a large den that could not be seen by those at ground level. Forty prominent citizens of Jotapata therein welcomed him. They had hidden there with enough provisions for many days. So, during the day Josephus hid from the Romans who had seized every inch of the city, but at night he climbed up out of the den and tried to find some avenue of escape, taking exact notes of the placement of Roman sentries. But with all escape routes guarded he realized there was no way of getting away unseen, so he went back down to the den where he hid for another two days. But on the third day, after the Romans captured a woman who had been hiding with him, Josephus was found. Vespasian quickly sent two tribunes to him, Paulinus and Gallicanus, ordering the tribunes to give Josephus their right hands as security for his life and asking him to come up.

2. The tribunes invited Josephus to the surface and assured him security

for his life. But the Jewish general did not respond immediately. He was suspicious, and reasoned that the person who made the Romans suffer so much must suffer torture in return, even though the courteous tribunes seemed to be in earnest. So, he was afraid to come out and be punished until Vespasian sent a third tribune, Nicanor. Josephus knew Nicanor well, having been his long-time acquaintance. When Nicanor arrived, he spoke of the natural pity the Romans had for their vanquished enemies and told Josephus that he had fought so valiantly he had invoked the general's admiration rather than his hatred. Vespasian greatly desired to have Josephus brought to his headquarters—not to punish him, for he could do that should Josephus not come voluntarily—but because he was determined to preserve a man of such courage. Vespasian added that if had resolved to punish Josephus, he would not have sent one of the general's friends to him. Nor would he have asked Nicanor to indulge in treachery against his old friend. Nor would Nicanor have agreed to go to Josephus with the intent of deceiving him.

3. As Josephus began to have second thoughts about Nicanor's proposal, angry soldiers came quickly to the den in order to set it ablaze. But the tribune waiting to take Josephus alive would not permit it. So Nicanor pressed Josephus to give him his response. He understood how threatened the Jewish leader must feel, surrounded by so many enemies. Josephus recalled frequent dreams where God spoke of the future destinies of Roman emperors and the coming Jewish calamities. But Josephus doubted such dreams had their source directly from God. Moreover, since he was a priest and the son of priests, he was well acquainted with the prophecies contained in the Jews' sacred books. Then, remembering the dreams and their vivid images, the Jewish commander grew silent and prayed. He said, "O God, since it pleases you to subdue the Jewish nation you created, and as all of its good fortunes has gone over to the Romans, and since you have given me the ability to foresee future events, I willingly surrender to them and choose to live. But I openly protest that I surrender not as a deserter, but as your minister."

4. When Josephus had finished praying, he surrendered to Nicanor. But when the Jews who had hidden with him heard of his decision they rushed at him as one man, crying, "No! May the God-ordained laws of our forefathers stop you! God has not created this race to fear death! O Josephus!

Are you still so in love with life? Can you truly bear the yoke of slavery? How quickly you have forgotten who you are. You, who have persuaded so many to die for their freedom. You have lied about being a real man or to be wise if you can now hope to be preserved by those against you have fought so zealously. The Romans have caused you to forget who you are! You are forsaking the glory of our forefathers. Here, take our right hand and a sword. If you willingly die, you will die as our general. But if you are unwilling to die you will be a traitor to the Jews." As soon as the Jews had finished speaking they leveled their swords at Josephus, threatening to take his life if he surrendered to the Romans.

5. Josephus feared that they would attack, but also feared that he would betray God's commands if he should die before fulfilling them. So, he began to speak to the Jews of the distress he was in, saying, "O my friends, why are we in such a hurry to kill ourselves? Why do we make such a distinction between our bodies and our souls? Can you pretend that I am not the same man as before? No. The Romans know that is not true. It is a brave thing to die in battle but only if it is at the hands of one's enemies. Therefore, if I avoid death by Roman swords, am I then worthy to be slain by my own hand and sword? But if they are merciful and willing to spare their enemies, how much more should we have mercy on ourselves and spare ourselves? It is certainly foolish to do to ourselves what we argue with them for doing to us. I freely confess that it is courageous to die for one's freedom. But only so long as it occurs in battle by those who would take our liberty away. But now we neither wage war against the Romans, nor are they killing us. The man is a coward not only when he will not die when he is obligated to die, but also when he will die when he is not obligated to do so. What do we fear when we go up to confront the Romans? Is it death? If so, of what are we afraid? But, when we merely suspect our enemies will kill us, shall we inflict death upon ourselves with certainty?

"Are we then currently at liberty to choose? Granted, I have heard it said that it is manly to kill oneself. But I believe suicide is most unmanly. Think about a cowardly ship's pilot, who, fearing an oncoming storm, sinks his ship of his own accord. Suicide is a crime unknown to all other animal life and is a sin against our Creator God. No animal chooses to die by his own will or his own means, for the animal's will to live is inbred in each of them. By the same token, we count as enemies those who would take our lives from us. Punishment awaits those that murder treacherously. Don't you think that God will be angry with anyone who injures what God has freely given him? God has given each of us our existence. It is up to Him alone to take it from us. Our bodies are subject to death, as created out of corruptible matter. But our souls are immortal and are part of the divinity that inhabits our bodies. Further, if someone destroys or abuses a trust given him by a man, he is considered wicked and perfidious. How much more wicked is the man who casts from his body this divine deposit?

"And we know that God sees it. Further, we know that slaves who run away from their masters must be punished, even though their masters may have been wicked men. So, shall we run away from God, who is the best of all masters and not be guilty of sin? Don't you know that those who depart this life naturally enjoy eternal glory as they repay their debt to God? Their children and posterity are established, and their souls are pure and obedient and have acquired the holiest place in heaven. From there, in the ages to come, they are to be united again in sinless bodies. But the souls of those who act rashly against themselves are received into the darkest places of Hades while their Father God punishes their posterity for their parents' wicked acts. God hates these things, and punishes them according to His wise counsel.

"Accordingly, Jewish law requires that the bodies of those who kill themselves are not to be buried but to be exposed until the sun has set, even though it is lawful to bury their enemies sooner. Other nations' laws order a suicide's hands cut off, since it was the suicide's hands that destroyed his body. They reckon that as the body is now alien from the soul, so a man's hands should be alien from his body. Therefore, my friends, it is proper to reason rightly and not to add sin against our Creator to the calamities brought upon us by other men. If we want to preserve ourselves, let us do it. To be spared by our enemies who have experienced our courage is in no way inglorious. But if we want to die, it is best to die by the hands of those who have been victorious over us. As for me, I will not go over to the enemy and become a traitor to myself. If I did that, I would be much more of a fool than those who have deserted to save themselves. I would desert to my own destruction. However, I heartily hope that the Romans prove treacherous, because if they kill me after they have given me their right hands as security, I will die cheerfully, carrying with me the knowledge of their treachery as a greater consolation than victory itself."

6. Josephus used these and other arguments to keep the remaining Jews from murdering themselves, but in their desperation the Jews had become deaf. They had decided long ago to die, and were angry with Josephus. Suddenly, they charged at Josephus brandishing their swords—one from this angle, one from another. They called Josephus a coward, and all appeared ready to kill him, but he called one of them by name, and another remembered Josephus's leadership. He took the hand of a third man and prayed for the fourth, making him ashamed. So even in his great distress, Josephus kept his friends from killing him. He was forced to do what the wild beasts do when surrounded. They always turn against the last to touch them. Finally the Jews' right hands hung limply out of reverence for their general in this fatal standoff. Their swords dropped to the ground. Though they all had been poised to kill him, but they now found themselves neither willing nor capable of the act.

7. But even in his extreme anxiety, Josephus held on to his wits. Trusting himself to God's providence, he put his life at risk saying, "Since you have made up your minds to die, come on, let us commit to die by lot. Let him on whom the first lot falls be killed by him on whom falls the second lot. And so fortune will determine the sequence of our deaths, so that no one dies by his own hand. When all of us are dead, it will be unfair that the last should change his mind and save himself." This proposal seemed just to the others, so Josephus proceeded to cast lots, casting one for himself as well. The Jew who drew the first lot bared his neck to the second, supposing that Josephus would die with them soon. They thought that to die with Josephus was sweeter than life itself. Yet Josephus had drawn next to last—whether that happened by chance or God's providence. He didn't want to be condemned by the lot, nor if he had been last in line, to shed the blood of a fellow countryman. So, he persuaded the last man to trust him and continue to live as well.

8. In this way, Josephus lived through the Roman war and the war with his friends. Then Nicanor led him to Vespasian. As he came up from the cave, the Roman soldiers rushed to get a glimpse of Josephus. Disturbances broke out as the crowd of soldiers pressed in upon the Jewish general. Some men were happy that Josephus was taken alive, while others threatened him. As some gathered close around him, other voices from farther away called for his execution. Those near Josephus thought of his

bold strategies against them and seemed deeply concerned for his welfare. Even the Roman officers, once they caught sight of Josephus, repented of their former rage against him. Above all, Titus's courage and Josephus's own patience under affliction made Titus pity him. Titus saw that they were the same age and recalled that his enemy, who was but a short time ago fighting against him, was now in Roman hands. Titus considered the power of chance and how quickly the tide of battle can change. No state of man is certain. As these thoughts poured out to the men around him, Titus brought the soldiers in the crowd to have pity on Josephus. Titus also brought weight to bear in persuading his father to keep Josephus alive. But Vespasian gave strict orders that the Jewish general be kept under heavy guard, as he would shortly be sent to Nero.

9. When Josephus heard Vespasian's order, he asked to speak privately to the Roman general. When all had withdrawn except Titus and two of their friends, Josephus said, "O Vespasian! Don't think that you have merely taken me captive. Rather, I have been sent to you by God as a messenger of greater news. I know how according to Jewish law it is best for generals to die. Truly, are you going to send me to Nero? Will Nero's successors all live until you are chosen? For you, O Vespasian, are to be both Caesar and emperor as is your son, Titus. Bind me tight but keep me for yourself, for, O Caesar, you are not only my lord, but lord of all the land and sea, over all mankind. I certainly deserve to be punished by far greater measure than these bonds if I ever give false testimony to what God has ordained." When Josephus finished, Vespasian didn't believe him, supposing Josephus was up to some trick in order to save his life. But shortly he became convinced and came to believe what Josephus had said.

God Himself must have caused Vespasian to think that this was a foretelling of his advancement to lead the Roman empire. He had also heard Josephus prophesy the future at other times. One of Vespasian's friends who was present at the initial conference said to Josephus, "I cannot but wonder why you could not predict the capture of Jotapata, or even of yourself being captured, unless what you say now is false and spoken to avoid the rage against you." Josephus replied, "I did tell the people of Jotapata that the city would be taken on the forty-seventh day of the siege and also that the Romans would take me alive." When Vespasian had privately asked the other prisoners if what Josephus had said was true, his prophesies were verified. Then the Roman general began to believe

Josephus's words concerning himself. Meanwhile, Josephus, while not set free, was given suits of clothes and other precious gifts. He was treated kindly by the Romans from that day forward. Titus also gave his approval to the honors bestowed on Josephus.

Chapter 9

How Joppa Was Taken and Tiberias also Delivered up to the Romans

1. Vespasian returned to Ptolemais on the fourth of Panemus [Tammuz] – (July, 67 AD), and from there went on to Caesarea by the sea. Caesarea was a large Judean city inhabited mainly by Greeks. The citizens welcomed Vespasian and his army with joyous shouts of acclamation. This they did partly out of good will toward the Romans but mostly out of their hatred of the Jews. For that reason, they came out in large numbers against Josephus demanding that he be executed. But Vespasian pretended he didn't hear the angry crowd and remained silent. He placed two legions at Caesarea in winter quarters, believing the city well suited for that purpose. The Roman general then placed the tenth and fifth legions at Scythopolis so that he wouldn't burden Caesarea with his entire army. Scythopolis was warm, even in the winter months, although suffocatingly hot in the summer, as it was situated on a plain near the Sea of Galilee.

2. While this was going on, Jews gathered in great numbers in Joppa. These had either escaped Roman captivity or had run from the cities demolished by the insurgents. Cestius destroyed Joppa, but the Jews began to reconstruct the seaside city. The region adjoining Joppa was incapable of supporting them as it had been wasted in the war, so the Jews determined to put out to sea. They constructed a number of ships and became pirates, roaming from Syria to Phoenicia to Egypt. The sea quickly became unnavigable because of their threat. As soon as Vespasian heard of these Jewish pirates he sent both infantry and cavalry units to Joppa, a city that lay unguarded at night. The Jews in Joppa caught wind of Vespasian's plans and feared for their lives. But they did not seek to defend the city. Instead, they fled in their ships at night to get out of range of the Roman missiles.

3. Joppa is not a natural port. While the shoreline on one side is rough the

rest is even, and the two ends bend toward one another in the shape of a crescent. High cliffs and great boulders jut out into the sea. This is where Andromeda's chains left impressions, attesting to the antiquity of that fable. But the wind from the north beats upon that shore and dashes huge waves against the rocks. The sea's haven there is more dangerous than the land the Jews had left. So, as the people of Joppa were floating about on the sea, a violent wind arose in the morning. The local residents call it "The Black North Wind." Their ships were dashed against one another and some against the rocks. Other ships were carried further out to sea, in the opposite direction, by the backlash force of the waves. They were afraid to make for land due to the rocks, which were covered with Romans. The waves rose to such a height that many Jewish ships swamped or capsized. There was no place to go, no way to save themselves. They were being thrust out of Joppa by the threat of Roman violence and out of the sea by the violent wind and waves. Loud screams went out as the ships collided with one another. Terrible noises erupted as the ships broke into pieces. Some of the Jews perished in the waves and many suffered shipwreck against the rocks. But some thought to die by their own swords was better than drowning, so they killed themselves before the sea could take them under. But the majority were pushed along by the wind and waves and dashed to pieces against the rock outcroppings. The sea far out was red with blood and full of floating corpses. Those coming alive onshore were set upon by Roman troops and massacred. The number of Jews that fell victim to the turbulent sea was 4,200. The Romans then took Joppa without opposition and demolished it once more.

4. Thus in a short time Joppa was taken twice by the Romans. But in order to keep the pirates out of Joppa permanently, Vespasian built an encampment where Joppa's fort had been located. He left a unit of cavalry there to despoil the surrounding countryside and a few infantry troops to guard the camp. The cavalry were ordered to destroy neighboring villages and towns. So, as they were ordered to do, the horsemen rode roughshod over the country, each day desolating portions of the whole surrounding area.

5. When Jerusalem heard of Jotapata's fate many didn't believe their ears. The calamity was too massive, and no eyewitnesses had appeared to confirm the truth. Not one resident of Jotapata was spared to carry the news

to Jerusalem. Instead, rumors were spread abroad about the city's doom. Bad news travels like wildfire. So, little by little the truth became known from the towns surrounding Jotapata; the news soon became convincing as fact. But much fiction was added to the facts. Reports arrived that Josephus was killed in action in the battle. This news filled the city with sorrow. Lamentations went up in every house for the citizens slain in the calamity. But the mourning for Josephus was very public. Some mourned for those individuals who had lived with them during the festivals, others for their relatives, still others for their friends or brothers. But everybody mourned for Josephus. The weeping went on for thirty days. Many hired professional mourners to play melancholy songs on their flutes.

6. When the whole truth about Jotapata became known, it became clear that earlier reports of Josephus's death were false. As the people came to understand that the Jewish general was alive and living in the Roman camp, and that the Roman officers treated him differently than their other captives, they became enraged. They were as angry with Josephus as they had been sorrowful at the report of his death. Some cursed him as a coward and others called him a deserter. The city was full of indignation at Josephus, and curses rained down upon him. The people's rage was aggravated by their sorrow for Jotapata, and even more inflamed by the terrible Roman victory. The afflictions that would have made wiser men more cautious became for the Jews of Jerusalem an occasion to press on toward future calamities. The end of one misery became the birth of another. The people vehemently resolved to step up the Roman conflict. They determined to avenge themselves on Josephus by their revenge upon the Romans. This was the mood in Jerusalem prior to the trouble that now threatened the city.

7. Vespasian then left Caesarea by the sea and went to Caesarea Philippi in order to visit Agrippa's kingdom. The king had extended invitations to him to do so. Agrippa wanted to host the Roman general privately in the finest manner possible, but to also have Vespasian help correct some problems in his own government. Vespasian rested his army there for twenty days while he feasted with King Agrippa. He also thanked God publically for the victories he had achieved. But as soon as he heard that certain disturbances were arising in Tiberias and that the people in Taricheae had revolted, both places ruled by King Agrippa, he decided to move against

these cities. Vespasian was now convinced that the entire Jewish nation had risen up against their Roman governors. So, for Agrippa's sake he moved to bring reason back to these two cities.

He sent his son, Titus, back to the other Caesarea to take the army that was there to Scythopolis, the largest city of Decapolis, near Tiberias. Then Vespasian went to Scythopolis to wait for Titus's arrival. He brought three legions with him and pitched camp some eleven miles from Tiberias at Sennabris, a site in full view of the revolutionaries. He also sent the decurion, Valerian, with fifty horsemen to speak peaceably to the Tiberians, encouraging them to give him assurances of their fidelity to Rome. He had heard that the people there wanted peace but were being coerced by the rebels to fight for them. When Valerian approached the city and was close to its wall, he jumped off of his horse and ordered the men with him to do the same. He didn't want the people to think they were there to fight before he had the opportunity to speak with them. But just then, a mob of the city's boldest rebels came out upon the Romans, armed to the teeth. Their leader was a man named Jesus, the son of Shaphat, the kingpin of a gang of thieves. Valerian did not think it safe to disobey his commanding general, even though he was certain he would win the fight. But he also knew that it would have been dangerous for so few to fight so many, or for those unprepared for battle to fight against those ready for it. Surprised at the Jews' unexpected attack, Valerian ran away on foot, followed by the rest of his fifty men, leaving their horses behind. Jesus then led the horses into Tiberias, rejoicing as if they were taken in battle and not by a surprise attack.

8. The elders of the people and those who held authority in Tiberias fled to the Roman camp fearing that no good could come from Jesus's treachery. They took their king along with them, falling at Vespasian's feet to ask for his mercy. They begged him to listen to their pleas and not attribute the madness of a few to the many. They asked Vespasian to spare the people who had always been civil and obliging to the Romans. Further, they requested that Vespasian justly punish those who had fomented insurrection and had long restrained the people, preventing them from seeking the right hand of peace from Vespasian. The general complied with their requests while expressing his rage at the loss of his fifty horses. He knew that Agrippa was also greatly concerned for the citizens of Tiberias. On hearing that, Vespasian and Agrippa gave the people the security of their

right hands; Jesus and his henchmen felt unsafe in Tiberias and escaped to Taricheae.

The next day, Vespasian sent Trajan with a detachment of cavalry to Taricheae to determine if the people were of the same mind as their elders and leaders. Finding the people wanting peace, Vespasian led his army to the city. The citizens opened the gates and greeted him with shouts of joy, calling him their savior and benefactor. But due to the narrowness of the city gates, the army took a long time entering. So, Vespasian ordered that the southern wall be breached to provide a broad passage for their entry. He commanded his troops to abstain from debauchery and to treat the citizens justly in order to please King Agrippa. It had been on Agrippa's account that Vespasian also spared what was left of the southern wall, for the king had promised the city's continuing loyalty to Rome. So Vespasian restored peace in Taricheae, which had been treated so severely by the insurgents.

Chapter 10

How Taricheae Was taken; a Description of the Jordan River and of the Country of Gennesareth

1. Vespasian made camp between Tiberias and Taricheae but fortified it more than usual, suspecting that the Romans could be forced to stay for a long battle. The insurgents had all gathered in Taricheae, relying on the strength of the city's fortifications and the lake next to it. The area's people call the lake Gennesaret. (We know it today as the Sea of Galilee.) Situated at the base of a mountain, the city resembles Tiberias. Taricheae had been strongly fortified by Josephus on those sides not facing the lake, but it was not as strong as Tiberias. The walls of Tiberias had been built at the onset of the Jewish revolt when plenty of money and power were available. Taricheae only got what was left over. However, many boats lay in wait upon the lake that might provide for a retreat should one be necessary. The boats were outfitted for war and prepared for battle on the water. But as the Romans were constructing a wall around their encampment, Jesus and his gang assaulted them. The insurgents were frightened neither by the Roman numbers nor their strict discipline and order. The builders scattered at the first sally and the Jews ripped apart their walls. But as soon as the revolutionaries saw the Roman troops gathering into formation, they returned to their main body. The Romans took after them in

swift pursuit, driving the Jews into their boats. They launched out beyond range of Roman missiles and lay at anchor, aligning their boats together as if in a line of battle. In this position, they fought the Romans from the lake. But Vespasian heard that a great number of Jews were congregating in the plain on front of the city. So he sent his son, Titus, with six hundred select horsemen to drive them away.

2. When Titus saw the multitude of rebels, he sent a message to his father asking for reinforcements. But he also realized that many of his cavalrymen were spoiling for a fight even before reinforcements could arrive, and even though some were concerned about the vast body of Jews they faced. Titus said to his men:

"My brave Romans! For it is proper for me to remind you of who you are at the beginning of my speech. You must not be ignorant of who you are or against whom you are to fight. No part of the known earth has been able to escape Roman hands. But as to the Jews, if I may also speak of them, they have already been beaten! And yet they still fight for their cause. If they, having already been defeated are standing fast, how can we, who have only tasted victory, be but disgraced by our fears? I rejoice at your obvious eagerness, yet I am afraid that their great numbers might expose fear in some of you. So let those who are afraid consider who we are as Romans and against whom we fight. These Jews, though they are brave and despise death, are disorderly and unskilled in battle. They ought to be called a mob rather than an army. I need say nothing about our skill and discipline. For this reason, we Romans are ready for war even in peacetime. We should not enter battle comparing our numbers against the enemy's numbers. For what advantage would we gain from our tactics if we must be equal to these undisciplined troops who oppose us? Also consider this: You are armed men ready to do battle against unarmed men. You are on horseback while they are on foot. They have no effective leader, while you have a general.

"These advantages will make you much more effective than mere numbers. For it is not the number of men who win wars, even though they are soldiers, but it is bravery that does it even though the numbers are but few. For a few are easily organized into battle formation and can easily come to each other's assistance. But armies that have too many troops hurt one another more than they hurt their enemies.

"The Jews are ruled by courage and rashness, the effects of insanity. Those passions are impressive when they succeed but vanish at the slightest setback. But we are led on by courage, obedience, and fortitude, which have been revealed by our good fortune. These never desert us in our misfortune. No, indeed, your motivation to fight exceeds that of the Jews, for even though they fight to liberate their country, what greater motivating force is there than glory?

"Let it never be said that the Jews are equal to us after we have conquered the entire known world. Also consider this: We have no fear of suffering an incurable disaster presently, as many are coming to assist us. In fact, they are almost at hand. But I think we ought to strike before my father's reinforcements arrive so that victory will be ours alone, and build our reputation. And I think this to be an opportunity where my father and I and all of you are on trial. Is my father worthy of his former glory? Am I truly his son and are you truly my soldiers? My father is usually on the winning side. As for me, I could not bear the thought of returning to him beaten by the enemy. How could you not be ashamed if you do not equal your commander's courage as he leads you into danger? For you well know that I will go into danger first and lead the first wave of the attack. Do not, therefore, desert me, but convince yourselves that you will assist my initial charge. Also know this before we begin: We shall have greater success in close combat than at a distance."

3. As Titus finished speaking, an extraordinary fury came over the men. They were uneasy that Trajan was now arriving with four hundred mounted cavalry, knowing that their own reputation in victory would be diminished by the addition of more troops. Vespasian had also sent Antonius and Silo with 2,000 archers and commanded that they occupy the mountain adjacent to the city, repelling those who were on the walls. The archers did as they were commanded, thwarting any assistance from other Jews within the city. Titus now spurred his horse into the attack, the others following him with loud shouts, deploying their line across the field to the width of the body of Jews facing them. By this means, they appeared more numerous than they were. The Jews, although surprised by the Roman attack and their good order, resisted for a little while, but when they felt the business end of the Roman lances and were overwhelmed by the noise of the horsemen, they were trampled under the horse's hooves. The slaughter was massive, causing many Jews to seek refuge quickly in the city. But Titus pressed upon them and killed them as they ran. Jews were

crowded into small groups and butchered. Others were met at the city gates and run through. Many tripped over one another and were trampled. The Romans encircled the Jews and cut off their retreat, turning them back into the open field. At last, some Jews forced a passage by their great numbers and escaped, running into Taricheae.

4. But now, terrible things began to happen to the inhabitants of the city. The people there with homes and possessions and to whom Taricheae belonged had wanted peace from the beginning. Now that the Romans had beaten the more warlike among them, they were even less willing to fight. But the many foreigners and tyrants among them continued to force them to fight. The two groups shouted at one another, and their frenzy intensified as the clamor arose within the city. When Titus heard the racket from his position just beyond the city wall, he cried out,

"Fellow soldiers, God is giving us the Jews! Why do we delay? Now is the time to attack! Don't you hear the noise they are making? Take the victory that is already in your hands. Those that have survived our attack are now in an uproar against one another. If we make haste, the city is ours. But we must also work and have courage, for no great thing is accomplished without danger. We must not only prevent their uniting again, which they will need to do to survive, but we must act before our reinforcements arrive. As just a few, we are poised to conquer a great city. May we alone take the city!"

5. Titus jumped on his horse immediately and rode down to the lake. [No city walls existed on the side of Taricheae bordered by Lake Gennesaret.] The Romans marched along the lakeshore with Titus in the lead and entered the city. The men on the walls were petrified with terror at the Roman's bold move. No one wanted to fight them or to stop them. So they stopped guarding the city while Jesus and some of his henchmen fled to the country. Other rebels ran down to Lake Gennesaret and were met by the Romans as they tried to enter their boats. A great slaughter began in Taricheae as some of the aliens who had not fled stood and fought. Even some natives of Taricheae died, even though they refused to fight. They had hoped for Titus's right hand as security since they had not consented to fight. But when Titus had killed all those who had joined the revolt, he stopped the slaughter out of pity for Taricheae's natural inhabitants. Those zealots who had reached the lake, seeing that the city had fallen,

quickly sailed away from the Romans.

6. So, Titus sent one of his horsemen to Vespasian to relate the news of what he had accomplished. His father was naturally joyful for both his son's courage and his glorious actions. He thought that the war was now close to over. Vespasian then rode to the city and set guards to encircle it. He ordered them to allow no one to leave secretly and to kill anyone who tried to escape. The next day the Roman general went down to the lake and ordered that the remaining boats be readied to pursue those who had fled on the water. The order was quickly carried out, due to the abundance of both materials and carpenters.

7. The Lake of Gennesaret (the Sea of Galilee), is named for the country that encompasses it. It is five miles wide and seventeen miles in length. Its waters are sweet and good to drink, and finer than the sediment-filled waters of typical estuaries. The lake water is also pure and is bordered with a sandy shore. When one draws water from its depths, its temperature is mild and much easier to swallow than that taken from a river or spring. Yet the water of Lake Gennesaret is also much cooler than one would expect of such a wide surface area. It seems as cold as ice when exposed to the air in the summer nights. The country folks are in the habit of making snow of it. Several species of fish live in the lake, distinct to the lake in both appearance and taste. The River Jordan divides the lake into two parts. Panium is believed to be the source of the Jordan, but the waters of the Jordan actually come from a place called Phiala, and in a strange way. Phiala lies just to the right of the road to Trachonitis fifteen miles north of Caesarea (Philippi). Its name, Phiala, means "bowl" and is a proper description of the place. It is round like a wheel, and its water is always pressing against its edges, the surface neither dipping lower nor overrunning its boundaries. This fountain of the Jordan was unknown until recently, when Philip was tetrarch of Trachonitis. He had chaff tossed into the Phiala spring, and it wound up in Panium where the people of old thought the headwaters of the Jordan were. The waters from Phiala had carried the chaff there, underground. Panium itself is a naturally beautiful place, and improved by Agrippa's liberality and expense. In its cave, Jordan's flow becomes visible, and flowing fifteen miles south, divides the marshes and fens of Lake Semechonitis. It flows first through the city called Julius then passes through Lake Gennesaret, then flows a long way

through the desert to end at Lake Asphaltites (the Dead Sea).

8. The countryside surrounding Gennesaret shares its name. Its natural composition is as marvelous as its beauty. The soil there is so fertile that many types of trees are grown in it, and the residents plant all sorts of trees. The varied air temperatures also agree with a variety of fruit trees. Walnut trees in particular flourish there in vast numbers, preferring the coldest air. Palm trees also grow there, but prefer hot temperatures. Olive and fig trees, requiring a more temperate climate, are planted next to those.

    One might call this place a feast of nature, seeing how so many varieties prosper in the same locale. These plants are rarely seen in the same vicinity anywhere else, yet here they all happily thrive in the varied climates of Lake Gennesaret. It's as if each species laid claim for itself upon the countryside. The climate not only nourishes different types of fall fruit beyond expectation, but it allows the fruit to ripen on the trees for a long time. Other fruit crops are also grown. Grapes and figs are in abundance and ripen throughout the year. In addition to its varied climates, the area is watered by a continuous spring the local people call Capharnaum. Some have thought it to be a tributary of the Nile because it produces Coracin fish that are also found in a lake near Alexandria. The country's length extends along the banks of the lake for almost four miles and is two and one-half miles wide. This is the nature of the place.

9. When Vespasian's boats were ready to sail, he placed as many of his forces on board as he thought sufficient to destroy the rebels who had escaped via the lake, and set sail for them. Those who had sailed from Taricheae couldn't land anywhere, for the shore was all in Roman hands. Nor could they fight on the water because their boats were small and useful only for piracy. They were too weak for Vespasian's boats and the sailors in them so few, they were afraid to venture near the Romans—who were to attack them in great numbers. However, as the thieves sailed around the lake, they threw stones at the Romans when far off, or fought hand-to-hand as they came closer. In either case, they suffered the most. The stones they threw merely made sharp noises bouncing off the soldier's armor or the boat, while Roman arrows reached the Jewish boats and inflicted much damage. When the Jewish boats ventured near the Roman boats, their vessels were sunk and all aboard drowned before the Jews could do

The Jewish Wars 229

any harm. Those crews that attempted to draw near the Romans to fight hand-to-hand were run through by the long Roman lances.

Occasionally, a Roman soldier would leap into the Jewish craft and kill with their swords. The Roman boats also rammed the Jews, sending the craft and all aboard to the bottom. Roman arrows killed some of those trying to swim, or Roman boats overran them as they tried to catch a breath. In desperation, some Jews tried to climb on enemy boats only to have their hands or their heads cut off. The Jews were annihilated by various means until those remaining crawled up on the shore with the Roman boats surrounding them. Those making it to shore died in the same way as those killed by arrows on the lake. The Romans leapt out of their vessels and killed a great many more rebels on land. The lake ran thick with blood and was littered with dead bodies, for not one Jew escaped. A terrible stench soon arose. The slaughter created a horrible and sad sight over the next few days. Shipwrecks and swollen bodies lined the shore. The Jewish corpses putrefied in the hot sun and fouled the air. Meanwhile, other Jews mourned over the misery inflicted on their brothers and the catastrophe that had been accomplished by those who hated them. This was the result of the battle on the lake. All those slain, including those at Taricheae, numbered 6,500.

10. When this fight ended, Vespasian set up his tribunal at Taricheae to separate the long-time inhabitants from the newly arrived aliens. Those foreign to the city had started the battle. He consulted with his commanders whether or not he should allow even the former citizens to survive. His officers advised the general that to dismiss them would not be to his advantage because when finding themselves free, they would not rest as they would be destitute and would be compelled to join the fight against Rome. Vespasian agreed that the citizens of Taricheae didn't deserve to live, and that if he let them go free they would turn against those who had allowed it. But he then considered how he might execute them all. If he had them killed publicly and in full view, he suspected that all of the country's people would rise against Roman rule. They would not tolerate such a thing—that those who only sought his mercy should then be massacred. And Vespasian could not bear to violate his oath to spare the people. But the Roman general's friends convinced him that no action taken against those Jews who remained would mean trouble. They argued that Vespasian should take the expedient and more profitable route rather

than being merciful, since a compromise was impossible. So Vespasian gave his officers liberty to do as they advised and permitted the Taricheaen prisoners to take the only road back to Tiberias. The Jews readily believed the freedom they desired was to be. Thinking they were safe, they gathered their belongings and headed up the road. But the Romans blockaded the road to Tiberias and corralled the Jews back into the city's stadium. He gave orders to execute the elderly and the useless. These numbered 1,200. He then chose 6,000 of the strongest young men and ordered them sent to Nero to help dig the canal at the isthmus in Corinth. He sold the rest of the people, 34,400 of them, as slaves. These were in addition to those he gave King Agrippa. Agrippa subsequently sold all these as slaves as well. The rest of the multitude was from Trachonitis and Gaulitis, Hippos, and Gadara. Most of them were rebels and fugitives—shameful characters who preferred war to peace. The Romans took these prisoners from Taricheae on the eighth of the month Gorpieus [Elul] (September 67 AD).

The Jewish Wars
Book IV

Containing the Timeframe of about One Year, from the Siege of Gamala to the Coming of Titus to Besiege Jerusalem

Chapter 1

The Siege and Taking of Gamala

1. Following Jotapata's fall, all those Galileans who had revolted against Rome came against the Romans after Taricheae's collapse. The Romans took over all the cities and fortresses except Gischala, and Mt. Tabor, that the Jews that had taken. Gamala, a city near Taricheae but on the other side of Lake Gennesaret, also held rebels who conspired against Rome. The city was situated on the border of King Agrippa's kingdom, as did Sogana and Seleucia. These cities were part of Gaulonitis: Sogana of Upper Gaulonitis and Gamala of Lower Gaulonitis. Seleucia sat next to Lake Semechonitis, a body of water almost four miles wide and eight miles long. Its marshes reach Daphne, which apart from the present war is a delightful place, having springs under the Temple of the Golden Calf that feed the river called the Little Jordan. It flows into the great Jordan River. Agrippa united Sogana and Seleucia to his kingdom at the start of the Roman war. Gamala did not join them, relying for safety instead on the difficult terrain surrounding it. Gamala is balanced on a ridge of a tall mountain that is rougher than the mountains around Jotapata. One must ascend to Gamala on a long neck-like ridge, and then descend into a valley before ascending again. The access is configured like a camel, hence the name Gamala, even though the locals don't pronounce it correctly. On two sides of the city are sheer cliffs that end in deep valleys. The side facing the mountain is less steep, but the people have cut a deep

ditch there, which also makes ascent difficult. Its houses lie on an acclivity that is straight, and the houses are thick and built close together. The city hangs on the cliffs as if defying gravity. It looks as if it should topple down the mountain. Gamala has a southern exposure with an extremely tall mountain to the south that act as a fortress for the city. Above that is a sharp precipice, not walled, but falling to a deep ravine. A spring also exists within Gamala's walls near the city limits.

2. Naturally fortified as it was, the city became even stronger when Josephus built a wall around it. He also dug ditches and tunnels under the ground. The siting gave courage to Gamala's people – more boldness than that of Jotapata's residents. But Gamala had fewer fighting men. They had such confidence in Gamala's strong fortifications that they thought no enemy could be strong enough to overwhelm them. Because of the city's defensive position, they had resisted a siege by Agrippa for seven months.

3. Vespasian now left where he had camped at Emmaus, which is near Tiberias. [Emmaus means "warm bath—for a spring of warm water used for healing purposes issues forth there.] Vespasian now came to Gamala. Seeing he could not surround the city, Vespasian nevertheless placed men to guard it and took an adjoining mountain that looked down upon it. In their normal pattern, Vespasian's legions fortified their camp on the mountain. They also began to build banks at the eastern lower level across from the highest tower of the city, and where the fifteenth legion had pitched their tents. Meanwhile, the tenth legion filled up the ditches and valleys. It was then that King Agrippa came near the walls and tried to convince those on the walls to surrender. Suddenly, a rock hurled from the wall hit him on the right elbow. The king's men surrounded him immediately. Indignant at this offense to the king, the Romans were enthusiastic to begin the siege. They also concluded that Gamala's defenders would stop at nothing in their barbarity against foreigners and enemies if they treated one of their own nation with such rage. The king only wanted to help them.

4. When the banks were finished—swiftly completed by many hands accustomed to the hard work, the soldiers brought up the war machines. Charles and Joseph, two of Gamala's most influential citizens, got their

armed defenders in position. The city was only afraid that they might lack a sufficient quantity of water and other necessities. They didn't believe they could hold out for long if they were short of supplies. But these two leaders encouraged the men and stationed them on the walls. Indeed, for a while they drove away those hauling up the machines. But when the engines began hurling darts and stones at them, they retreated into the city. Then the Romans brought battering rams to three positions, causing the walls to shudder and fall. The Romans poured into the city through the newly broken down walls with loud trumpet blasts and the noise of their armor. The incoming soldiers shouted and fell by force upon the citizens. But the men of Gamala counterattacked and held off the Romans, keeping them from proceeding further. With great courage, they beat the Romans back. The Romans were overpowered on every side by a great multitude, and were forced to run to the city's upper parts. Seeing their retreat, the people turned and followed the Romans, thrusting them down to Gamala's lower city. The Romans were greatly distressed by the steep and narrow lanes and were slaughtered wholesale. They couldn't defend against those above them nor escape their own men who were pressing upward.

The soldiers sought shelter in Gamala's low rooftops. But the roofs of the houses, full of soldiers, could not bear their weight and suddenly fell. When one house fell, those below it also fell like in a cascade of crumbling houses. A vast number of Roman troops were crushed to death in this way. As more Romans saw the houses toppling over, their distress caused them to leap higher up, with the result that those under the tonnage were ground to powder. Some did escape with their lives but lost arms or legs. But an even greater number of Romans suffocated in the dust that arose from the ruins.

Gamala's people saw these events as God's providential care for them. Disregarding any injuries they suffered, the people pressed forward, throwing the enemy from the housetops. The Romans were continually slipping and falling in the steep and narrow streets, while the citizens threw stones and darts at them, killing many. The ruined buildings also afforded the people many stones to hurl. The bodies of fallen Romans gave up their swords. They were used by the Gamalans to dispatch those trapped or merely unconscious. A number of Romans caught in the ruins killed themselves with their own daggers. Being unacquainted with the city, the going was difficult in the thick dust for those trying to escape. The soldiers wandered about like blind men, falling dead amongst their slain brothers.

5. Those Romans who could find a way out of the city retired to their camps. Vespasian, as always, gathered with the survivors of the debacle. Deeply saddened about the falling houses that had crushed so many of his troops, the general forgot to care for his own life. Accompanied by a few others, Vespasian climbed slowly up to the city's highest reaches, seemingly unaware of the danger. Titus was not with him, having been sent to Mucianus in Syria.

Vespasian thought it not fit to turn and run, but recalled his courage and the heroic deeds of his youth. Then, as if he had been strengthened by a divine fury, he and those with him covered themselves with their shields and formed a testudo over their armored bodies. They withstood the enemy attacks that came running down from the city's highest reaches. Without showing any fear of the enemy or their missiles, Vespasian kept moving.

Witnessing the general's supernatural courage, the citizens stopped their assault. When they held back, Vespasian got away. He did not show his back to the enemy until he was out of the city walls. A large number of Romans died in the battle, among whom was Ebutius the decurion, a man who in every battle not only showed great courage, but also put many Jews in misery. Gamalan citizens surrounded another man, a centurion named Gallus. He and ten other Romans secretly crept into a certain house and heard people conversing around the supper table. They were Syrians planning what they were going to do against the Romans. So Gallus got up in the dark of night and cut all of their throats, then escaped with those accompanying him to the Roman encampment.

6. Vespasian now comforted his troops, who were depressed by their misfortune. They had never been overwhelmed by such a disaster and were terribly ashamed. The soldiers were also remorseful that they had left their general alone in the midst of such dangers. Vespasian said little, not wanting to complain about his circumstances. But then he said to his men, "We ought to bear up under the usual circumstances of warfare. Such is the nature of war. We can never conquer without shedding some of our own blood. For our fortunes are in no way immutable. We have slain tens of thousands of Jews and have now paid a small share of fate's reckoning. People who become puffed up at their successes are impotent. In the same

way, cowards are much too fearful after suffering defeat.

"The best warrior is he who bears both circumstances soberly, continuing in that mindset in order to joyously recover precious losses. What has now taken place was not caused by any lack of manliness on your part nor Jewish courage. But it was because of the difficult terrain on which you fought that was to their advantage and to our disappointment. Upon reflection one might blame the defeat on your ungovernable zeal. While your enemy had retreated to the utmost parts of Gamala, you followed them to the top and exposed yourselves to danger. If when the enemy retired to the upper reaches and you were to have waited in the lower city, you would have provoked them to return and the battle would have been easier. But in rushing for a quick victory you forgot your own safety. This unconsciousness in battle and this zealous madness are not Roman values. Barbarians and Jews do this, not Romans. We do all we attempt with skill and good order. We are going to remember what we are trained to do and not to feel dejected because of this one failure. So let us each seek to console ourselves, for by doing so we will avenge those who have been killed and punish their killers. As for me, I will seek to be first among you to attack our enemies and the last to retire from the battle."

7. So Vespasian encouraged his army by his speech. Meanwhile, the people of Gamala took courage from the unaccountable victory they had won. But when they realized they now had no hope of achieving terms of surrender, and knowing that escape was impossible and that their provisions were scarce, they became depressed and lost their courage. They continued to plan for further enemy engagements with what little they had. The more emboldened men guarded those parts of the broken down wall, while the wounded and the sick manned the rest of the city walls. The Romans then raised their siege embankments and attempted to get into Gamala a second time. Meanwhile, many residents fled from the city through steep valleys and subterranean caverns unguarded by the Romans. Those fearful of being caught while escaping stayed in the city and died of starvation. All the city's food was gathered in one place and reserved for Gamala's fighting men.

8. The people of Gamala faced a difficult situation. During the siege, Vespasian went to work against the rebels who had taken Mt. Tabor. This mountain lies between Scythopolis and the Great Plain and tops out at

approximately 1,850 feet in elevation. Its northern face is almost impossible to climb. Mt. Tabor's top is as flat as a table and is approximately 3,000 feet long by 1,800 feet wide, and is encompassed by a wall built by Josephus in forty days. He brought up materials and water from below, as its inhabitants only had rainwater collected in cisterns. A multitude of Jews had gathered on Mt. Tabor, so Vespasian sent Placidus with six hundred cavalry troops.

Realizing any ascent of the mountain to be impossible, Placidus gave the Jews his right hand as security, interceding for them and the people and asking them to come to terms of peace. Accordingly, the people descended, yet with a deceitful intent. Placidus also came with the same intent to deceive the Jews. He had spoken mildly all the while, wanting to get the Jews off of the mountain and onto the lower plain.

So, complying with his proposition the multitude descended—not to surrender, but to attack. But Placidus was too smart for them. When the Jews began to attack, he pretended to retreat. The Jews raced after the Romans, part of Placidus's plan to entice them away from the mountain onto the plain. Suddenly, the Romans turned and counterattacked. The stunned Jews were hacked to pieces by Placidus's cavalry. He cut off any Jewish retreat, forbidding their return to the mountain. So the Jews left Mt. Tabor and fled to Jerusalem. The remaining Jews needed water and surrendered to Placitas, giving themselves and their mountain over to the Romans.

9. Meanwhile, those people of Gamala who could do so fled from the city and hid themselves. The weaker citizens died of starvation while the fighting men of the city defended it until the 22nd of Hyperbereteus [Tisri] – (November, 67 AD). Three Roman soldiers of the fifteenth legion had secretly undermined a high tower. One night, after they had mined under the tower and had rolled away five large stones on which the tower rested, the diggers were discovered. They made their escape just as the tower came crashing down, carrying the guards stationed on its heights. Frightened, the Jews stationed throughout Gamala ran away. The Romans then killed many Jews who ran forward to attack them—among these was Joseph, who was killed by a dart as he retreated over the city's damaged wall. The citizens of Gamala were also terrified by the din of the falling tower and ran to and fro thinking the enemy was already upon them. One of the citizens, a sick man named Chares, was so frightened at the noise that he

died of fear in his bed. The Romans, meanwhile, hesitated to venture into the city until the next day, remembering their former defeat.

10. In the meantime, Titus had returned to Gamala after learning of the Roman embarrassment in his absence. He took two hundred select horsemen and a few foot troops and silently entered the city. As Gamala's watchmen perceived his approach they cried out and took up arms, alerting the town to the new Roman assault. Some of the defenders grabbed their wives and children and with great anguish and tears carried them to the citadel at the highest point of the city. Others were quickly dispatched as they went forward to meet Titus. Most of the citizens did not know which way to turn. They tried to get to the fortress but were killed by the Roman troops. Heard were the agonized groans of those dying throughout Gamala, and their blood ran down from the upper to the lower city. Now Vespasian brought his entire army up against those who had fled to the citadel.

This highest part of the city was very high, rocky, and difficult to climb. Precipices surrounded the citadel, which was filled with Gamala's citizens. The Jews slew many Romans who came up after them, either by their darts or by rolling large stones down upon the ascending soldiers. The Jews were too high up for the Roman arrows to be effective. But just then a supernatural storm came upon the city that paved the way for the Jewish destruction. The wind grew to such proportions that the Jewish darts were turned back against them or blown aside from the enemy. The Jews couldn't stand against the furious winds nor could they see the Romans advancing against them. The Romans were able to climb to the citadel and surround the defenders, killing some who tried to surrender before they knew what had happened. For the Romans remembered their former assault into Gamala and now let out their rage against its people. Many of those surrounded and in despair threw their children, their wives, and then themselves off of the cliffs into the deep valleys below the fortress that had been dug even deeper by the Gamalans. The Romans' rage was nothing compared with the madness of the citizens as they killed themselves. For the Romans killed 4,000 Jews within the city while 5,000 died jumping or being thrown from the ledges. Only two women escaped with their lives. They were Philip's daughters. He was a prominent Jew and the son of Jacimus, who had been one of King Agrippa's generals. The women hid from the Roman siege and survived, unlike the many infants who were thrown down from the citadel. So the Romans took Gamala on the 23rd day of Hyperbereteus [Tisri] – (November, 67 AD). The city had first revolted on the 24th of Gorpieus [Elul] – (October, 67 AD).

Chapter 2

The Small City of Gischala Surrenders; John Flees to Jerusalem

1. Gischala was the only place in Galilee remaining unconquered. The citizens there wanted peace, for they were mostly farmers who tilled the soil. However, among them were also a great many thieves joined by some of the more powerful citizens. A man named John ben Levi had drawn these men into rebellion against Rome. He also continued to encourage them to revolt. John was a crafty knave who was of a changeable disposition. He was very optimistic in his fanciful goals but also very capable of achieving them. Everyone knew John loved to fight and used this love to enhance his leadership. The rebels in Gischala followed John and by the rest of the people who were ready to sue for surrender waited for the arrival of the Roman troops in their full battle regalia. Vespasian sent Titus with 1,000 cavalries to Gischala. Titus then withdrew the Tenth Legion to Scythopolis while Vespasian returned to Caesarea with two other legions to refresh themselves after their rather long and difficult campaign. He thought rest in Caesarea and Scythopolis would help them prepare for future trials. Vespasian knew that the siege of Jerusalem would be very difficult because it was a royal city and the leading metropolis of the entire nation. Furthermore, all the rebels who had escaped previous engagements had fled there.

Jerusalem was a natural stronghold and its walls gravely concerned the Roman general. He also believed that the holy city's defenders would be courageous and bold and difficult to conquer, even without their formidable walls. For that reason, Vespasian prepared his men for the hard work, in the same way that wrestlers train prior to beginning a match.

2. Titus thought Gischala would fall easily as he rode out to the city. But he knew that if he took it by force his soldiers would not leave one person alive. [Titus had seen enough bloodshed and felt compassion upon the innocent victims of the slaughter to come, who would die along with the guilty.] So, he was hoping the city would come to terms of surrender. But when Titus spotted Gischala's walls full of rebels he said to them that he could not help wonder what they relied upon, seeing that they now stood

alone to face the Romans. Every other city, some much better fortified than Gischala, had been overthrown on the first attempt, while others saved themselves by trusting the security of the Roman right hand. He now offered his right hand to them, forgetting their former insolence in order that they might enjoy their possessions and their safety. He said he could forgive them for wanting to recover their freedom, but to continue their rebellion was inexcusable because victory was impossible. If they failed to comply with his humane offers and to accept his right hand of security, the Romans would not spare one person in Gischala. They would soon be aware that their wall would crumble immediately on the impact of the Roman battering rams. All they would be demonstrating was their arrogance, although they were no better than slaves or captives.

3. No one in the city could reply to Titus. The thieves so filled the walls and guarded the city gates that no citizen could get out to propose terms of surrender or to invite Titus's cavalry into the city. But then John spoke to Titus from the wall. He said that he was happy to hear the Roman general's proposition, and he would personally persuade or force those that refused to listen. But he asked Titus to please regard the Jewish Sabbath day and let them celebrate. It was unlawful for them to meet with him to negotiate peace or even to take off their armament on the seventh day. The Romans knew that the Jews cease from all work on the Sabbath and that those that compel them to transgress their law are equally guilty before God with those whom they compel.

One day could not be disadvantageous to the Romans, for why would anyone think of doing anything unless it was to escape from the city? Titus could prevent that by encircling Gischala with his troops. The Romans would bring honor to themselves if they chose to make the Jews not break the law. If Titus truly wanted peace, he should preserve not only the people's lives but also their laws. But John wanted only to trick Titus. He didn't care about the Sabbath day but only about his own safety. He had hopes of preserving his life by fleeing the city that night. Allowing John to escape was God's work. He preserved John so that Jerusalem might ultimately be destroyed. Titus listened to John and was prevailed upon to delay the siege. He then pitched his camp away from Gischala, in Cydessa. Cydessa was a strong Tyrian village on the Mediterranean Sea. The city had always hated the Jews and had warred against them. It was also a well-populated place and well fortified—which was fitting, as it was a town where the Jews were enemies.

4. When night came, John observed no Roman troops around Gischala. He took the opportunity to flee to Jerusalem, taking with him not only his own armed henchmen but also a considerable number of innocent men and their families. As John fled in haste, he was tormented with the thought of losing either his life or his freedom, even as he took this multitude of men, women, and children two and a half miles into the countryside. He then left them there and headed for Jerusalem.

Those left behind wailed plaintively. Far away from their own people and Gischala, they believed themselves to be near the enemy and feared the Romans would take them prisoner. Still, they turned back toward Gischala as each sound they heard made them think the Romans were almost upon them. Many strayed off track while others, in their hurry to return to Gischala, trampled women and children to death. Some wives called to their husbands who had left with John for Jerusalem with loud weeping and moaning. But John's voice had won out as he and his followers flew toward the holy city. He shouted for those left behind to save themselves. He also told them that if captured by the Romans he would wreak revenge for them. So this multitude of Gischala's citizens became broadly dispersed. Some ran faster while others moved more slowly.

5. The next day Titus went again to the wall to seek the city's surrender. The people opened the city gates to him and came out: men, women, and children, with shouts of joy to their benefactor who had delivered the city from bondage to John and his thieves. They also told Titus of John's flight and begged him to spare them and to come into the city. They also hoped he would bring the rebels to justice. But Titus, not fully regarding the requests of the people, sent some of his cavalry to capture John before he and his men could get to Jerusalem. Though the detachment could not overtake the rebel leader, the Romans did kill 6,000 of his followers and brought back 3,000 of the women and children John had abducted and left in the countryside.

Titus was not pleased that John was able to escape. He had enough captives and those residents whom John had corrupted to ameliorate his anger. So Titus entered Gischala amidst shouts of joy. He ordered his men to destroy part of the city walls as if they have been overwhelmed in battle. Titus did not execute those guilty of disturbances in the city because he thought if he tried to determine who was guilty and who was innocent, false testimony arising from former animosities and previous disputes might condemn the innocent. It was better to let the guilty live in fear of punishment than to execute anyone who did not deserve to die. The guilty would learn prudence by the fear of deserved punishment and for the shame of being forgiven their prior offenses. But to execute someone was an irremediable punishment. Titus then placed a garrison inside Gischala for its security, in order to restrain any thoughts of rebellion that remained and to leave those peaceful citizens in greater security. And so all Galilee was conquered, but not until the Romans had suffered greatly for their achievement.

Chapter 3

Concerning John of Gischala and the Zealots; the High Priest Ananus; and Also How the Jews Practiced Rebellion Against One Another in Jerusalem.

1. When John arrived in Jerusalem, the masses of people there were in an uproar and ten thousand crowded around him and the fugitives who had arrived from Gischala. The people asked John and his men what miseries had happened to them there. But the new arrivals' hot and quick breath spoke of their distress without words. They preferred not to speak of any difficulties but pretended they had not fled from the Romans but had retreated to Jerusalem to fight Rome with much less danger. It would have been unreasonable and fruitless to expose themselves to the utmost danger of fighting in Gischala or any other such weak city. It would be wiser to save themselves for Jerusalem. But when they spoke of the capture of Gischala and their orderly departure from the city, the people understood it to be no less than abject flight, especially when they learned of those captured.

The people in Jerusalem were then greatly concerned, thinking that these events were but a forewarning of what would happen to them. But John had little concern for those he had left behind. Instead, he went through all the people in Jerusalem, giving them hope and persuading them to resist the Romans. He said that the Roman army's condition was weak and at the same time affirmed his own power. He ridiculed the Romans as ignorant and inexperienced. Even if they sprouted wings, he said, they could never fly over Jerusalem's walls. They had suffered enough problems taking the Galilean villages. The Romans had broken their battering rams against weaker walls.

2. John's continuous harangue and bravado brought many young men into the rebel camp and gave them enthusiasm for a fight. But the more prudent young men and all the older men saw precisely what was coming and mourned over the city as if it had already fallen. The people were in great confusion. But the whole country was enveloped in partisan strife even prior to the sedition being encouraged in Jerusalem.

Titus now went from Gischala to Cesates, and Vespasian from Caesarea to Jamnia and Azotos, taking those cities. Vespasian installed garrisons and brought back a large number of people who had surrendered. He had vowed to preserve their lives. In addition to these places, disorder and civil war broke out in every city. Those not fighting Rome fought amongst themselves. A very bitter struggle developed between those who wanted war and those who wanted peace. The struggle began in private families who could not agree amongst themselves. These families, where love for one another should have restrained tempers, split apart in their diverse opinions as brothers rose up against each other. The rebellion became universal. The young men desired war and were too bold and too strong for the older and more prudent men. Rapine was everywhere as men gathered together in gangs to rob their fellow countrymen. The Jews practiced the same barbarity and iniquity against one another as did their Roman enemies. It seemed to be more palatable to be scourged by the Romans than by one's fellow countrymen.

3. The Roman garrisons in the cities did little or nothing to stop the Jews from pillaging and destroying their own country. Their lethargy was due partly out of not wanting to take the trouble to help and partly out of their hatred for the Jews. But then, the leaders of the gangs of thieves throughout the country—growing tired of raping and pillaging their own people, joined in one massive gang of wickedness and crept into Jerusalem. The city had no civil leadership, nor any government, and received everyone without distinction into it gates. They believed that all who came into the city came to help them out of kindness. But these men, setting aside their rebellion against the established authority, came to assist in the city's destruction. For they were unprofitable and useless and had wasted those provisions that could have helped the city survive. They not

only started the Roman war, they now brought new internal trouble and famine upon Jerusalem.

4. But men even more depraved than before were to enter the city gates. No barbarity was beyond their means. They didn't just measure their boldness by pillaging or rape, but by murder. And they didn't wait for nightfall to practice their butchery, but killed in broad daylight. Their victims were not ordinary citizens but were among the most prominent people of Jerusalem. The first man they took was Antipas, of royal lineage, who as city treasurer was the most powerful man in the city. They kidnapped this Antipas and hid him away. They then kidnapped Levias, an important person, then Sophos and another man of royal lineage, Raguel, were abducted. They did the same thing with many more important men of the city. The people were terribly concerned and frightened. Everyone looked to his own safety. It was as if the entire city had fallen by siege to Rome.

5. The kidnappers were not satisfied with the captives' chains, nor did they think it was safe to keep them for long. The men they had taken were very powerful and had strong families of their own that were able to avenge them. They thought that the very people involved would be so concerned by the kidnapping of their loved ones, they would rise against them.

It was, therefore, resolved to have the captives executed. So they sent for a man named John, who was the most bloodthirsty of them all, to proceed with the execution. The man was also called "the son of Dorcas" in the Hebrew language. John took ten more men with him into the prison with their swords drawn, and so cut the throats of the men in irons. They created a preposterous lie to cover up such flagrant wickedness, stating that the captives had held secret conferences with the Romans to surrender Jerusalem. They said that they had only executed men who had been traitors to their common cause. With this bold hoax, they grew even more insolent, pretending they were the benefactors and saviors of Jerusalem.

6. The people had fallen to such a degree of callousness and fear, and these thieves to such a degree of lunacy, that the robbers took it upon themselves to appoint high priests. They cast aside the high priestly line accord-

ed to the families from which the high priests came. Then they ordained unknown and contemptible men for that office, so that these shameful high priests might assist them in their wickedness. These new high priests were obliged to comply with everything their thieving bosses required of them. The rebels also set Jerusalem's prominent men against one another by all sorts of maneuvers and deceits. By ordaining their own high priests, the thieves were able to do what they pleased, as those who might have stopped them were silenced by their mutual quarrels. Shortly, the thieves were so filled with unjust behavior against others; they transferred their repulsiveness to God and entered his sanctuary with unclean feet.

7. Meanwhile, Ananus, the oldest of the high priests, persuaded the people of Jerusalem to rise against the thieves. Ananus was a very judicious man and would have saved the city had he escaped those who plotted against him. The robbers had made a fortress of God's temple, a place where they might hide to avoid the people they feared. God's sanctuary had become a sanctuary for them, a place that housed tyranny. The rebels laughed at the miseries they had caused the people. Their laughter was even more intolerable than their wicked activities. The rebels tested the extent of their own power by always keeping the people off balance. They began to cast lots for the high priesthood that, as we've said, had otherwise been handed down for centuries through families. They fabricated a lie saying that from ancient times lot had determined the high priesthood, but in actuality it was nothing but a crafty attempt to seize control of the government. They wanted to appoint governors as they pleased – governors who would do what they were told to do.

8. So the tyrants sent for one of the high priestly class called Eniachim, and cast lots amongst them to see who would be the high priest. The lot fell upon a man named Phannias ben Samuel, of a village named Aptha. The action clearly demonstrated their wickedness, for this man Phannias was not only unworthy of the high priesthood, he did not even know what a high priest was. Nevertheless, they snatched this man out of the country without his consent as if they were putting on a stage play. They put the sacred garments upon him and told him what he was supposed to do. They did it all for sport and as if it was all a big joke. But the true high priests who saw their law being made into a joke shed tears and lamented over the destruction of their sacred calling.

9. The people of Jerusalem finally had enough of this insolence and gathered together to overthrow the tyrants. Leading the people were Gorion ben Josephus and Symeon ben Gamaliel. These two men addressed both groups and individuals, encouraging them to punish these vermin, these destroyers of their freedom, and to purge the holy temple of its bloody polluters. The most esteemed of the high priests, Jesus ben Gamalas and Ananus ben Ananus, also criticized the people for their apathy and stimulated them to rise against the Zealots. The tyrants called themselves "Zealots" as if they were zealous for practicing virtue. But these thieves and tyrants were zealous in the worst possible way.

10. So, the people of Jerusalem gathered together in a great assembly. Every one of them was indignant at the Zealots' capture of the holy sanctuary and their looting and murdering. This assembly was the first time the people had cried out against these men, thinking beforehand that it would be very difficult to suppress them. Ananus stood in the midst of the assembly with tear-filled eyes. He frequently glanced at the temple, then said, "It would've been better for me to die than to have seen God's house filled with so many horrors. These sacred places have now been trampled on by the feet of these blood-shedding villains. But I still live, and am clothed with the high priestly vestments and am called by the name 'High Priest.' I am very fond of living and hope not to endure the death that would bring glory to my old age. If I were the only person concerned in these matters as if I lived in the desert, I would give up my life for God's sake. For what purpose is it to live among people who are oblivious to their misfortunes – people who have no clue of any remedy for the miseries that afflict them? When men are kidnapped you allow it! When you are beaten you remain silent! And when the tyrants commit murder you won't even shed a tear in public! Oh, the bitter tyranny! But why do I complain about these tyrants? Was it not you and your toleration of them that allowed them to do this? Was it not you who overlooked their assembly when they were but few and by your silence allowed the few to become many? Was it not you who took no action when they were first arming themselves against you? You ought to have stopped them when this thing first started – when they began insulting your relatives. But by neglecting what you could have done you encouraged these wretches to plunder and loot. When they robbed houses, nobody said a word. When

they carried off the owners of these houses and dragged them through the streets of Jerusalem, nobody came to their aid. Then they put these same men who you betrayed in irons. I don't know how many men there were and know nothing of their characters, but I know there were no accusations against them, and no one had condemned them. When put in chains, no one helped them. The tyrants subsequently executed these men.

"I have also seen this; it was as if these men were merely the best of a herd of brute animals being led to sacrifice. No one said a word or raised his right hand to preserve them. Will you also allow them to trample your sanctuary underfoot? Will you lay out our red carpet for these profane wretches upon which they may walk to higher degrees of disrespect? Or will you pluck them down from their exalted state? For if they have been able to overthrow the sanctuary will they not plan even greater evil? They now occupy the strongest place in the city of Jerusalem. You call it your temple although it's more like a citadel or fortress. Now that tyranny has penned you in and your enemies are in the upper city over your head, what purpose does it serve to ask for anyone's counsel? What will embolden you in this misfortune? Perhaps you're waiting for the Romans to protect our holy places. Has our situation come to that? Have our agonies become so great that we expect our enemies to have pity upon us? O, wretched creatures! Will you not rise and turn against those who harm you? Even the wild beasts will take revenge upon those who strike them.

"Can you not remember the calamities that you have suffered? Can you not visualize the inflictions that you have undergone? Will not such things embolden you to avenge yourselves? Have you, therefore, lost that most honorable and natural passion for being free – to live in freedom? In truth, we are in love with slavery and in love with those who lord it over us. It is as if the desire for slavery had been handed down from our ancestors. Yet did not they undergo many great wars for liberty's sake? Did they not do what was right even when overcome by the power of the Egyptians or the Medes? And why is there now war against Rome? I'm not asking whether war with Rome is beneficial or rewarding. But what is the reason for it? Is it not that we may enjoy our liberty? Shall we not bear those who have conquered the known world to be lords over us and yet allow tyrants from our own country to rule us? I must say that we must endure submission to foreign powers because fortune has already submitted us to them. But our submission to these tyrants is cowardly and has been brought upon us by our own consent. However, since I have mentioned the Romans, I want to reveal something that comes to my mind as I'm speaking, and that affects me deeply. It is this. Even though we should be overthrown by the Romans, and God forbid this to happen, yet nothing would be more difficult than what these tyrants have already brought upon us.

"How then can we avoid shedding tears when we see what the Romans have given to our temple? When we see the Zealots from our own nation plundering our glorious city and slaughtering our own men—evils that even the Romans would not have committed? The Romans have never gone beyond that which is allowed the Gentiles, nor have they interrupted any of our sacred customs. They respect our sacred walls even when they view them from a distance. And yet some men, who have been born in this country and brought up according to our customs and called Jews, walk around in the middle of these holy places with hands covered with their own countrymen's blood. How can we be afraid of war with foreigners when those foreigners will behave with greater moderation than our own people? In actuality, we will probably find the Romans support our laws while the tyrants are rebellious against them. I am fully persuaded that every one of you here is satisfied, even before I have spoken, that those who will destroy our liberty deserve destruction. The punishment for the evil they have done has yet to be invented.

"And you who have suffered so greatly are now angry with them because of their wicked actions. But perhaps many of you are afraid of the great number of the Zealots and their fearlessness. Perhaps their more elevated position causes you to be afraid. These are all circumstances caused by your carelessness, and they will become even greater by your continuing negligence. For each day, more thieves are coming to their side as men from all over the countryside are rushing to join those who are like themselves. The tyrants' boldness is, therefore, elevated because no one will challenge them. If we don't act soon, they will make use of the higher ground to employ war engines. But be assured of this, if we go up to fight them, their own consciences will witness against them. For any advantage they might have on higher ground will be lost by the opposition of their own reason. Perhaps God himself, whom they have also offended, will make what they throw at us return upon themselves so that these impious rascals will be killed by their own arrows. Let us just rise against them, and they will come to nothing. If we face any danger in our attempt, it is good to die before these holy gates. Let us spend our lives not just for our wives and children, but also for God's sake and the sake of His sanctuary. I will help you both with my counsel and my hand, and I will spare neither my energy nor my body for your support."

11. With these words Ananus encouraged the citizens of Jerusalem to fight against the Zealots, although he knew how difficult it would be because of their numbers, their youth, and their boldness. But mostly, Ananus knew that the Zealots would fight bravely because there was no hope of pardon for the crimes they had committed. But Ananus determined to undergo whatever sufferings might come upon him rather than to allow matters to continue in such confusion. So the citizens shouted to him to lead them against those who opposed them. Every citizen was ready to run any hazard in order to destroy the Zealots.

12. While Ananus began enlisting men and placing those chosen by him in battle formation, the Zealots became aware of what was happening because some informers ran to them. They were angered by Ananus's speech and leapt out of the temple en mass, killing everyone who crossed their path. Ananus immediately gathered the citizens, who were more in number than the Zealots but not as skilled with weaponry. They also had little experience in fighting, but the enthusiasm they showed made up for both shortcomings. The citizens' passion was stronger than any weapon. They also derived a great deal of courage from their defense of the temple, more than any other foe. If they could not break the hold the robbers had on Jerusalem, the citizens would give up all hope of dwelling in the city. On the other hand, the Zealots knew that unless they won there would be no end to the punishments inflicted upon them. So, passions prevailed on each side of the conflict. At first the two sides threw stones at each other in the city and before the temple. They also cast their javelins from a distance. But when one side began to win they met hand-to-hand with swords. The slaughter was great on both sides, and the survivors suffered many wounds. Relatives of the dead carried their bodies to their houses. But when one of the Zealots was wounded he went up to the temple and defiled the sacred floor with his blood. One might say that it was Zealot blood alone that polluted our sanctuary.

In these conflicts, the Zealots always sallied quickly out of the temple and came down upon the citizens. But as the citizenry grew in anger and numbers they began to reprimand those who retreated, making them return to the battle. At last, the entire body of citizens turned against the Zealots. The tyrants had no choice but to seek refuge in the temple. When Ananus and his men followed the Zealots into the temple, it frightened them because the move chased the tyrants out of the first court and immediately into the inner court, where the Zealots shut the gates. Ananus was hesitant to attack the holy gates to the inner court, even as the Zealots threw stones and shot arrows at them from above. The high priest also believed that to allow the unpurified citizens into the inner court would be unlawful. He, therefore, chose 6,000 men by lot and placed them as guards in the cloisters. The guards rotated one after another, with each man forced to keep watch. Many of those in leadership were dismissed to attend to the government, but even these hired substitutes to do guard duty for them.

13. Now this John, who had escaped from Gischala, became the cause of all the citizens being annihilated. He was a very crafty man whose soul was driven to oppression. John directed the Zealots' actions from a distance, but during this conflict he pretended to be one of Jerusalem's citizens and sat each day with Ananus as he planned strategy with the other citizen leaders. John also went with the high priest as he visited the watch at night. So, the secrets he obtained were divulged to the Zealots. The enemy discovered every question that the people deliberated even before they had settled the issue. To avoid suspicion John cultivated a close friendship with Ananus and the other leaders. He began to be suspected due to his excess flattery, and the plot began to unravel. Further, he was always in their presence, even when uninvited. They began strongly to suspect that he betrayed their secrets to the Zealots. They figured out that the rebels knew in advance every move the people made. No one else could be the spy – it had to be John. But it wasn't easy to rid themselves of him as he was so accomplished in wickedness. Several prominent men always consulted on important matters also liked and supported John. These men thought it wise for John to pledge his loyalty under oath. John took the oath readily, and pledged his loyalty to the people, saying he would never betray any of their secrets or practices to the enemy. He also pledged to assist them in overthrowing the Zealots both by his guidance and his hand. Ananus and the rest believed John's oath and received him into their consultations without further suspicion. They believed him so

strongly they sent him to the Zealots in the temple as their representative, where he transmitted offers of accommodation. The citizens wanted to avoid the temple's pollution as much as possible, particularly that no Jew would be slain there and so defile the holy place.

14. So John, as if his oath had been made to the Zealots and not to the citizens, went into the temple and stood among his friends. He said that he had encountered many dangers on their behalf in order to make plain to them everything Ananus and the citizens planned against them. But both he and the Zealots may be placed in imminent peril unless some supernatural help was forthcoming. For Ananus did not delay any longer but had sent couriers to Vespasian to come and take Jerusalem. Ananus had appointed the next day as a day of fasting, desiring that the people might be admitted into the temple for religious purposes. If not, John said they would take it by force and fight them at the temple. John did not see how they could both endure siege and fight against so many citizens. He added that he was sent by God's providence as an emissary to them to negotiate a truce because Ananus made these propositions in order to throw them off guard. Then, when they were unarmed, Ananus would attack.

John said that if the Zealots valued their lives, they should choose one of two avenues: either they should sue for peace with their enemies or they should look for help outside Jerusalem. If captured and they still held any hope for pardon, then they had forgotten what dreadful things they had done. John said they should not believe that their repentance would be received by the citizens or followed by any reconciliation. On the contrary, wrongdoers are often even more hated by their enemies for this sort of repentance. Those who have suffered, when power comes into their hands, are more severe upon those who have opposed them. The friends and relatives of those who have suffered would continue to seek revenge on those who have slain their loved ones. He said a great number of citizens are very angry with them because of their gross breaches of the law. They could not hope to survive against such a very bitter majority.

Chapter 4

The Idumeans Are Called by the Zealots and Come Immediately to Jerusalem and When They are Excluded from the City They Spend the Night There. Jesus, One of the High Priests Speaks to Them, and Simon the Idumean Replies to It

1. John put fear in the Zealots' hearts by his clever speech. He did not directly identify the outside help he meant, but covertly intimated he meant the Idumeans. (Note: Idumea is a Jewish region just south and east of Judea near the Dead Sea.) Now wanting particularly to enrage the Zealot leaders, John began to accuse Ananus of atrocities, saying that the high priest had threatened them in an exceptional way. The leaders were Eleazar ben Simon, who was the most capable of all the Zealot leaders both for his planning and the execution of those plans. Zacharias ben Phalek was another Zealot who had, with Eleazar, come from a family of high priests. When these two men heard the threats against the rebels and themselves, and particularly how Ananus and his people had invited the Romans to occupy Jerusalem in order to secure their rule, they hesitated, wondering what they should do next. [This subterfuge was also one of John's lies.] They feared the people would attack soon, and that their plot against them had destroyed any hope of getting foreign assistance. They might even be on the brink of disaster before any notice for outside help could be sent.

However, they then decided to invite the Idumeans into the city and wrote a letter to this effect: That Ananus had assumed dictatorial powers over the people and had betrayed Jerusalem to the Romans. They had separated from the rest of the citizens and had gained control of the temple in order to preserve their freedom. Only a few days remained for hope of their deliverance, and unless the Idumeans came quickly, they would soon be in Ananus's hands and the city would be under Roman rule. They also charged the messengers to tell the Idumeans many more of their circumstances. Two dynamic Zealots were chosen to carry the message to Idumea. They each had the ability to speak effectively and persuade the Idumeans of the letter's truth. They were also very fast runners, a major qualifying factor behind their selection. The Zealots knew that the Idumeans were a warlike and disorderly nation and would respond quickly to their requests as they were always ready to rise to any fight or innovative exploit. If they were flattered a little and merely asked to come, they would rally to their arms and rush headlong into action as they were always happy to go into battle. It was as if going to a festival. The Zealots did need to hurry in delivering the letter, and the messengers were very capable and fast. They were both named Ananias, and they ran quickly to the Idumean leaders.

2. The Idumeans were quite surprised by the letter's content and the extra detail the runners gave them. They began to run around the area like madmen, proclaiming that their people should gather for war. A great many Idumeans came together quickly, even sooner than their leaders had requested. Everyone took up his arms in order to save Jerusalem's freedom. In all, 20,000 were placed in battle formation and marched to Jerusalem under four commanders: John and Jacob, the sons of Sosas, Simon ben Cathlas, and Phineas ben Closuthus.

3. Ananus and the other citizens fighting for the city knew nothing of the messengers' mission, but it soon became known as the Idumeans approached Jerusalem. Ananus ordered the gates closed, and that men should guard the city walls. He had no thought of fighting against his countrymen. He wanted to negotiate with the newcomers before things came to blows. Jesus, the next in authority to Ananus as high priest, stood up on the tower near the Idumeans and said,

"Many and various troubles have fallen upon Jerusalem yet none have made me so discouraged as now. You have come to help damnable men in a very astonishing way. You have enthusiastically come to support the most depraved men against your friends. You would have hardly been as enthusiastic if we had called you to help us to fight barbarians. If I knew that your army consisted of the same type of men who invited you, I would not have been surprised, for nothing so cements men's friendship with other men as their common values. But if you were to examine the men who have invited you, one by one, you will find that every one of them deserves 10,000 deaths. They are rascals who've scoured the whole countryside spending themselves on debauchery and wildly plundering our neighboring villages and towns. Now they have all secretly run into the holy city. These bandits have now profaned the holy temple in their utter wickedness. Even now they are drinking themselves drunk in our sanctuary and eating the food of those whom they have slaughtered, gulping it into their insatiable bellies. All of your men are so decently fitted out in their armor one would think they have been called by a large city to defend against foreigners. What can one call this march of yours, but a freak of fortune when he sees your whole nation of the Idumeans come to protect those who are the filth of the earth?

"I cannot imagine what could motivate you to come here so suddenly, because certainly you would never don your armor on behalf of bandits or against your own flesh and blood without some very important reason. We have heard it said that we had invited the Romans in order to turn the city over to them. Now you have raised an uproar about this lie and have come to set city free. One can only admire the wretches that created this fiction, for they knew there was no better way get men enraged than to appeal to their inborn desire for freedom. On that account, the best men are prepared to fight enemies from abroad. But the Zealots have devised this propaganda saying that we are ready to betray that which is most coveted and precious to us – our liberty. But you should have considered what kind of people have invited you here and against what sort of people have circulated this lie. You need the truth, not fictitious speeches. But consider first the actions and character of both parties. For why would we ever sell ourselves to the Romans when it was in our power not to have rebelled against them in the first place? Or when we first revolted why didn't we return under their rule? And this was a time when our neighboring countries had not yet been devastated. It isn't exactly easy to be reconciled to the Romans.

"If we wanted reconciliation now that they have conquered Galilee and have become proud and insolent, to try and please them now when they are so close to attacking us would bring reproach upon us far worse than death itself. As for me, I would have preferred peace with Rome rather than death. But now that we are at war with them and have fought against them I prefer death with my reputation intact rather than living as a slave under them. Who do the Zealots think we are? As the people's leaders would we have sent secret messages to the Romans? Shouldn't this have been done by common vote of the citizens? If it was only us who did this thing, let them name who among us was sent to accomplish this treachery. Was one of them caught because they went on this errand or captured on his return? Are the Zealots in possession of our letters? How could we few conceal such an act from our fellow citizens? We speak with them constantly! It seems that the Zealots know all our secrets, even though the Zealots are few and confined to the temple. Are they finally sensible to what punishment they should receive for their abominable crimes? When these men were free from the fear that now haunts them, they never spoke to us as traitors.

"But if they charge the people with this abomination, the issue must have been debated in a public forum where not one dissenting voice spoke. If it were debated openly, such knowledge of any such treason would have been made known to you sooner than any private communication. But how could this be? Where are the ambassadors who confirmed these agreements? Who was it that was appointed to be an ambassador? This charge of treason is only a lie promulgated by men who do not want to die and are doing whatever they can to avoid the punishment due them. If they had determined that Jerusalem was to be betrayed into Roman hands, no other than these false accusers could have had the impudence to do it, as they will not stop to advance their traitorous intentions.

"And now you Idumeans have arrived armed for battle. Is it not your first duty to come to your city's assistance, joining with us to kill the tyrants who have broken our city's policies and trampled on our holy laws? They have made their swords the judges between right and wrong. For they have kidnapped important innocent men as they stood peaceably in the marketplace, then tortured them and put them in chains without bothering to hear what they had to say. Then, deaf to their pleas for mercy, they murdered them. You may come into the city if you please, yet not to wage war but to see for yourselves the evidence that what I say is true. Look at the houses that now stand vacant because of their rapacious hands. See the wives and children who are wearing black, mourning for their dead husbands and fathers. Listen to their groans and lamentations all over Jerusalem. Not one soul in Jerusalem has not felt the weight of these profane wretches who have escalated their madness and who have brought their wicked banditry out of the countryside and from the distant cities into Jerusalem. This city is the center of the entire nation, and they have taken its very heart — God's temple. They have made the holy place their refuge and fortress and the headquarters from which they make their preparations against us. Now this holy place, loved by the entire known world and honored by many throughout the earth who have never even seen it, is trampled upon by these wild beasts that have been raised up among us. They now exult in their desperation when they hear people fighting other people, city against city, and that you Idumeans have gathered an army to fight against your own flesh and blood.

"Instead of that, and as I said before, is it not more highly fit and reasonable for you to join with us to kill these miserable wretches? To avenge yourselves for being tricked by them? I mean for you to take revenge upon them for having the impudence to invite you here to help them. Rather, they should have stood in fear of your coming to punish them. But if you still think their invitation worthy you may put down your weapons and come into the city as brothers, not as auxiliaries or enemies, and judge the case for yourselves. However, think of what the Zealots will gain by being tried by you for such undeniable and flagrant crimes. No citizen will come to speak in their defense. You'll hear only the Zealots. But if you will neither share our outrage at these men nor desire to act as judges, a third option exists: Leave us both alone. Neither condone the atrocities we have suffered nor join with the Zealots against our city. For even though you still think we have parlayed with the Romans, you may watch the approaches to Jerusalem in case the accusations were found to be true. Then come and defend your city and punish those found guilty. For the Romans can't stop you now that you are so close to us. But if none of these proposals seem good to you, don't wonder that Jerusalem's gates bar you outside while you bear your weapons against us."

4. So Jesus finished his speech. But not one of the Idumeans had listened to it, for they were enraged at being kept out of the city. The Idumean generals were also incensed at the notion of laying down their weapons, seeing such a suggestion as surrender, a casting of their arms aside at some enemy's command. But Simon ben Cathlas, one of the Idumean officers, silenced his men's uproar and stood where the high priests could hear him. He said,

"I am now certain why those who want freedom are in irons in the temple since you have closed the gates of our common city to your nation's citizens. At the same time, you were willing to admit the Romans and are probably ready to festoon the gates with flowers on their arrival. But you speak to us from your towers and ask us to throw down our arms that we have taken up in the cause of freedom. While you will not trust your brothers to guard our common city, you profess to make us judges of the differences between you and these other men. You accuse some men of having executed innocent men without a fair trial, but you still condemn us, the entire nation, of acting in a disgraceful manner. Now you have shut up Jerusalem from citizens of your own country where they used to be open, even to foreigners who came to worship here. We have indeed hurried to get here into war against our own countrymen. The reason for our haste is so that we may preserve the freedom that you so sadly betray. You have probably been guilty of the same crimes as those you have imprisoned. You have probably collected plausible falsehoods against them that you will now use against us. These lies have been the source of your imprisonment of those in the temple while they were merely taking care of the public's interests. You have shut the gates against people who are most closely related to you. You make injurious demands of others while you complain that you are the victims of tyranny. Then you blame those tyrannized as being the tyrants. Who can listen to your abusive words? Unless, of course, you mean that as you exclude Idumeans from the sacred offices of our country, so also may we exclude you from your own metropolis. One may justly complain that those men you now hold captive in the temple found the courage to punish those tyrants whom you call 'innocent and important men' because they were your partners in crime. It is odd that they did not begin with you and nip in the bud the most treasonous of you all! But if these men have been more merciful than normal civility requires, we Idumeans will serve God's house and fight for our common country. We will also war against those who attack us from abroad as well as those internal enemies. We will stay in front of your walls in our armor until either the Romans grow tired of waiting for you or that you become liberty's friends and repent of your actions against it."

5. The Idumeans cheered as Simon finished his speech. But Jesus left sorrowfully seeing the Idumeans were against all reason and that the city was now besieged both internally and externally. The Idumeans were also dismayed, enraged that they had not been allowed to set foot within the city. They had thought the Zealots were strong, but seeing none of them they began to doubt their strength. Some Idumeans began to be sorry they came to Jerusalem at all. But they knew that if they returned home without doing anything they would be the objects of shame. Their fear overcame their sorrow. So they camped that night alongside Jerusalem's wall. But a violent storm broke out during the night with heavy rain and continuous lightning and thunder—epic concussions. It seemed like the earth would split open as the ground trembled. This violent storm bore witness to the destruction to come. When chaos rules nature, anyone would understand that its violence foreshadowed the horrific catastrophe ahead.

6. Both the Idumeans and the citizens of Jerusalem were of the same opinion. The Idumeans believed that God was angry with them for taking up arms and that punishment would come upon them for making war

on Jerusalem. Ananus and those around him believed they had won the victory without a fight and that God was acting on their behalf. But both these conjectures proved wrong. The destiny the storm predicted seemed to each side to be ominous to their enemies, but it was the citizens of Jerusalem who would ultimately not escape its ill effects. The Idumeans gathered as one unit, holding their shields over their bodies as protection from the rain and cold winds.

Meanwhile, the Zealots became increasingly worried about the Idumeans, more concerned than they were about themselves. The leaders got their heads together and came up with a way they could help their friends from the south. The more warlike among the Zealots were in favor of storming the citizens' guards with their weapons and forcing entry into the middle of Jerusalem. Once there, they could open the city gates to the Idumeans. They supposed the guards to be under the same disarray and would fall for their surprise attack, particularly since the majority was either unarmed or unskilled in battle.

The vast number of Jerusalem's citizens were hiding from the storm in their homes and could not be quickly assembled. If there were any danger in the attempt, such danger would only accrue to the Zealots glory. But if nothing were done to help the Idumeans who were suffering, the Zealots' inaction would be seen as cowardly. But the more sensible Zealots disapproved of the plan. They knew the guards holding them in the temple were numerous and that the city walls were carefully watched due to the Idumean threat. They also believed that Ananus would be both on the walls and visiting the guards, as he did every night. But Ananus's visits were not accomplished that night, not due to any laziness on his part, but due to Fate's overruling appointment that both Ananus and his many guards might perish. Late that night when the storm was at its worst, Ananus allowed the guards manning the cloisters to go to sleep. Meanwhile, the Zealots thought to make use of the temple saws to cut the bars of the city gates. The great clamor of the violent wind, the rain, and the thunder masked the sound of the saws, and collaborated with the Zealots.

7. So a few Zealots secretly stole out of the temple and onto the city walls. Arriving at the gate next to the Idumean camp, they took the saws and cut through its bars. At first The Idumeans feared that Ananus's men were coming to attack. Every man had his hand on his sword to defend him-

self. But as soon as this they saw who had come to help them they went into Jerusalem. The Idumeans were so enraged that had they immediately began to attack; nothing could've stopped them from killing every man in the city. But the men with the saws had encouraged them to first quickly go to the temple to release the Zealots. They had begged the Idumeans not to overlook those they came to save from their distress nor to put them in further danger. The remainder of the city could be taken readily once they had captured or killed the guards. But if the city were alarmed, overcoming guards would not be easy. Once the guards realized the Idumeans were in the city they would gather to fight and would prevent the Idumean advance to the temple.

Chapter 5

The Idumeans and Zealots Display Their Cruelty When They Arrive in the Temple During the Storm. Also, this Chapter Concerns the Slaughter of Ananus, Jesus, and Zacharias, and How the Idumeans Returned Home.

1. The Zealots' advice pleased the Idumeans, and they ascended through the city to the temple where the Zealots eagerly awaited. As the Idumeans were entering the temple, the Zealots came boldly out to greet them. Mixing in with the new arrivals, the Zealots attacked the citizens' guard. Some guards were killed instantly who were on watch but had fallen asleep. But those awakened cried out in amazement, whereupon the entire guard unit arose and took up their weapons in defense. They thought only the Zealots had attacked them but then saw the Idumeans and realized they had gotten into the city. When they saw these new arrivals pressing in, some dropped both their weapons and their courage and took to weeping. Other, younger guards covered up with their armor and valiantly fought the Idumeans, protecting the older men for a time. Others shouted to those in the city of the disaster, but when the guards realized that the Idumeans had come in, no one came to help them. The city only echoed back the horrible wail of their misfortunes. As all of the guards were in danger of being slaughtered another painful howl erupted from the city's women. Zealots added their shouts to the others while the storm itself made the noise even more dreadful. The Idumeans did not give any quarter. They are a naturally very bloody and barbarous nation, and having been greatly upset by the storm they hacked away at the Jews who had shut the city

gates. They did the same to those who begged to spare their lives as well as to those who fought back. Those were treated no differently who tried to remind the Idumeans they were fellow Jews and begged them to honor the holy temple. There was no place to escape, nor hope of survival. The citizens were thrown on top of one another and killed. Seeing no way of escape and the murderers upon them, some guards threw themselves headlong into the lower city. In my opinion, these suicides underwent a more miserable death than that which they avoided because it was voluntary. Now the temple's outer court overflowed with blood. Later, 8,500 dead bodies lay there.

2. But the Idumeans' rage was not satiated by the slaughter. They now went down into the city and killed everyone they met. The Zealots, meanwhile, did not go with the Idumeans but went after the high priests and those with them. All who were captured they immediately executed. The Zealots then stood upon the dead bodies as a joke, laughing at Ananus because of his kindness to the people, and Jesus, for his speech to the Idumeans from the wall. They were so impious as to throw away the high priests' dead bodies without their accustomed burial. The Jews always bury their dead. Even crucified criminals were taken down from their crosses before sunset. It is no mistake to say that Ananus's death marked the beginning of Jerusalem's destruction. One may date from that very day the ultimately overthrown wall and the ruination of all Jerusalem. For the people began to lose hope when they saw their high priest killed in the midst of their city, the man who had championed their survival. Ananus was very revered and very just. Besides the splendor of the nobility, dignity, and honor he possessed, Ananus's love for the equality of the people even extended to its lowest citizen. He loved liberty with all his soul and admired democracy in government. He always sought the public's welfare over his own benefit and preferred peace above all. He knew that the Jews could never conquer the Romans. He also saw that the Jews would be annihilated unless they quickly pressed the Romans for terms of surrender. In a word, if Ananus had survived, events would have transpired differently, for he was wise in speaking and in persuading the people. He had previously won over those who were for the war and those few who opposed his plans. Under Ananus's leadership as a general officer, the Jews would have delayed the Romans' advance. Jesus was his ally, and although inferior to Ananus by comparison, and he was a superior person. I cannot

think that it was because God doomed the city to destruction. God had decided to purge the pollution from his sanctuary by fire. He allowed those to die who clung to Him with tender affection. So, those who had recently worn the sacred garments, who had led ceremonies of world renown, and who were held in reverence by visitors from every corner of the globe, were now cast out naked to be eaten by dogs and beasts of prey. I think Virtue herself mourned for them, sorrowful at such a total downfall at the hands of Vice. Ananus and Jesus had met their end.

3. Having disposed of the high priests, the Zealots and Idumeans now turned upon and butchered the people of Jerusalem like they were a herd of pigs. Ordinary citizens were murdered on the spot. The youth and nobility were captured and slapped in irons and thrown into prison. Their executions were delayed in the hope that some of them might join the Zealot ranks. But not one would comply. All preferred death to being counted among such dreadful scalawags who made war on their own country. But their refusal brought them painful torments. Their bodies were so tortured and afflicted they collapsed in pain. At length, and with great striving, the Zealot swords brought death. Those who were captured in daylight were executed that night. Their corpses were carried out and thrown away like trash to make room for other prisoners. The people were terror stricken. No one had the courage to either weep for those slain or the victims' relatives, or even to bury them. The people shut themselves in their homes and shed secret tears, even groaning with great caution. They feared if their enemies heard their weeping they would suffer the same fate as those for whom they wept. Only in the dark of night would the people take up dust and throw it upon their heads, although some did this in the daytime, exposing themselves to great danger. The number of citizens who perished in this way was 12,000.

4. The Zealots and the Idumeans soon became tired of merely killing men. Impudently, they began to set up a fictitious court in order to have some fun. They sought to execute Zacharias ben Baruch, one of Jerusalem's most prominent citizens. He had provoked the Zealots by his vehement hatred of iniquity and his love of liberty. Zacharias was a rich man as well, and so by killing him they also hoped to seize his money and rid themselves of a powerful enemy. So, by public proclamation, the Zealots called seventy of the most prominent men of Jerusalem for a sham trial, as

if the seventy men were actual judges or had proper authority. Zacharias was brought before them on trumped-up charges of betraying Jerusalem to the Romans by sending traitorous messengers to Vespasian. There was whatsoever no proof of the charges, but the Zealots believed that the charges themselves were sufficient evidence for his conviction. Zacharias understood that all avenues of escape were closed to him. He had been taken by treachery and put in prison without benefit of a trial and was despairing now even of life itself. Accordingly, Zacharias stood up and laughed at the Zealots' false accusations. In a few words, he refuted the crimes charged against him. Then he turned to his accusers and began to summarize all their transgressions, groaning for the distress and bewilderment they had brought upon the city. Hearing Zacharias's words, the Zealots were enraged and had to restrain themselves from drawing their swords. They wanted to maintain at least a semblance of proper jurisprudence. They also desired, for other reasons, to bring the seventy judges to trial. They wanted to see which of the judges would play along with their mock trial. But the judges brought back a verdict of not guilty. They had chosen to die with Zachariah ben Baruch rather than be accomplices to his death. Hearing the acquittal, the Zealots rose up in great anger and were indignant at the judges for not seeing their authority was given only as a ruse. So, two bold Zealots fell upon Zachariah in the middle of the temple and killed him. As he fell to the floor dead, they mocked him, saying, "That's our verdict! Your death will prove a more certain acquittal than the other." They then immediately took his body and threw it from the temple into the valley. The Zealots then struck the judges on their backs with the flats of their swords, abusing them. They threw the seventy judges out of the temple, sparing their lives for the purpose that now dispersed among the people, that the judges might become their messengers, letting the people know that they were no better than slaves.

5. By this time, the Idumeans were sorry they had come to Jerusalem. Neither were they happy with what they had done while in the city. One of the Zealots privately assembled them together and reminded them of the wicked deeds they had done in conjunction with Zealots who had invited them. He enumerated one-by-one the trouble they had brought against the city. He said that though they had taken up arms believing that the high priests had betrayed Jerusalem to the Romans, they had found no evidence of any such treachery. Instead, they had befriended

those who had pretended to believe the lie and insolently joined them in acts of war and tyranny. He accused the Idumeans of not checking to see if the actual treachery ever took place. But then once they had taken part in shedding their own countrymen's blood, it was now time to stop such crimes and not offer assistance to those who were undermining their forefathers' laws. Even though still angry at not being allowed to set foot in the city, those who shut the gates were now dead. Ananus was dead, and almost all the others had died that first night. These actions now produced repentance even among members of their own party. But those who had invited the Idumeans to come were still full of horrible brutality and had no regard for those who had saved their lives. The Zealot continued:

"The Zealots continue to commit atrocious crimes under your very eyes, which crimes they are attributing to you. They will continue to do this until someone stops them or separates themselves from this vile barbarity. You should return home since the original charge of high priestly treason is now known to be a lie and no Roman arrival is expected. The city's walls are protection enough for us and cannot be easily battered down. So avoiding any further partnership with these vile men in order to make restitution for being tricked into taking part in these crimes, it would be best for you to return home."

Chapter 6

Being Released from the Idumeans The Zealots Killed Many More Citizens; Vespasian Dissuades the Roman Officers from Marching Against the Jews at the Present Time

1. The Idumeans soon complied with this advice. But before they left the city they freed about 2,000 citizens from the prisons, who immediately fled Jerusalem to go to Simon, of whom we shall soon speak. The Idumeans retired from the city and returned home. Their departure was a surprise both to the Zealots and to the citizens. The people had no knowledge of the Idumean repentance and so their courage was invigorated for a while, after losing so many enemies. But the Zealots grew more audacious, not from being deserted by their allies, but by being freed from those who might hinder their evil plans and plot to end their wickedness. Accordingly, the Zealots delayed no longer nor took time to further deliberate their horrible practices, but executed their plans faster than anyone could have imagined. In particular, their thirst was to be slaked by the

blood of heroic men from prominent families. One type of victim was to be destroyed from envy and the other from fear. The Zealots thought they were only secure if all Jerusalem's influential men were dead.

Because of this they killed Gorion, a very dignified and important gentlemen of high birth. He championed democracy and had as bold and free of a spirit as any Jew. The principal capability that ruined Gorion, in addition to his other qualities, was his candor. Nor did Niger of Peres escape the Zealot clutches. Niger had shown great courage in his battles against the Romans. The Zealots dragged him through the middle of Jerusalem. As he went, he frequently cried out and showed his battle scars. When he was taken outside the gates he despaired of being kept alive, and asked the Zealots to allow him burial. But they had threatened beforehand not to grant that one desire of his heart. So they killed Niger and did not allow him Jewish burial. While they were dispatching Niger he placed a curse upon them, calling on the Deity to make them undergo famine and pestilence in the war, and that the Zealots might slaughter each other. All these curses God justly confirmed against these sinful men. It wasn't long before the Zealots tasted their own lunacy in actual treasons against one another. So when Niger was killed, the Zealots' fear of losing their power diminished.

Indeed, not one pretext was overlooked to see that people from all sections of Jerusalem were slaughtered. Some were slain because of differences between them and the Zealots. Others died because they had opposed the Zealots in peacetime. The Zealots looked for any opportunity to make accusations. If someone refused to come near them he became under suspicion of being prideful. If someone approached them more openly, he stood in condemnation. If anyone approached the Zealots desiring to oblige them, he was seen as planning treachery. The only punishment for these crimes, whether real or imagined, was death. No one could escape unless he was very poor, either because of his low birth or some other misfortune.

2. The commanders of the Roman army thought this sedition of their enemies to be to their great advantage, and urged Vespasian, as their lord and general, to march with haste to Jerusalem. They said to him, "God's providence is on our side as he has set our enemies against one another. But change can come quickly and the Jews may reunite with one another. They may grow weary of their civil warfare or repent of such goings-on."

But Vespasian replied that they were greatly mistaken in their proposal. The officers were interested only in theatrics, wanting to make a show of their hand and arm strength without regard for their safety or timing. Were he to attack Jerusalem immediately, "the Jews would unite against us. Now their strength is at its height and they fight amongst themselves. But if we wait a while we shall have fewer enemies because they are all consumed in this domestic struggle. God is a better general than I and is giving the Jews over to us without any struggle on our part, granting our armies victory without any danger. The best way is to sit back and watch the enemy destroy itself by its own hands. Better that the enemy annihilate itself through civil strife while we sit as spectators of their calamities, than to interfere and fight hand-to-hand with men who love to murder one another and who are enraged against each other. But if anyone imagines that victory's glory is more colorless without fighting, let him know this: A glorious success, quickly obtained, is more profitable than facing the hazards of battle. Those who act with moderation and caution are to be esteemed more than those who have gained a reputation from actions in war. He shall lead a much stronger army when the enemy forces are weakened and his own army refreshed after the continuous work they have undergone. However, now is not the right time to propose victory for ourselves. The Jews are not currently at work on their armor or on their walls, nor in gathering auxiliary troops. It will accrue to our advantage to delay. Every day brings them more trouble in civil disturbances and discords. They are under much greater misery now than we could bring upon them. So if one cares about our army's safety he should let the Jews go on destroying themselves. If one has greater regard for the glory of battle, neither should we attack them now that they are under such affliction. Then it would be said that our victory was not due to our bravery but to their agitation."

3. The Roman officers joined in approval of what Vespasian had said. The true wisdom of his words was soon discovered. Many Jews were deserting Jerusalem daily, fleeing from the Zealots, although it was a formidable task. Each escape route out of the city was heavily guarded and many escapees were killed trying to get out. Zealots took it for granted that anyone trying to leave the city was deserting to the Romans. Every Jew who greased their palms with money was given safe travel while those with nothing to bribe the guards were slain. Vast numbers of dead bodies

lay in heaps alongside the roadways. Many who dearly wanted to leave Jerusalem chose instead to die in the city in the hope of at least getting buried there. They were forced to choose between the lesser of two evils. But the Zealots soon revealed their ultimate barbarism. They neither allowed burial within the city nor along the roads. It was as if they had decided to eliminate not only the country's laws but nature's laws as well. As they continued to defile the citizens with their evil deeds, it seemed the Zealots would pollute God himself. They left their victims' corpses to putrefy under the hot sun.

The same punishment that came to those trying to desert also came to those who attempted to bury the dead. They were both slaughtered. The man who granted burial to another man soon stood in need of a grave himself. In a word, the gentle passion of mercy was completely lost as those most deserving of pity only enraged the Zealot wretches. They transferred their rage from the living to the dead and then from the dead to the living. The terror was so overwhelming that those who were still alive called the dead happy, as they were already at rest. Those who lay unburied were declared happier in comparison to those suffering torture in the prisons. The Zealots trampled upon all mankind's laws and laughed at God's law. They ridiculed the prophetic oracles as the deceits of magicians. Yet these same prophets had foretold many things about the rewards of Virtue versus the punishments of Vice. In violating the prophets' words the Zealots fulfilled prophecies in Judea in their own time. A certain prophecy of old stated that Jerusalem would fall when uprisings would occur among the Jews and the sanctuary burned as in warfare. By the Jews' own hands God's temple would be polluted. While they may not have disbelieved the prophecy, the Zealots made themselves the instruments of its fulfillment.

Chapter 7

John Tyrannizes the Rebels; the Sicarii Wreak Havoc Near Masada; Vespasian Takes Gadara; and Placidus Chases the Fugitives

1. By this time, John had joined in the subversion but thought it beneath his stature not to be given the highest honors. He joined with one of the wickedest bands of Zealots and broke off from the rest. He could not agree with others' opinions, rather choosing to assert his own orders in a most imperialistic way. It became evident that John was setting himself

up as a sort of king. Some Zealots submitted to his authority out of fear, while others submitted out of their comradeship. He had a very clever way of enticing men to himself, both by deceiving them and by his many lies. Some men thought they would be safer under his command and that the blame for any past crimes might rest on his shoulders alone rather than on several others. Many were drawn to John's amazing energy and insight, both in his undertakings and his direction. Many of his gang left him and became enemies, mostly out of envy. They thought it unbearable to be subjected to one who had formerly been their equal. But the central reason that drove men from John was their fear of his authoritarianism. For once in power, John would not be easily unseated. They knew John would ultimately have this against them: that they had opposed him before he rose into power. They would rather suffer anything in wartime than to be in servitude for a time and then die. So the sedition was divided into two factions, John reigning over one of them and opposing the other. The faction leaders closely watched one another while not coming to blows. They were too busy fighting the citizens and contending with one another over who would get the biggest target. Jerusalem struggled with three great misfortunes: war, tyranny, and sedition. It looked to the citizens like war was the least troublesome of all three. So the citizens tried to run away from their homes to foreigners, finding safety with the Romans that their own people had prohibited.

2. But a fourth misfortune now arose that also threatened the country with destruction. A very strong fortress existed near Jerusalem that had been built by ancient monarchs. Historically, valuables were stored there, and it served as a personal retreat in wartime. It was called Masada. Men called Sicarii had gained possession of Masada and began to scour the neighboring countryside trying to import sufficient supplies necessary for their survival. They feared they would be prevented in the future from securing those necessities. But once they heard that the Roman army was encamped and not on the move and that the Jewish factions of tyrants and seditionist factions were battling amongst themselves, they began to undertake bolder exploits.

At the Feast of Unleavened Bread [in which the Jews celebrate the memory of their deliverance from slavery in Egypt and their return into the land promised to their forefathers], the Sicarii left Masada secretly by night and overran a town called Engaddi. They arrived in a surprise attack before the citizens could arm themselves, and then took the town. The Sicarii drove the citizens out of the city and scattered them into the countryside. Of those who could not escape, mainly women and children, they slaughtered over seven hundred. Following the fight, when the Sicarii had plundered everything in Engaddi's houses and had seized all of the city's fresh fruit and vegetables, they carried their loot back to Masada. The Sicarii then began to lay waste to all the towns and villages around the area, making the country desolate.

Meanwhile, corrupt men like themselves gathered from all parts of the country. At this time, the bandits began to raid other areas of Judea that had previously enjoyed peace. When these robbers had plundered their own villages, they fled to the desert. They gathered together and conspired as raiding parties, too small to be called armies and too large be called gangs of thieves. They began to attack Jewish synagogues and towns with unabated violence, taking prisoners of the townspeople as if in war. Unable to retaliate, the people were hauled away by their captors. No part of Judea escaped the misery that had also befallen Jerusalem.

3. Deserters from Jerusalem told Vespasian of these events. Although the Zealots watched all the escape routes from the city and killed all they discovered on them, a few disguised themselves and got away. They had fled to the Romans and tried to persuade the general to come and help Jerusalem's citizens, advising him that many had been slaughtered due to the people's goodwill to the Romans. The survivors were in danger of the same treatment. Vespasian already felt sorry for the calamities in which the citizens found themselves and seemed as if he were ready to lay siege to the city. But in fact, he wanted to deliver the people from a far worse siege than they were already suffering. However, other areas of the country were also threatened, and the general believed he could leave nothing outside Jerusalem that might cause him to interrupt the city's ultimate siege. Accordingly, he marched his army to Gadara, the capital of Perea, a strongly fortified city. He entered Gadara on the fourth of Dystrus [Adar].

Unbeknownst to the seditious men there, the Gadarenes in power had sent messengers to Vespasian asking him for terms of surrender. Many of Gadara's citizens were wealthy and had a strong desire for peace because they wanted to salvage their homes and belongings. The rebels only discovered the proposed deal with the Romans as Vespasian's army approached the city. But, being fewer in number than their enemies within Gadara, and seeing the Roman army approaching, the rebels decided to flee.

But they also thought it would be dishonorable to leave the city without shedding some blood and taking revenge on those who had asked the Romans for help. So they captured Dolesus, a very prominent man whom they supposed had sent for the Romans, and killed him. The rage against Dolesus was violent. They treated his body in a vicious way. They then ran out of the city as the Roman army was just outside the gates. The people of Gadara welcomed Vespasian into the city with joyous acclamations, receiving from him his right-hand of security. Vespasian also garrisoned the city with both cavalry and infantry to guard against a return of the renegades. The Gadarenes had already pulled down the city walls before the Romans could order them destroyed. In this way, they assured the Romans of their love of peace. Even if they had wanted war against the Romans it would now be impossible.

4. So Vespasian sent Placidus with a detachment of five hundred cavalries and 2,000 infantry to run down the renegades who had fled Gadara. The general then returned to Caesarea with the rest of his army. As soon as the Gadarene fugitives saw the Romans about to overtake them, and before any fighting began, they ran into a village called Bethennabris. They found a great many young men whom they armed, partly by their consent and partly by force.

In a sudden and rash move the Jews assaulted Placidus's troops. The cavalry gave way a bit at the onset, hoping to draw the fugitives and Jews away from the village walls. When successful, they surrounded the fugitives and cast their darts at them. In this way, they cut off the fugitives' flight while the foot troops destroyed the rest of the men who fought against them. The Jews were annihilated who did no more than show their courage. When they tried to attack the Roman troops, encased as they were in their armor and shields, no place could be found to throw a dart. Nor were the Jews able to break the Roman ranks, while they were run through by Roman darts. Like the wildest of beasts, the Jews rushed upon the Roman swords. Some were killed by slashes of Roman steel across their faces while Placidus's cavalry scattered others.

5. Placidus was concerned that the retreating Jews would reenter Bethennabris. Accordingly, he placed his cavalry between the Jewish throng and

the village. Placidus then turned back upon them as his men unleashed their darts at those nearest. This attack caused the Jews farthest away to turn back in terror. Ultimately, the most courageous among the Jews broke through Placidus's cavalry and fled toward the village wall. Those Jews manning the wall did not know what to do. They couldn't bear the thought of keeping their countrymen out of the village, and yet if they let them in, they could expect to die with them. This prospect was exactly what happened. As the Jews were crowding together at the wall, the Roman horsemen were almost upon them. But the guards on the wall stopped them by shutting the gates. When Placidus attacked, fighting courageously until it was dark, he took the wall and the village and the inept multitude was destroyed. But the strongest and the fastest men got away. The Roman soldiers then plundered the houses of Bethennabris and then set the village on fire. Those who escaped ran into the countryside, stirring up trouble by exaggerating their own misfortunes and shouting that the entire Roman army was upon them. Fear arose on every side. So the people gathered together in great numbers and fled to Jericho, for they knew of no other place that could afford them the slightest hope of escaping.

Jericho was a city with a strong wall and was well populated. So Placitas, relying on his cavalry and his former success, followed the Jews. He killed all he caught as far as the Jordan River, where the swollen river's powerful current halted them. He then placed his troops into battle formation against the Jews who had their backs to the river. The Jews had no choice but to fight. They extended themselves in a long line along the riverbank and received the Roman darts as well as cavalry attacks. The Romans beat back the Jews and forced them into the fast-moving current. In hand-to-hand combat, 15,000 Jews lost their lives. A great many more were forced into the Jordan and drowned. Another 2,200 were captured. In addition, the Romans garnered a huge bounty of donkeys, sheep, camels, and oxen.

6. This defeat was the most horrendous that had befallen the Jews to date, although it did appear greater than it was. This appearance was because corpses littered the whole country through which the Jews fled. The Jordan River could not be crossed because of the dead bodies that were in it. Lake Asphaltites (the Dead Sea) was also full of human corpses that fell into it, washing down from the river. Then, following up his victory,

Placidus fell violently upon the neighboring villages and small towns. The Romans took Abila, Julius, and Bezemoth and all those towns that lay north of Lake Asphaltites. He put his prisoners into each of these towns as he thought best. Then Placidus put his soldiers into boats to kill those who had fled into the lake. The result was that all Perea as far as Macherus had either surrendered or was overrun by the Romans.

Chapter 8

Vespasian, Hearing of New troubles in Gaul, Hastens to End the Jewish Uprising. A Description of Jericho and of the Great Plain, Together with a Description of Lake Asphaltites

1. While all this was taking place, a report came to the Roman camp regarding uprisings in Gaul. Vendex, along with other powerful men in that country, had revolted against Nero. [The affair is more accurately described elsewhere.] When Vespasian heard the report, he was encouraged to move quickly to wrap up the Jewish war. He foresaw that civil wars in other places were coming upon Rome and that its very government was in danger. He thought if he could just bring peace to the eastern parts of the empire, he would ease the anxiety for Italy. The winter hindered him from taking the field, so he placed garrisons in many villages and small cities for their security. He also placed decurions in the villages and centurions in the towns and rebuilt many destroyed cities.

When spring arrived, Vespasian took most of his army and led it from Caesarea to Antipatris where he spent two days restoring order in that city. On the third day, he marched out, destroying and burning all the neighboring villages. When he had wasted the entire area around the Thamnus toparchy, he went on to Lydda and Jamnia. He placed a number of captives within these already suppressed cities. Vespasian then marched to Emmaus, took control of the passage to the city and erected his fortified camp there, leaving the fifth legion in it. The general then went on to torch the Bethletephon toparchy and its neighboring districts. He fortified Roman strongholds all over in Idumea. When he had taken Betaris and Caphartobas, two villages in the midst of Idumea, he killed more than 10,000 other Jews there and carried more than 1,000 away captive. The rest of the citizens were driven away, chased by many of his men, who then laid waste to the entire hill country. Vespasian and the rest of his forces then returned to Emmaus via Samaria, marching close to the city some call Neapolis [Shechem], or which the natives call Mabartha, then on to Corea where he made camp on the second day of Decius [Sivan]. The following day he marched to Jericho and was joined there by Trajan, one of his higher officers, with all his forces from Perea. Vespasian had subdued the places east of the Jordan.

2. Masses of people evacuated Jericho and fled to the mountainous region toward Jerusalem. The Romans found the city almost deserted, and most of those left were put to death. Jericho is situated on a wide plain with a long, bare mountain standing over it. The mountain extends from Scythopolis in the north as far as Sodom, and in the south to the southern shores of Lake Asphaltites. The mountain is very rough and unoccupied as it is completely barren of vegetation. Another mountain sits next to it on the Jordan River's west side. This mountain rises at Julias in the north end extends southward to Somorrhon on the border to Petra in Arabia. In this range of mountains is one called the Iron Mountain that runs lengthwise as far as Moab. The region that lies between these mountain ridges is called the Great Plain.

The plain extends from the village called Ginnabris to Lake Asphaltites. It is twenty-nine miles in length and fifteen miles wide. Flowing north and south, the Jordan divides the great plain into two parts. The Great Plain has two lakes within its boundaries, Lake Asphaltites and Lake Tiberias, which are completely different from one another. Lake Asphaltites contains salt water while the waters of Tiberias are sweet and fruitful. The plain burns with heat in the summer and because of the extraordinary heat its air is very disagreeable. No other potable water exists in the Great Plain save that of the Jordan. But the Jordan waters plantations of date palm trees near its banks. The closer to the water, the more fruitful the palm trees.

3. However, a large spring exists near Jericho that brings forth much fresh water that is suitable for irrigation. It comes up near the ancient city which Joshua ben Nun, the Hebrew general, took in warfare. Of all the Canaanite cities, Jericho was the first to fall. Tradition has it that in the beginning the large spring poisoned the earth and trees and even sickened the local women and children. It was completely corrupt and sickening to every plant and person. But it became very wholesome and fruitful by the hand of the prophet Elisha. Elisha was a protégé of the prophet Elijah and

succeeded him. At one time when Elisha stayed in Jericho the residents treated him with great kindness. He returned the favor by going out to the spring and casting a clay pot full of salt into it. Then, stretching out his right hand up to heaven and pouring out a drink offering, he prayed that God would calm the waters and that other channels of fresh, sweet water might gush forth and that God would bring a healthy temperature to the spring. He prayed that God would also bestow upon the people there the fruits of the earth in abundance and give them many children and that the prolific spring would never fail them as long as they lived lives pleasing to God. To these prayers Elisha added certain religious ceremonies and the spring changed from foul to sweet.

From that time on the people prospered in both children and abundant fruit, where previously only unfruitfulness and starvation reigned. The water there is so powerful in irrigating the ground that if taken to other areas and allowed to penetrate the ground over long periods, it gives greater nourishment than any other water. Other water, even if it is plentiful, cannot compare with this water from Jericho, even if only in small quantities. Accordingly, it waters a much larger area than most other waters would cover effectively, passing through the plain a distance of nine miles and two-and-a-half miles wide. In its path rich gardens thick with fruit trees are irrigated by its nourishment. Many types of palm trees are watered by it, different in both name and taste from any other. When pressed, the best dates yield a delicious juice that tastes like honey.

The country also produces honey from bees and grows the prized balsam fruit, the most precious of any region. Cypress trees also grow there, producing myrobanalum. Anyone who considers this place to be divine would not be in error because its fruit trees are rare and of the finest varieties. One could never find another climate in the entire earth that would compare to it. Everything planted here comes up in bunches, it seems, caused by the warm air and the fertile waters. Warmth brings sprouts from the seeds and causes the plants to grow. The moisture ensures firm root structure and allows the plants to thrive in the hot summer sun.

In the summertime, the country is so hot that no one wants to go there. But when the water is drawn before sunrise and exposed to the air, it becomes very cool, quite contrary to the surrounding air. In the winter, the water becomes warm, and if you go into it, it is very soothing. The air in the winter is so very mild that the populace dresses only in linen clothing, even when snow covers the rest of Judea. The place is almost nineteen miles south and west of Jerusalem and almost eight miles from the Jordan River. The countryside all the way to Jerusalem is wilderness and rocky. It is also wasteland and desolate as far as the Jordan and Lake Asphaltites. But I've said enough about Jericho and the sublime bliss of its location.

4. The nature of Lake Asphaltites is also worth mentioning. As I have said, its waters are bitter and worthless. The water is so dense that even heavy things float when cast into the lake. It is not easy to make things sink to the bottom if one should seek to do so. And so, when Vespasian went to see the lake he commanded that some who couldn't swim have their hands tied behind their backs and thrown into its depths. They floated to the surface as if the current had driven them there. What's more, the lake changes color three times each day as the sun's rays strike it at different times and the sun's light is reflected from the various angles.

On the other hand, gigantic black globs of bitumen, or tar, float to the surface in many areas, resembling headless bulls both in their shape and size. The workers who gather the stuff grab hold of these big blogs of tar and haul them into their ships. When the ship is full, it is difficult to offload the bitumen as it hangs tenaciously to the ship's sides. The only materials that will cut it and make it separate are women's menstrual blood and urine. Bitumen will yield only to those. The pitch is useful in caulking boats and for healing the body, by formulation into several medicines.

The lake is over seventy-two miles long and almost nineteen miles wide, bordered by the country around Sodom. In ancient times, this was a most delightful land, both for its rich agricultural products and its wealthy citizens. But it is now all burned up. Its destruction was apparently the consequence of the inhabitants' iniquity. Divine lightning bolts burned the cities and the entire area. Reminders of the divine fire remain as the traces of the five cities are still in evidence, as are the ashes of its ancient fruits. These fruits are colored just as if they were ripe and ready to eat, but if you pick one up it dissolves in smoke and ashes. With our own eyes, we have confirmed the ancient writings concerning this land around Sodom.

Chapter 9

Vespasian, After He Had Taken Gadara, Made Preparation for

the Siege of Jerusalem; but Hearing of Nero's Death Changed His Mind. Also Concerning Simon of Gerasa

1. Vespasian now fortified all the places around Jerusalem and also built fortresses at Jericho and Adida, placing garrisons in both. His own forces and auxiliary troops manned the garrisons. He also sent Lucius Annius to Gerasa with the unit of cavalry and a large body of infantry. After Ananias had taken Gerasa easily, he executed 1,000 of its young men, whom he captured before they could take flight. He also made their families captives and allowed his men to plunder their property before setting their homes on fire and leaving for the surrounding villages. The more prominent men fled the city, but Ananias killed the poorer people and burned everything else to cinders. So, the war had gone through all of the mountainous region and throughout the plains as well. The people in Jerusalem were trapped there, having no place to escape. The Zealots closely watched any that wanted to leave the city. On the other hand, the Romans had surrounded Jerusalem, keeping the people inside its walls.

2. So Vespasian had now set his sights on Jerusalem. As he was preparing his entire army to march there, he learned that Emperor Nero was dead. Nero had reigned for thirteen years and eight days, and had greatly abused his power while in office. Nero had committed the management of Roman affairs to two wretches, Nymphidius and Tigellinus, his unworthy freedmen. They had conspired against Nero, who when deserted by his guards, escaped with four of his most trusted freedmen but committed suicide just outside Rome. Meanwhile, the men who had prompted Nero's death were brought quickly to justice and punished. The war in Gaul ended. Galba became emperor and returned to Rome. But Galba was shortly accused by his soldiers of spinelessness and was covertly assassinated in the middle of a Roman marketplace. Otho then became emperor, but then died in combat against Vitellius's officers. Vitellius then caused much turmoil in the fighting around Rome until Antonius Primus and Mucianus killed Vitellius and defeated his German legions, thereby putting an end to that civil war.

All of these issues I have chosen but to skim over here because they are well known by everyone. They are described in detail by a number of Greek and Roman authors. [I have merely touched upon these events here so that my history is positioned in context.] So on hearing of these things, Vespasian delayed his march to Jerusalem and waited to hear whom would be crowned Rome's emperor following Nero's death. When he heard that Galba was the emperor, he decided to await further instructions from him regarding future war plans. But he did send his son, Titus, to Rome to salute Galba and to receive his orders concerning the Jews. King Agrippa sailed to Rome with Titus on exactly the same mission. As they were en route in their long ships near the Achaean coast, for it was winter, they got word of Galba's assassination after he had reigned seven months and seven days. Otho succeeded him in the government and undertook the affairs of Rome. Agrippa wasn't afraid to continue to Rome on account of the change in governments, but Titus, by Divine impulse, sailed back to Syria from Greece and hurried to his father in Caesarea. Both were now in doubt about the situation in Rome, being as it was in flux, and did not proceed to march to Jerusalem. They thought it best to hesitate to attack the Jewish city until they received orders from Rome.

3. Meanwhile, another war erupted in Judea. A young man from Gerasa named Simon was the son of Giora. He wasn't as shrewd as John of Gischala, who had previously captured the city. But Simon was superior to John in both physical strength and courage. When Simon had been driven out of the Acrabattene toparchy by the high priest Ananus, he went to the bandits at Masada. At first the robbers didn't trust Simon. They only allowed him to come with the women he brought with him into the lower part of the fortress, while they lived in its upper parts. But his ways so agreed with theirs, and he seemed so trustworthy that he soon rode out with the bandits, looting and burning the country around Masada. When Simon tried to convince the bandits to undertake larger targets, they backed off. They had grown accustomed to living in their fortress and were afraid of going too far afield of their hiding place. When Simon heard that Ananus was dead, and being ambitious in his tyranny, he left Masada and went up into the mountains. There he proclaimed all of the enslaved to be free and to reward all the free, then began to gather an army of wicked men from around the countryside.

4. Organizing a substantial body of men, Simon overran the mountain villages. When still more and more men joined him, Simon ventured down into the country's lower regions. Since he had now become powerful enough to terrorize the larger cities, he deceived many powerful men.

Eventually, he no longer led only slaves and bandits, but a great many of the populace now looked upon Simon as their king. He began to overrun the Acrabattene toparchy up to and including places as far as Great Idumea. He built a wall around a village called Nain and used the village as his own private fortress. Simon enlarged many of the caves in a valley called Paran. Along with other already useful caves he stored his treasures. These caves became the hiding places for his prey and the fruits he had gathered through his plundering, and many of his followers made their homes in the caverns. Simon made no secret of his plans to prepare his men for their ultimate assault on Jerusalem.

5. Meanwhile, the Zealots were afraid Simon would attack them. Unwilling to allow Simon's force to become large enough to oppose them, the armed Zealots went out of Jerusalem to fight him. Simon met the Zealots, killing many of them and driving the rest back into the city. But he didn't think his men yet capable of storming Jerusalem's walls. First he decided to conquer Idumea, as he now had 20,000 armed men. So Simon marched to the Idumean border. The Idumean leaders quickly gathered 25,000 of their fiercest warriors while the rest of their men were assigned as guards to protect against this Sicarii raiding parties from Masada.

The Idumeans met Simon's army at their border and fought against them for an entire day. The battle was considered a draw, as neither the Idumeans nor Simon's men won a clear victory. So Simon returned to Nain while the Idumeans returned home. But it wasn't long before Simon went back to Idumea and resumed his attack. He pitched camp at a village called Thecae and sent one of his companions, Eleazar, to the garrison at Herodium, demanding the Romans surrender the fortress to him. The garrison received Eleazar not knowing the object of his mission. But as soon as he spoke of their surrender they drew their swords and attacked. Having no place to retreat, Eleazar threw himself off the wall into the valley below, dying instantly. Meanwhile, the Idumeans were still afraid of Simon's powerful army and decided to spy upon them before another battle erupted.

6. One of the Idumean officers, a man named Jacob, readily volunteered to become a spy. In actuality, he wanted to betray his own people. He left the village Alurus near the Idumeans' encampment and went to Simon. Jacob agreed to betray Idumea to him, and received assurances that Si-

mon would always consider him his ally. Jacob also promised Simon he would always help in the subjugation of Idumea. On this basis, Simon treated Jacob royally and was elevated for his many promises. When Jacob returned to his own people, he lied about Simon's army, saying that it was far larger than it was. Following that, he deviously persuaded the Idumean officers, and gradually the entire army, to render the country to Simon without a fight. As this ruse was going on, Jacob sent a message to Simon inviting him to come back, and promising to disperse the Idumean army, which he did. As soon as Simon's army came into view, Jacob jumped on his horse and fled along with all of those he had misled. Great fear descended upon the rest of the Idumeans, who on seeing their officers flee, broke ranks and retired to their own homes.

7. In this way, Simon marched into Idumea without a drop of blood being spilled. He made a quick assault upon Hebron and captured it. He garnered much plunder and commandeered a large supply of fruit. The people of Hebron say it is a much older city than any in the country, even older than Memphis in Egypt. They reckon it to be 2,300 years old. They also tell that Hebron was home to Abram, the father of the Jews, after he had moved from Mesopotamia, and that his progeny left Hebron to go to Egypt. Excellent and elegant marble monuments are in Hebron that attest to that period. Additionally, less than a mile from the city, a large turpentine tree grows said to have been there since the creation of the world.

So Simon went all through Idumea ravaging the towns and villages and laying waste to the entire countryside. In addition to those armed men who had followed him, Simon now had 40,000 more. His problem was that he didn't have enough provisions to supply such a multitude. Besides that problem, Simon was a barbarous man and bore great animosity toward the Idumeans. Idumea was seriously depopulated by the time he finished. Like trees defoliated following a locust attack, nothing was left of Idumea but desert. Some places were burned to the ground while others demolished. Whatever grew in the country was either trampled down or eaten by Simon's army. The ground over which they ran was packed down and more difficult to plow than when it had been barren. In short, the nation was left in such a wasted state as to make one believe it had never been a nation at all.

8. Meanwhile, the Zealots became even more nervous, learning of Simon's

successes. They were afraid to engage him in open combat, preferring instead to set ambushes in the mountain passes. They kidnapped Simon's wife along with a host of her attendants, then returned to Jerusalem with joy as if they had captured Simon himself. They thought Simon would lay down his arms and plead with them for his wife's return. But Simon did no such thing. He was enraged at the Zealots for taking his beloved wife. He came to Jerusalem's wall and like a wounded wild beast that cannot take revenge on its foe, bellowed at all within earshot. He grabbed everyone coming out of the city gates to gather herbs or sticks, even those unarmed and elderly, torturing them before executing them. Such was Simon's horrendous fury. He was almost ready to eat the flesh of their dead bodies. He also cut off a great many hands and sent the amputees back into Jerusalem to put fear in his enemies' hearts, and to provoke the citizens to rise against those who had kidnapped his wife. He ordered the amputees to tell the people that Simon, swearing by the God of the universe, that unless his wife were restored to him he would break down their walls and punish everyone. He would neither spare infants nor the elderly, nor make any distinction between who was innocent and who was guilty. Simon's threats cast a pall of fear upon not only the populace but on the Zealots as well—so much so that they sent his wife back to him. Welcoming his wife, Simon's anger ameliorated, and he ceased his bloody rampage.

9. So uprisings and civil war prevailed not only in Judea but Italy as well. Galba was murdered in the midst of Rome's marketplace. Then Otho was made emperor and fought against Vitellius, who saw himself as emperor on the acclamation from his German legions. But when Otho fought Vitellius's generals, Valens and Cecinna, at Betriacum in Gaul, Otho prevailed on the first day. But on the second day the Vitellius's soldiers emerged victoriously. After much bloodshed Otho committed suicide when at Brexia he heard of the defeat. He had managed Rome's affairs for three months and two days. Otho's army then deserted to Valens and Cecinna, and Vitellius rode with his army back to Rome.

Meanwhile, Vespasian left Caesarea on the fifth of Decius [Sivan] and proceeded to march against any place in Judea that remained unconquered. He went up to the mountains and took the Gophnitica and Acrabattene toparchies. He then took the small cities of Bethel and Ephraim and placed garrisons in them. He then rode toward Jerusalem taking prisoners as he marched. But one of his officers, Cerealis, took a constituent of cavalry and infantry and laid waste to Upper Idumea. He attacked Caphethra, taking the town quickly and lighting it ablaze. He also assaulted Caphatabira and laid siege to its formidable walls. While he was preparing for a long siege, the residents suddenly opened the city gates and ask for pardon, surrendering the town. The Roman officer then went on to Hebron, that ancient city. I have already told you of how that city is located in the mountains not far from Jerusalem. When he had forced his way into Hebron, he killed everyone there and burned the city to the ground. Now all Judea was taken except Herodlum, Masada, and Macherus, which were held by the bandits. So now the Romans looked to Jerusalem.

10. As soon as Simon had freed his wife and recovered her from the Zealots, he returned to what remained of Idumea. He drove the people out of every part of the country, prompting many to go to Jerusalem. Simon then followed them back to the city and once again surrounded it. When he caught anyone coming from the country into the city, he killed them. Simon now roamed free without the protection of the wall, and aroused more fear among the people of Jerusalem than did the Romans. The people also feared Simon even more than they feared the Zealots, who were a more formidable enemy than either Simon or Rome.

While Simon's plundering was taking place, John's criminal maneuvers tricked the people of Galilee. The Galileans had promoted John to high office among them. In return, John gave them what they wanted: to ride roughshod over the people. The Galileans' lust to rob and to steal became insatiable, as did their eagerness to ransack rich mens' homes, murdering the men they found and abusing their women. It was like sport to them. The Galileans consumed whatever spoils they got their hands on, slaking their thirsts with blood. Without compunction, they indulged in cross-dressing until satiated. They braided their hair, put on women's clothes, and drenched themselves in perfume. They imitated women's affectations and lusts. The Galileans bore the shame of insufferable sinfulness, inventing all sorts of impurity. They sashayed up and down the streets like harlots. Their faces bore women's makeup as they walked imitating women's swaying mannerisms. They killed with the right hands, unsheathing swords from beneath their finely decorated cloaks. Instantly, they transformed into warriors, running through everyone crossing their paths. But anyone in Jerusalem who tried to escape John's Galileans ran into a bloodier foe awaiting them outside the walls—Simon. All attempts were quashed to flee the city and desert to the Romans.

11. But John was not without his troubles. His army turned against him, and the Idumeans who remained broke away from the tyrant and tried to kill him. They were envious of his power and hated his brutality. So they joined forces and killed many of the Zealots, driving the rest into the palace that was built by Grapte, a relative of Izas, king of Adiabene. The Idumeans drove the Zealots out of the palace into the temple and began to loot John's treasures obtained through his tyranny. Meanwhile, the rest of the Zealots dispersed throughout the city ran to the temple, where John organized them anew, preparing to attack both the citizens and the Idumeans. The Idumeans were not as afraid of the coming attack as much as they feared the Zealots' lunacy, for they were better soldiers than the Zealots. They feared the Zealots might steal secretly out of the temple and mingle with the crowds, where they might kill wantonly and perhaps set the city on fire. So the people and Idumeans gathered with the high priests and decided what to do to avoid any Zealot assault. But God intervened and allowed them to heed the absolute worst advice. The medicine they devised for freeing themselves was worse than the disease itself.

They decided that the best way to overthrow the Zealot, John, was to bring Simon into the city. They wanted to bring a second tyrant into Jerusalem. So, resolving to do this, they sent the high priest, Mathias, to invite the man whom they had been afraid to ask to help them. Those also who had previously fled Jerusalem joined in the request, wanting as they did to keep their homes and property safe while away. Accordingly, Simon haughtily granted their request for his benevolent protection and went into Jerusalem to save it from the Zealots. The people shouted joyfully upon his arrival, seeing Simon as their deliverer and protector. But once inside with his army, Simon took pains to consolidate his authority. He looked upon those who had invited him as his enemies, just as those they intended to invite.

12. In this way, Simon took possession of Jerusalem in the third year of the war, in the month Xanthicus [Nisan]. Meanwhile, John, with a great many Zealots, was locked up in the temple. Additionally, Simon's army had stolen all their possessions. The Zealots began to lose all hope and

despaired of escape. Simon then attacked the temple with the people's assistance. The Zealots successfully defended themselves from the cloisters and the battlements, as a large part of Simon's men had died, and many others were carried off wounded. For the Zealots threw their darts from higher positions and rarely missed. Earlier, the Zealots had erected four high towers from which to launch their missiles. One was at the northeast corner of the court, one above Xystus, the third at another corner near the lower city, and the last built above the top of Pastophoria. [It was on Pastophoria where one of the priests historically stood in the twilight and gave the signal for the start of the Sabbath day. The trumpets also signaled when the Sabbath was over. The signal gave the people notice when to stop working and when to begin work again.] The Zealots also placed their war engines on the towers along with their archers and slingers. Simon's assault diminished inasmuch as his men had grown tired. He did not withdraw from opposing the Zealots as he had a superior force. However, the darts and stones thrown by the engines carried a great distance, and killed many of his men.

Chapter 10

How the Roman Soldiers, Both in Judea and Egypt, Proclaimed Vespasian Emperor, and How Vespasian Released Josephus from His Bonds

1. Around this time, troubles began coming upon Rome from every side. Vitellius had arrived from Germany with his legions, accompanied by a multitude of foreigners. When the soldiers' normal encampments could not hold them all, he made all Rome their camp. He filled the citizens' homes with armed men, men who had never before seen such riches. They found themselves surrounded by objects of silver and gold and found it difficult to restrain their covetousness. They stood ready to steal whatever they wanted and kill anyone who stood in their way. These troubles describe the state of affairs in Italy at the time.

2. When Vespasian had subdued all the countryside around Jerusalem, he returned to Caesarea. There, he was advised of the troubles plaguing Rome and of Vitellius's promotion. He was indignant. Although the general knew what it was to be governed as well as to govern, he could not abide anyone as lord over him who acted like a madman and who had

seized this opportunity to take over the Roman government. Vespasian came under great sorrow to the degree that he became depressed and unable to apply himself to warfare in Judea while his own country was being laid waste. But as much as he wanted to avenge his native land, he was physically far removed from taking any action. It was a long way to Italy from Syria, and it was winter. Even if he could sail for Italy, other misfortunes might keep him from it. So Vespasian was forced to contain his intense anger.

3. But Vespasian's officers and soldiers began meeting together in groups and speaking openly about the changing situation in Rome. In their indignation, they cried out, "In Rome our soldiers live in comfortable circumstances, not even hearing a rumor of warfare. Meanwhile, they are ordaining whomever they please to be our leaders in hopes of making them emperor, while we who have gone through such toil and have spent years in our armor, allow them such power. We have a man more worthy to rule than any other they have put forth. How can we ever repay our general if we do not use this opportunity to act? Do not more just reasons exist why Vespasian should be emperor instead of Vitellius? Are those reasons not more compelling than those that elevated other emperors? Have we not undergone greater wars than those troops who fought in Germany? Nor are we inferior in any way to those who have accompanied this tyrant, Vitellius, to Rome, nor have we undergone lesser labors. Neither the people of Rome nor the senate will bear the lecherous Vitellius, in comparison to our virtuous Vespasian. Nor will the people and the senate stay silent in the face of this most brutal tyrant, in place of a just and noble governor. Nor would they choose a childless governor to rule over them instead of one who has a son, because the succession of one's own children is the greatest security kings can provide for themselves. If we believe the ability to govern is best located in those who have years of experience, then we'll have to have Vespasian. If, on the other hand, youth and strength are determined to be best, we ought to have Titus. By having both Vespasian and Titus, we will have both advantages. They have the strength of emperors, having already the acclamation of three legions plus the auxiliaries from our neighboring kings. Additionally, they will have all the armies of the east to support them, as well as those in Europe. They are far away from Vitellius and do not fear him. Besides, Vespasian and Titus have other allies in Italy itself. Don't forget Vespasian's brother,

Flavius Sabinus, and his second son, Domitian. Domitian will bring with him many young men of dignity. Sabinus is already the mayor of Rome, which office itself will play no small role in bringing Vespasian to power. But if we delay any longer the senate may choose another emperor whom we, the saviors of the Roman Empire, will hold in contempt."

4. These were the conversations the soldiers had amongst themselves in their various companies. Following these meetings, they gathered together as one, encouraging one another and declaring Vespasian emperor. They encouraged him to save the Roman government, which was in danger. Vespasian, meanwhile, had been greatly concerned about civil affairs in Rome, but did not intend to set himself up as emperor. His credentials revealed that he was very qualified to rule the empire, but he preferred the security of private life to the dangers of a public profile. When he refused to listen to his troops, their officers insisted more earnestly on his acceptance. The soldiers gathered around their general with their swords drawn, threatening to kill him unless he would now live according to his immense capabilities. Reluctantly, Vespasian thought a great while, hoping to deny the power being thrust upon him. But ultimately, not being able to subdue his men's enthusiasm, Vespasian yielded to their pleas to become emperor.

5. So, upon the encouragement of Mucianus and the other commanders, the general accepted leadership of the empire. The rest of the army voiced their willingness to follow Vespasian against all who opposed him. So the general relented and agreed. He wanted, first of all, to gain dominion over Alexandria, knowing that Egypt was of utmost importance in any quest to the office.

Egypt supplied corn to Rome. If Vespasian could control the Roman supply of corn he could depose Vitellius, who would be forced to resign if the population of Rome had nothing to eat. Vespasian also wanted to join the two legions in Alexandria to the legions that were with him in the east. He also considered that if fortune frowned against them, Egypt would be a place of security. Egypt would be difficult to attack by land, and has no good ports on the Mediterranean Sea. To the west lie the arid deserts of Libya; to the south the Syene and the unnavigable wild cataracts that separate Egypt from Ethiopia. The Red Sea protects Egypt on the east, reaching as far as Coptus, and is fortified to the north by land that reaches Syria and the Egyptian Sea, where no ports exist.

Egypt is fortified on every side. From north to south, Egypt runs from Pelusium to Syene and is six hundred fifty miles in length. The sea voyage from Plinthine to Pelusium is about four hundred fourteen miles. The Nile is navigable as far as the city called Elephantine where the aforementioned cataracts begin, preventing ships from going farther. The port of Alexandria presents its own problems even in peacetime, for the channel is narrow and full of subterranean rocks that force the sailor to weave between them. Man-made levees barricade its port side and to the starboard lies the island call Pharus, just before the entrance. Pharus supports a tall tower from which its fire may be observed from a distance of thirty-eight miles, allowing ships to anchor well off its dangerous shore at night. Around the island, large piers have been constructed, against which the sea waves crash. Navigation in these waters is very troublesome, and the entrance to its narrow passageway is very dangerous. Once inside the port the danger vanishes. The port itself is four miles in size, into which arrive Egypt's imports and from which are distributed the country's abundant resources throughout the known world.

6. So, it is obvious why Vespasian desired to control Alexandria, in order for it to be the foundation of his attempt to control the empire. Accordingly, he immediately sent a message to Tiberius Alexander, then governor of Alexandria and Egypt, informing him what his army had placed upon his shoulders. He wrote of how he had been forced to accept the governmental burden, and how he hoped to have Tiberias as his colleague and supporter. As soon as Alexander had read the letter, he ordered his legions and the citizens to take an oath of loyalty to Vespasian, both of which were willing to do so, knowing of Vespasian's courage from his earlier conduct in their area. And so Vespasian, believing himself already entrusted with the Roman government, began preparations for his journey to Rome. His fame preceded him. The news traveled fast that Vespasian was already the eastern empire's head. Every city held festivals in his honor and celebrated sacrifices and oblations, upon hearing the news of the general's ascendency. The Roman legions in Mysia and Pannonia [on the Danube], which had been in upheaval because of Vitellius's insolence, happily took the oath to Vespasian upon his succession as emperor. So Vespasian left Caesarea for Berytus, where many ambassadors arrived from Syria to see him, as well as many from the provinces, bringing with them their city seals

and the people's congratulations. Mucianus also arrived, and as president of the province, told him of how the people joyously received the news of Vespasian's advancement and that the people in every city had taken the oath of loyalty to him.

7. Vespasian's good fortune exceeded his hopes on every front, as the civil affairs were, for the most part, already in his hands. He knew that he had received imperial power by the hand of Divine Providence—a righteous kind of fate. As he recalled the many and varied signals that foretold of his ascension as emperor, Vespasian also recalled that Josephus had accurately predicted his future while Nero was still alive. He then became very concerned, realizing that the Jewish general was still being held in irons. So he called Mucianus and his other commanders and friends, informing them how brave a man Josephus had been and what great hardships he had been forced to suffer in the siege of Jotapata. He then told them of Josephus's predictions which he had first suspected as lies, conjured up because of the trouble the Jew was in, but which ultimately had been demonstrated to be by Divine inspiration.

Vespasian said, "It is a shame that this man who foretold of my becoming emperor and had been the Minister of the Divine message to me should still be our prisoner." So he called for Josephus and commanded that he be set free. The officers thought that if Vespasian treated a prisoner in this way, what even larger honors were in store for them as well. Standing next to his father, Titus said, "O father, we should remove Josephus's chains, but also the shame of ever having been a prisoner. We should treat him as if he was never in irons at all. So let us not just loose his chains but let us cut them into pieces." That was the usual practice for someone imprisoned without just cause. Vespasian agreed with Titus's advice, and the man came in and cut Josephus's chains into small pieces. Josephus received this as a rewarding testimony to his integrity. Henceforth, Josephus was regarded as a person who could be relied upon to foresee future events.

Chapter 11

How Titus Marched to Jerusalem and the Danger He Faced as He Overlooked the City. A Description of His Encampment

1. When Vespasian had responded to the ambassadors and had justly dis-

tributed government offices according to everyone's own merit, he went to Antioch. There he deliberated as to what to do next. He preferred to go to Rome rather than to march to Alexandria, seeing that Alexandria was already in his camp. On the other hand, Rome's affairs were in complete disorder due to Vitellius. So he sent Mucianus to Italy, along with a large force of cavalry and infantry. Mucianus was fearful of sea travel for it was in the middle of winter, so he led his army overland through Cappadocia and Phrygia.

2. Meanwhile, Antonius Primus took a third of the legions in Mysia [for he was president of that province], and quickly went to Rome to attack Vitellius. But Vitellius sent a great army led by Cecinna to meet him. Vitellius had full confidence in Cecinna because of his victory over Otho. So Cecinna hurriedly marched out of Rome and found Antonius near Cremona, a city near the northern border of Italy. But when Cecinna saw that Antonius's troops were numerous and well disciplined, he chose not to fight. Also believing an attempt to retreat would be dangerous, he began thinking about surrendering his army to Antonius. By doing so, he would diminish Vitellius's power and enhance Vespasian's. He told his men that with Vitellius came only the bare name of power, but with Vespasian lay the actual power. It was better for them to gain favor with the real power. They would likely be overwhelmed if they fought, so it was better to avoid that danger in advance by willingly going over to Antonius. Vespasian was already able to conquer those who remained with Vitellius without their help, while Vitellius couldn't save what he had with it.

3. Cecinna, by these words and more like them, persuaded his army to comply. Both he and his entire army then deserted to Antonius. But that very night the soldiers regretted what they had done, and were seized with fear at the prospect that Vitellius might be the most powerful. So, drawing their swords they attacked Cecinna, attempting to kill him. They would've been successful if their tribunes had not fallen upon their knees and begged them not to do it. So, the soldiers spared Cecinna, but put him in irons as a traitor. They were about to send him to Vitellius. When Antonius Primus got word of this, he awakened his men immediately and had them don their armor, leading them against those who had revolted. The men loyal to Vitellius aligned themselves for battle and resisted for a little while, but were soon overcome and fled to Cremona. So, Antonius

quickly ran ahead of them with his cavalry and cut off their entrance into the city, surrounding them and destroying a great many of them. He then followed those who had gotten into Cremona and gave his soldiers leave to plunder the city. Many citizens of Cremona also perished, many of whom were merchants. Among the dead was Vitellius's entire army of 30,200 men. Antonius lost more no more than 4,500—those who had joined him in Mysia. He then set Cecinna free and sent him to Vespasian with the good news. So, he was received by Vespasian, who covered up the discredit of Cesena's treachery by giving him unexpected honors.

4. In Rome, Sabinus took courage at the news that Antonius was approaching. He assembled the cohorts assigned to the night watch and in the darkness seized the capitol. When the sun rose, many men of character came over to him with Domitian, Vespasian's other son, whose encouragement carried great weight in the coup. Meanwhile, Vitellius wasn't as much concerned with Antonius Primus as he was enraged at those who had revolted with Sabinus. Thirsting after noble blood according to his own innate barbarity, Vitellius sent out the army that had come with him from Germany to fight against those holding the capitol building. A formidable fight ensued, with heroic actions on both sides. But finally the soldiers from Germany proved to be too numerous and took the temple hill. Domitian, with many other prominent Romans, were able to escape while the rest of the defenders were cut to pieces. The soldiers brought Sabinus before Vitellius and executed him. They then plundered the temple's ornamentation and set it on fire. Within a day, Antonius arrived with his army and were met by Vitellius and his troops. The battle took place in three separate locations. The result was the destruction of Vitellius's army.

    Vitellius then came out of his palace, drunk as a dog and full of a luxurious and extravagant dinner, as it was to be his last meal. The multitude that greeted him threw him around, Vitellius suffering all kinds of torments. Then they cut off his head in the middle of Rome. He had reigned for eight months and five days. Had he lived much longer, I do not think the empire would've survived his lust. The others killed numbered over 50,000. The battle was fought on the third of the month of Apelleus [Kislev] (roughly November). On the following day, Mucianus arrived in Rome with his army and ordered Antonius and his men to stop the slaughter. For, they were still searching houses and killing many of

Vitellius's soldiers and many of the populace, supposing them to be allies. The soldiers' rage prevented any accurate distinction between them and others. He then brought forth Domitian and recommended him to the multitude of Romans until his father, Vespasian, arrived in the capital city. Freed from their fears, the people shouted for joy for Vespasian as their emperor and held a festival in confirmation of that, as well as for Vitellius's death.

5. When Vespasian arrived in Alexandria, the good news had already come from Rome. At the same time, ambassadors from all over the earth came to congratulate Vespasian on his ascension as emperor. Though Alexandria was the largest of all the world's cities next to Rome, it proved too small to hold the multitude that came into it. So as Vespasian's ascension to emperor was confirmed in Rome—which had been delivered unexpectedly from utter ruin—Vespasian turned his thoughts to subduing other uprisings in Judea. But he needed to make haste in sailing to Rome, as the winter was almost over, and he had set affairs in Alexandria in order.

So, he sent his son, Titus, to destroy Jerusalem with a specially chosen portion of his army. Titus and his men marched by foot to Nicopolis, two-and-a-half miles from Alexandria. There, he put his army on board some long ships and sailed on the river along the Mendesian Nomus as far as Tumuis. He got out of the ships in Tumuis and walked to a small city called Tames where he spent the night. He then went to Heracleopolis, then Pelusium, where he rested for two days. On the third day, Titus crossed over the mouth of the Nile at Pelusium. He then proceeded into the desert, making camp at the temple of the Cassian Jupiter, and then went on the next day to Ostracine. That place had no water. The people there had to import it from other places. After Ostracine, Titus rested at Rhinocorura and then on to Raphia, his fourth station. The city is at the southern border of Syria. For his fifth station, Titus made Gaza, then on to Ascalon, then from there to Jamnia, and after that Joppa. From Joppa, Titus marched to Caesarea, having decided to concentrate all his forces at that one place.

The Jewish Wars
Book V

Containing the Interval of about Six Months From Titus's Arrival to Lay Siege to Jerusalem to the Great Extremity to which the Jews were Reduced.

Chapter 1

Concerning the Seditions in Jerusalem and what Terrible Miseries they Inflicted upon the People

1. Therefore, when Titus had marched over the desert between Egypt and Syria, as I've mentioned before, he arrived in Caesarea having decided to coordinate his forces there before beginning the war. While he had been in Alexandria helping his father settle the government that God had recently conferred upon them, the subversion in Jerusalem revived and divided into three factions. Each faction fought against the other two. It may be said that this division was a good thing, even in the cause of evil, and the cause of Divine justice. We've already accurately described the attack the Zealots made upon the people, which I believe was the beginning of Jerusalem's destruction, as when it began and how it greatly enlarged. This present agitation maybe called a sedition caused by sedition, and likened to a wild beast becoming mad, which lacking other food, began to feast upon itself.

2. Eleazar ben Simon, who first separated the Zealots from the people and who ordered the Zealots' retreat into the temple, became enraged at John's audacious daily attacks upon the people. John never stopped murdering. The real reason was that Eleazar could not bear the thought

of submitting himself to this tyrant who had taken over his command. Eleazar wanted to keep all power and dominion for himself. So he broke off from John and took as his assistants Judas ben Chelsias and Simon ben Ezron, who were among the most powerful citizens. Also joining Eleazar was Hezekiah ben Chober, also an eminent person. Each of these men brought along a great many Zealots, seized the temple's inner court, and stacked their weapons against the holy gates and the temples' fascia. They had plenty of provisions and were of good courage. An abundance of food consecrated to sacred use existed in the temple. The Zealots had no scruples about eating it. Yet they feared their numerical inferiority, so they had to be content to lay still and hold what ground they had. Meanwhile John, although having the advantage in superior numbers, nevertheless was at the disadvantage of having his enemies right above his head. He could not assault them without fearing heavy losses, yet his anger was so fierce as to not let him be at peace. John also suffered more damage from Eleazar than Eleazar did from him. Yet he would not stop assaulting the other faction. So, continuous sallies went forth against one another and numerous missiles were thrown from each side. The temple was defiled everywhere with their blood.

3. But now, the tyrant Simon ben Gioras held the upper city in his power, as well as a large part of the lower city. People under the mistaken impression that he could help them fight the Zealots invited him into Jerusalem. He now made vicious assaults upon John's Zealots, catching them in the crossfire from Eleazar's Zealots, who held the temple mount. So, John both received and inflicted great damage as his enemies attacked from both sides. He had the same advantage over Simon that Eleazar had against him, since Simon was physically lower than John. Because of this, John easily fought back the attacks made from below using only handheld weapons. For those above him, John used his war engines to throw many darts, javelins, and stones. By these methods, he not only defended himself from his enemies but also killed many of the temple priests as they went about their sacred daily functions.

 For although the Zealots in the temple courts were irrational and practiced every sort of irreverence, they continued to admit those desiring to offer sacrifices. They always searched their own countrymen beforehand; suspecting them of treachery they watched them carefully. They weren't so much afraid of the foreigners who, when admitted to the temple court, often became victims of the Zealots' missiles. For John's engines threw darts with great force toward all the buildings, often reaching as far as the altar in the temple itself, hitting the priests who engaged in their sacred duties. People who had come to Jerusalem from every nation to offer sacrifices at this famous place, believed holy by everyone on earth, fell dead upon their oblations. The altar was sprinkled with that which was honored among all men—Greeks and Barbarians alike—their own blood. The remains of foreigners lay side-by-side with those of Jews, those of Gentiles with priests, and the blood of various animal carcasses all pooled together in God's courts. And now, "Oh wretched city, would you had but suffered at Roman hands who came to purge you of your internal hatreds. You are no longer God's chosen city nor will you continue to exist after becoming the tomb for your own people's cadavers. You have made the holy temple itself a graveyard in your civil war! But perhaps there is a chance for you: if you will appease the anger of God, who is the instigator of your devastation." But I must restrain myself by rules guiding historians. Now is not the proper time from my personal lamentations. I must continue with the historical narrative and so shall return to what these agitations have brought upon Jerusalem.

4. So, three treacherous factions fought each other in the city, led by John, Eleazar, and Simon. Eleazar's men seized the sacred temple wine and came against John in a drunken fury. Those with John plundered Jerusalem's citizens and fanatically attacked Simon. Meanwhile, Simon gathered his necessary supplies from the citizens as he fought John. Because John fought on two fronts, he had his men turn and throw their darts from the cloisters he controlled while he attacked Eleazar's men in the temple courts with his war engines. Whenever the Zealots above him frequently stopped their assault, mainly for being tired and drunk, he sent many of his men out against Simon. John entered many areas of the city setting fire to those houses full of corn and other foodstuffs. Simon did the same thing. When John's men retreated, Simon turned upon the citizens. It was as if these men had done it all on purpose, to serve the Romans by destroying that which Jerusalem had laid up as provisions for the coming Roman siege. In doing so, the three factions cut off their own ability to survive. And so it came to pass that all the buildings surrounding the temple were burned to the ground. The area had become like a desert. Almost all the city's corn burned up, corn that would have been sufficient

to withstand a multiyear siege. So, unwary of what the Zealots had done, the city was destined to fall due to starvation. Famine would have been impossible unless the Zealots had by their own actions made it so.

5. Now was the city totally encompassed in warfare from these large gangs of wicked men. The people of Jerusalem were like a body torn to pieces. The older men and the women were in great distress due to the internal strife. They even hoped the Romans would arrive and begin an external war to deliver them from their internal miseries. All the citizens were under a terrible dread and fear. They had no hope of gathering together, of altering their own conduct, nor of making peace with their enemies. Nor could anyone who wanted to flee the city do so, for guards were everywhere. The robber chieftains, while at war with one another in most respects, agreed on this: to kill anyone who wanted to make peace with the Romans, or who were suspected of attempting to escape. They saw these potential escapees as their common enemy. They agreed upon nothing but this—to kill the innocent. The clamor created by the continuous fighting went on day and night, but the lamentations of the mourners even exceeded that din. The mourning never ceased because disasters came one after another. However, in their deep consternation the mourners could not cry and wail outwardly. Stifled by the fear of outwardly revealing their inner passions, they dared not even open their lips to groan.

The people still living were treated as nonexistent even by their relatives and friends. Nor was any care shown for burying the dead. The result was that every person mourned within himself. Those who took no part in the warfare lost interest in everything, expecting to be dead shortly. Meanwhile, the seditious continued to fight each other as they walked over the corpses that lay heaped in piles. The dead bodies underfoot only increased their rage and made them even fight more fiercely. The factions continued to plot maneuvers against one another and carried out their wicked plans without mercy. They omitted no form of torment or barbarity. John took sacred materials and used them to build his war engines. [The people and priests had at one time determined to raise the temple sixty feet higher, and the materials were there for that purpose. At very great expense and difficulty, King Agrippa had brought in the materials. Large and very straight timbers lay unused because of the coming threat of war. So the work was interrupted.] But now John had these timbers cut and prepared for the construction of towers. They were long enough to raise his engines up to the level of the temple above him. He also had them carried in and set up behind the inner court, the only place the towers could be erected. The other sides of the court had steps that would not allow John's towers to rise as high as the temple cloisters.

6. So, John hoped to defeat his enemies by these towers, constructed in his irreverence. But God proved, by bringing the Romans upon the city, wasted John's energies before any of his towers could be erected. When Titus had gathered a portion of his forces about him and ordered the rest to meet him in Jerusalem, he marched from Caesarea. He had with him the three legions that accompanied his father when he had laid waste to Judea, along with the twelfth legion, which was defeated in Cestius's debacle. That legion was otherwise well known for its valor and now marched with great eagerness to avenge themselves upon the Jews, remembering their former suffering at Jewish hands. Titus ordered the fifth legion to meet him by taking the route near Emmaus while the tenth legion was to march through Jericho. Titus then set out his remaining forces together with a large contingent of troops from the kings, and a considerable force of Syrian auxiliaries. Another 2,000 men chosen out of the Alexandrian army filled the ranks vacated from the four legions sent to Italy with Mucianus. He drew another 3,000 men from the River Euphrates guards. A friend of Titus, Tiberius Alexander, also came along. He was the most valuable ally both for his friendship and his wisdom. Tiberius had been Alexandria's former governor, but now found himself worthy to be next in command to Titus. This was the reason that Tiberius had been the first to encourage Vespasian to accept the role of Emperor. He had stood by Vespasian when events were uncertain, and the new emperor's fortune was not yet clarified. He also followed Titus as his counselor, very useful in this war both due to his age and his vast experience.

Chapter 2

How Titus Marched to Jerusalem, and how He Exposed Himself to Danger as He was Viewing the City. And of the Place where He Pitched His Camp

1. As Titus began his march through enemy territory, the troops sent by the kings marched first, in order, accompanied by all the rest of the auxiliaries. After these came those Romans whose duties were to grade the

roads and measure out the encampments. Then came the officers' baggage and their armed guards in support. After these marched Titus surrounded by his select troops. Then came the pike-men, and following them the legion's cavalry. All of these marched before the war engines. Following the engines came the tribunes and cohort leaders with their select units. After these marched the Roman flags with the golden eagle, which preceded their trumpeters. Then came the main body of soldiers and their ranks — six abreast. The servants of every legion followed them, and then finally came the mercenaries and the rear guard. So Titus marched his legions in good Roman order through Samaria to Gophna, the city formerly conquered by his father that now held a Roman garrison. He lodged there one night and then marched on the next morning. When he had marched a full day, Titus pitched camp in the valley which the Jews call in Hebrew, "The Valley of Thorns." Located near a village called Gabaothsath, which means "The Hill of Saul," the valley is about four miles from Jerusalem. There, Titus chose six hundred horsemen to observe the city: what strength its walls had, how courageous the Jews were, and whether when they saw him, and before the battle began, the Jews would be frightened and surrender. He had heard the truth: that the people of Jerusalem had come under the power of the Zealots and the seditious, and truly wanted peace. But, being too weak to rise against their internal enemies, they couldn't move.

2. As Titus rode along the highway that led straight to the city wall, nobody came out of the gates. But when he left the road with a few of his men and marched toward the tower called Psephinus leading a small group of men, an immense number of Jews charged out of the so-called "Women's Towers." The Jews passed through the gate next to the Queen Helena monuments and intercepted him. They got between the main cavalry forces on the highway and those few that had left it and intercepted Titus with his small unit. Titus could not move forward due to the many trenches protecting the gardens in front of him. He also saw that it was impossible to return to his men on the highway since a multitude of Jews were between them. Many of the cavalrymen on the road did not even realize that Titus was not among them. So, Titus saw that his life would be spared only by his own audacity. He wheeled his horse around and cried out to those with him to follow him. Then he galloped violently into the midst of the Jewish crowd in order to force his way through to

his own men. Here we may learn an important principle: both successes in warfare and the peril of kings are under God's providential hand. For a while, a great many darts were launched at Titus when he had neither his helmet nor his breast armor on. [As I said, he had just gone out to observe the city and not to fight.] But not one arrow touched his body. All were shunted aside as if the Jews missed on purpose. The arrows merely hissed as they streaked past.

With his sword in hand, Titus averted those who came close to him and ran over anyone in his path, trodden under by his horse's hooves. The Jews cried out at Caesar's boldness and encouraged one another to attack Titus, yet those in his path gave way and left in great numbers while his men rode close by his side, though receiving arrows in their backs and shoulders. They all had just one hope of escaping – if they could help Titus make a way through the multitude and that the Jews might not be able to surround him before he got away. Two Romans were at some distance from Titus. One man was surrounded, and arrows killed both him and his horse. The other Roman died as he leapt off his horse, which the Jews then carried off. But Titus escaped with the rest of his men and made it back safely to camp. This, the first Jewish success, served to increase their courage and gave them reason to hope. This brief instance of good fortune made them even bolder as the days progressed.

3. When the legion that had taken the Emmaus route drew near at night, Titus broke camp that next morning and went to a place called Scopus, from where Jerusalem and its temple were in plain sight. This place, opposite the northern quarter of the city, was flat and very appropriately named Scopus—"The Prospect." It was less than a mile from the city gates. Titus ordered the encampment fortified for the two legions that would camp together, but ordered yet another fortified camp less than half a mile behind that for the fifth legion. He thought that the fifth legion had been marching all night and was probably worn out. They deserved to be further separated from the enemy and there fortify themselves with less fear of attack. As these fortifications were under construction, the tenth legion, which had marched through Jericho, had already arrived at a certain place where a gang of armed men had formerly guarded the pass into the city. Vespasian had routed them previously. The tenth legion was ordered to camp closer to the city on the Mount of Olives, located three-quarters of a mile east of Jerusalem. The deep valley called Cedron

separates the Mount of Olives from the city.

4. As the Romans now threatened the city, the three Zealot and robber factions suddenly put a stop to their internal conflict against one another. They could clearly see the Romans making camp in three separate locations and began to consider an awkward peace agreement. They said to one another, "What are we doing, allowing these three fortified camps to wall us in so that we shall not be able to breathe free? The enemy is building a city to oppose us while we sit here within our own walls as mere spectators of what they are up to. Our hands are idle, and our armor laid aside as if the Romans were doing this for our benefit. Apparently we are courageous only in fighting amongst ourselves. Will we allow the Romans to conquer Jerusalem without shedding blood while we go on killing each other?" In this way they encouraged one another; they gathered together, immediately donned their armor, and ran out of the city, loudly assaulting the tenth Legion as they fortified their camp.

Meanwhile, the Romans had gathered into separate groups to carry on their work. For the most part, they had laid their weapons aside. They thought the Jews wouldn't dare sally out against them, distracted as they were by their own internal struggles. But, unexpectedly, they were put into confusion. Some Romans left their work and retreated, while others ran to pick up their weapons, but were struck and killed before they could turn back upon the enemy. More Jews joined the attack, encouraged by the initial surprise assault. Enjoying their good fortune, the Jews seemed to themselves and to the Romans more numerous than they were. The confusion of the Jewish attack puzzled the Roman defenders who had always been taught to fight skillfully and in good order, keeping within their ranks and obeying their officers' orders. So, they were surprised and had to retreat from the Jewish assault. As the Romans were in flight, many turned back upon the Jews and stopped their pursuit. But many were wounded by the vehement attack as still more and more Jews sallied out of the city.

Ultimately, the Romans were totally confused, and fled running away from their camp. The entire tenth legion would have been in danger had not Titus been informed of their situation and immediately sent help. He admonished the fleeing men for their cowardice and turned around those in flight. He then personally fell upon the enemy flank, and with his special forces killed many Jews and wounded even more. The rest fled, running quickly down into the Cedron Valley. Driven into the valley, the Jews suffered many casualties, but once on the rise near the city they turned, faced the Romans, and fought them hand-to-hand. The battle raged until noon. Shortly thereafter, Titus sent his other legions and his auxiliary troops against the Jews, attempting to stop further Jewish sallies. He then sent the tenth legion to the height of the Mount of Olives to resume fortification of their camp.

5. As the tenth legion left the battle, the Jews thought they were retreating. The watchman on the city walls shook his robe as a signal that sent a fresh crowd of Jews running violently into the fray, like a herd of wild animals. I tell you the truth; the Romans could not withstand the fury with which the Jews attacked them. It was as if they had been thrown out of a war engine. The Jews decimated the Roman ranks and put them to flight, back to the Mount of Olives. Only Titus himself and a few of his friends were left standing halfway up the hill. Paying no heed to the danger facing them, and ashamed to leave their general, the men earnestly begged Titus to get away from these Jews, who were so fond of death. They implored Titus not to risk danger for the sake of those who had already fled the battle and not to assume the role of an ordinary soldier. The entire success of the war depended upon their general, and the lord of all the habitable earth, on whom everything depended. Titus seemed not to hear their words, but continued to face the onrushing Jews. He sliced away at the enemy with his sword. He forced the Jews to retreat, and killed many as they fell back down the slope. The Jews were amazed at his courage and strength but did not immediately rush back into the city. Rather, they stepped aside and pursued those Romans still fleeing up the slope. Seeing their counterattack, Titus struck their flank, trying to halt the fury of their maneuver.

Meanwhile, confusion and fear fell upon those fortifying their camp on the mountaintop. As they saw those below them running away, they thought the entire legion was fleeing and that the Jewish sallies were unstoppable. They thought Titus himself was now running away, because they took for granted that if Titus had stood and fought, the rest would have never fled. Panic prevailed on every side as some fled one way, and some another way, until some observed Titus fighting in the very middle of the battle. Being greatly afraid for him, they shouted of the peril he faced to the retreating legion, shaming them to turn back. They scolded one another that they had done far worse than merely run away – they had deserted their general. So they came again in utmost force against the Jews. Running down the hillside, they drove them down to the valley floor. The Jews turned to fight, but as the Romans had the high ground and the Jews were running downhill, they were all driven further downward. Titus continued to press against those Jews near him and ordered the men of the tenth legion back to fortify their camp. He and the others with him continued the fight in order to keep the Jews from attacking further. Allow me here to add nothing to flatter Titus, nor to take away anything out of envy, but to speak only the plain truth. Caesar twice saved the entire tenth legion when it was in danger, giving them the opportunity to fortify their camp.

Chapter 3

How the Sedition Was Revived in Jerusalem, and the Jews Devise Snares for the Romans. How Titus Threatened His Soldiers for their Ungovernable Rashness

1. Now, as the war with the Romans subsided for a while, the sedition in Jerusalem revived. The Feast of Unleavened Bread arrived on the fourteenth day of Xanthicus [Nisan] - (April). It is a day when the Jews celebrate being freed from bondage in Egypt. Eleazar and his Zealots opened the gates to the temple's innermost court to allow in those people who wanted to worship God. But John used the festival opportunity to mask his treacherous plans. He secretly armed the most inconspicuous of his followers, many of whom were not purified. He ordered their weapons hidden under their robes and sent them as worshippers into the temple to seize it. When they arrived at the temple, they cast off their garments, revealing their armor and their weapons. This caused great confusion and distress in the holy place, as the worshipers who had no part in the uprisings believed John's men to be targeting everyone without distinction. The Zealots had only wanted to attack other Zealots. Eleazar's men, therefore, quit guarding the gates to the inner court and leapt down from their battlements before they could be attacked, fleeing to the subterranean caverns under the temple.

The people at the altar and throughout the temple trembled as they were thrown down, trampled upon, and beaten without mercy with wooden and iron weapons. Some Zealots who had prior enmity and hatred of others killed their innocent victims in the melee as if they were their Zealot opponents. Every citizen who had offended any of the Zealots was carried off and murdered. When John's men had done all these horrible things to the innocent people, they made a truce with Eleazar's tribe and let them come out of the caves in safety. John's followers seized the inner temple along with Eleazar's war engines and then moved out to oppose Simon. And so, this sedition that had been divided into three factions was now reduced to only two.

2. Meanwhile, Titus revised his plans and began to pitch the Roman camp nearer to the city than Scopus. He wanted to place many of his best cavalry and infantry units near the walls to prevent further Jewish sallies upon his men. And so he gave orders for the entire army to clear all the land between him and the city walls. The Romans destroyed all the hedges and walls that inhabitants had used to mark off their gardens and fruit groves. Fruit trees were also cut down, and all the ditches and chasms were filled up. They leveled the rocky outcroppings with iron hammers. The entire area between Scopus and Herod's monuments that adjoined the Serpent's Pool was leveled.

3. But the Jews, meanwhile, had contrived the following plot against the Romans: the bolder Jews went out of the city at the Women's Tower as if they had been thrown out of the city by those desiring peace. They stumbled about as if in fear of a Roman attack or afraid of one another. Those remaining on the walls that seemed to be on the people's side cried out to the Romans for peace and asked for security for their lives. They promised to open the gates to the Romans, while throwing rocks as if to drive away those Zealots who had gone outside of the city. The men outside pretended that they had been driven out by force, and cried out to be allowed back. They first rushed madly at the Romans and then fell back, and seemed to be in total confusion. Many Roman soldiers believed this cunning ploy was real and thinking that those on the wall would open the gates to them, they began to move toward the gates. But Titus was suspicious of this new Jewish maneuver. Only the day before had he sent Josephus to Jerusalem to invite the Jews to come to terms of accommodation.

But Josephus had been received discourteously and returned with no answer. So, Titus ordered his men to stay put. But some Roman troops who were in front of the works could not hear him. They took up their weapons and ran to the gates. Seeing this, the Zealots who had pretended to be thrown out of the city at first retreated, but as soon as the soldiers had advanced between the towers on either side of the gate, the Jews ran out and surrounded them. The Zealots then fell on them from behind, while the many Jews standing on the walls threw all kinds of stones and darts at them and killed a considerable number of soldiers, wounding many more. Trapped between the Zealots who pressed them against the wall, escape for the Roman troops proved difficult. They were also ashamed of themselves and feared what their officers would do to them. So, realizing their mistake, they stayed and fought with their spears for a long time, receiving many strikes from the Jews and delivering the same in return. When at last the soldiers managed to fight off those surrounding them, they were able to escape. But the Jews followed the retreating soldiers closely all the way to the Queen Helena monuments, throwing their darts.

4. The Jews, delighted by their sudden victory, grew insolent and taunted the Romans for being deceived. They beat their shields loudly and leapt with joy, mocking the Romans with loud shouts. Meanwhile, their commanders angrily received the deceived soldiers, and by Caesar himself, who said to them: "These Jews are not ruled merely by their madness. Everything they do is done with careful planning. They devise clever strategies and set ambushes whereby they gain good fortune and success. And they are obedient, honoring their faithfulness and goodwill to one another. On the other hand, we Romans rule Fortune by our obedience to orders and always move in proper discipline and complete submission to our officers. But you have acted contrary to the Roman way. You were caught by your own inability to restrain yourselves from action. You went off without your commanders and did it in Caesar's very presence. Truly," Titus continued, "the rules of Roman warfare groan heavily, as will my father when he hears of this debacle. In all his days in battle, he never made a mistake as grievous as this.

"Our rules of warfare also call for those who break the least of them to forfeit their lives. Here we have seen an entire army run into disorder. However, those of you who have acted insolently will be made quickly sensible. Even if victory is obtained by such disorder it would be a disgrace." When Titus had finished speaking, it appeared evident that he would execute all who were concerned with this breach of discipline. The soldiers sunk in utter despair expecting to be put to death immediately as their just punishment. But the other legions gathered around Titus and asked him to spare their fellow soldiers. They begged him to pardon the rash actions of the few in order to ensure proper obedience of the rest. They promised that the men who had disobeyed would make amends for this mistake and prove to be better soldiers in the days ahead.

5. So, Caesar complied with their requests. It also seemed prudent to him to only execute individuals for their misdeeds. Punishment inflicted on whole groups of men should be only in the form of reproof. So Titus was reconciled to his men, warning them that they should act with more wisdom in the future. He also thought of ways to get even with the Jews for their cunning hoax. Now, the Romans had leveled all the land between them and the city—all accomplished in four days. Titus wanted to bring forward the army's baggage and bring the entire multitude that had followed him safely to camp. He set the strongest part of his army against the northern and western walls of the city. He then positioned the army in seven ranks. The infantry was in the first position in three ranks, with the cavalry behind them, also in three ranks. Titus placed his archers in the center of the seven ranks. With this placement of such a large body of Roman troops, the Jews were restricted from making sallies against them. The three Roman legions' beasts of burden and the rest of the multitude then marched in without fear. Titus stationed himself about four hundred yards from the wall near the tower called Psephinus. From that spot, the northern wall bent around to the west. The remainder of the Roman army built fortifications at the tower called Hippicus, also a distance of about four hundred yards from Jerusalem. The tenth legion, meanwhile, continued to occupy its place on the Mount of Olives.

Chapter 4

A Description Of Jerusalem

1. Three walls fortified the city of Jerusalem, covering those areas of the city not protected by impenetrable valleys. Near such valleys, or ravines, it had only one wall. The city sat on two hills opposite one another, with a valley dividing them. The rows of houses on each hill ended at that valley. Of the two hills, the hill containing the upper city is much higher and has a straighter ridgeline. King David called this hill "The Citadel". David

was the father of Solomon, who built the first temple. We now call this high hill "The Upper Marketplace." The lower hill on which the lower city sits is called "Acra." Acra is shaped with a hump like a hog's back. A third hill next to Acra is lower still, and is separated by another broad valley. During the era when the Asamoneans reigned, they filled that valley with dirt and stones, wanting to join the city to the temple. They also lowered Acra's height to make the temple sit much higher relative to it. Called "The Valley of Cheesemongers," it separated the upper city from the lower and extended all the way to Siloam, the name of the fountain from which sprang copious sweet water. Outside the city, these hills are encompassed by deep valleys, the steep sides of which render them impassable.

2. Of the three city walls, the oldest one was the most difficult to be taken because of the steep valleys leading to the hill above, on which the wall is built. Besides having that great geographical advantage, it was also built very strong. David and Solomon were very passionate about making it so. It begins on the north side of the tower called "Hippicus" and extends to the place called "Xystus," and then, joining the council house, ends at the west cloister of the temple. If one begins at the same point and goes westward, the wall extends through a place called "Bethso" to the Essene gate. After the Essene gate, the wall heads south, bending around the Siloam fountain, then bending again at Solomon's Pool, and toward a place called "Ophlas," where it adjoins the temple's eastern cloister.

The second wall originated at the "Gennath" gate, which belonged to the first wall. The second wall surrounded only the city's northern quarter, and reached as far as the Antonia Tower. The third wall began at the Hippicus Tower and reached as far as the city's north quarter and the Psephinus tower, and continued to rise against the Helena monuments, named for Helena, queen of Adiabene and daughter of Izates. This third wall then extended a great distance, passing by the royal caverns of the kings until ending at the corner tower at the "Monument of the Fuller," and then joining the old wall at the Cedron Valley. Agrippa had encompassed the newer parts of Jerusalem and added the old city within this wall. The newer city had lain unprotected before that. As Jerusalem became more populated, the city limits gradually crept outward. Those sections north of the temple joined the temple mount to the city, making Jerusalem considerably larger. The fourth hill, called Bezetha is also inhabited. It sits against the Antonia Tower, separated by a deep chasm—

dug out in order to keep Antonia's foundations from joining this hill, and to increase its security, because of its heightened elevation. Otherwise, the Antonia Tower might be taken with ease. The depth of the man-made ravine made the elevation of the tower seem even higher. This newly built section of the wall we called "Bezetha," which in Hebrew means "The New City."

Since its inhabitants stood in need of protection, Agrippa, the present king's father by the same name, built it. So, he began to build the wall but stopped construction after the foundation was laid. He feared that Claudius Caesar would suspect the construction of a major wall was being done in order to start a rebellion. If Agrippa had been able to finish it in the way that he had planned, there would have been no way the city could be overrun. Its parts were connected by stones forty-five feet long and fifteen feet wide. The stones could have never been undermined by any iron tools, or shaken by a battering ram. The wall was fifteen feet wide when completed, and would have had a greater height than that had not his enthusiasm wavered. The wall was never completed. The wall was ultimately completed by the Jews to a height of thirty feet, above which were battlements of three feet and turrets rising four-and-a-half feet. So, its entire height equaled almost thirty-eight feet. (Editor's note: This is using the Roman measurement of one cubit = 18 inches.)

3. The towers upon the wall were each thirty feet wide and thirty feet tall. They were square and as solid as the wall itself. The tight joints and the stone's beauty were in no way inferior to those of the holy temple itself. Above the solid thirty foot rise of the towers were magnificent chambers, with upper rooms above those, and cisterns to catch the rainwater. There were many such chambers, and the steps by which one ascended to them were wide. The third wall had ninety of these towers, each four hundred feet from the other. A total of forty towers existed in the center of that wall, while the old wall boasted sixty. The perimeter of the city walls ran to over four miles. The third wall was wonderful, yet the Psephinus Tower rose high above it on its northeast corner. It was there that Titus made his headquarters, for being one hundred five feet high, one could see the sun rising in Arabia. One could also see the utmost limits of Hebrew possessions, westward to the sea.

Psephinus was octagonal in shape and was situated near the Hippicus Tower and near two other towers built by King Herod on the old wall. These towers were larger, more beautiful, and stronger than any others in the known world. Herod was of a generous nature and a great benefactor to the city on other occasions. But he built these extraordinary structures to gratify his own affections, and dedicated the towers to the memory of those who had been most dear to him. He named them for his brother, his friend, and his wife. He had executed his wife out of love and jealousy, as has already been said. The other two men were lost to Herod in battle, in which both fought fearlessly. Hippicus was named for Herod's friend. It was a square tower; its length and breadth were each thirty-eight feet, and it rose to a height of forty-five feet. It was comprised of solid rock, with huge stones bonded together. It held a reservoir thirty feet deep over which a two-story house rose another forty feet. The house was divided into several rooms. Over this were battlements of three feet high and turrets all around of almost five feet, so that the entire height of the tower was about one hundred twenty feet.

The second tower, which Herod named for his brother, Phasaelus, was thirty feet square. Its solid stone base rose to thirty feet in height, over which a fifteen-foot high cloister was shielded from enemy missiles by breastworks and walls. Over the cloister was another tower, divided into magnificent rooms with a place for bathing. The Phasaelus tower lacked nothing that any royal palace might offer. It too was adorned with battlements and turrets, and more than the others. Its entire height reached almost one hundred twenty feet. It resembled the Tower of Pharus—which held the fire to warn sailors as they approached Alexandria—but was much larger in diameter than Pharus. At this time, it had been connected to living quarters where Simon exercised his dictatorial authority.

The third tower was called Miriamne, Herod's queen's name. Its first story was also solid rock thirty feet wide, deep, and high. Its upper buildings were more magnificent and had greater variety than the others of Herod's towers. He thought it best to adorn this tower after a more feminine manner, while the other towers he deemed best fashioned after male interests. These other towers were also built stronger than Miriamne. The entire height of this tower was seventy-five feet.

4. While these towers were all very tall, they appeared much taller seen from the ground from which they rose. They were constructed on the old wall, which was itself built on a high hill. The old wall was forty-five feet tall, over which the towers rose, making them appear much higher.

The gigantic stones were marvelous for they were not made from smaller, common stones, or even of stones that a man might be able to carry. But these were of white marble cut out of the quarry. Each stone was thirty feet long by fifteen feet wide by seven-and-a-half feet deep. They were so perfectly fitted together that each tower looked like it was formed by one gigantic stone, so perfectly were they cut and polished into shape by the skilled hands of stonecutters. The stones were so tightly joined together their joints were almost imperceptible. The king had his palace joined on the inside of the towers, which is beyond my ability to describe. No skill or cost was spared in its construction. A wall forty-five feet high surrounded the palace, which was adorned with towers at equal intervals from each other. It contained many large bedrooms, each of which would sleep one hundred guests. The variety of stones was beyond number, for many rare types were collected together. The ceilings also were breathtaking, both for the length of their exposed beams and their ornamentation.

A great number of these large rooms existed within the palace, with a prodigious variety of statues adorning them. The apartments were fully furnished and boasted plates and bowls of silver and gold. The palace also contained many porticos, one after the other in every direction. Each portico contained curious pillars, and each of the outdoor courtyards was planted with grass. Additionally, there were several groves of trees with long walkways through them, as well as deep canals and cisterns. Several areas contained brazen statues out of which water flowed. Around the canals stood many cages containing tame pigeons. It is impossible to completely describe the palace. The very remembrance of it is tormenting, as I think about how these vastly imposing buildings were consumed by the fire kindled by the thieves. For the Romans did not burn them. Jerusalem's internal enemies destroyed them as I've already written. It took place at the beginning of this rebellion. The fire began in the Antonia Tower, and then spread to the palace, finally consuming the upper parts of the three towers.

Chapter 5

A Description of the Temple

1. The temple, as I have said, was built upon a very strong hill. The flat area on top of the hill was insufficient for the holy place and the altar. The

ground there was also very uneven and precipitous. But King Solomon, in planning and building the first temple, constructed a wall on the east side of the hill. He then added a cloister, founded on a bank he filled in behind the wall. The temple stood isolated and unprotected. But in the future, others would add new fill dirt, so the hill became a much wider and larger plot. Additionally, they later tore down the north wall to create an area more than large enough to construct the entire temple. When they had surrounded the temple with walls on three sides at the base of the hill and had done more than could ever have been hoped, they surrounded the upper courts with cloisters and did the same to the temple's lower courts. [This work took many years and exhausted the sacred treasuries, which in time were replenished by the gifts sent to God by people the world over.] The lowest part they built up from a depth of four hundred fifty feet – even more in some places. The foundation footings could not be seen, for they brought in dirt and rocks and filled up the ravines in order to make them the same level as the city's narrow streets. In this endeavor, they use stones as large as sixty feet, for they had great wealth and the people gave generously, making the entire project a great success. What had been unimaginable in scope was with time and perseverance brought to fulfillment.

2. Let me now describe the work carried on above these foundations, which were completely worthy on their own merit. The cloisters, or porticos, were in double rows with pillars almost forty feet high supporting them. Each pillar was made entirely from one piece of white marble. Overhead, the roofing was made of interestingly carved cedar. The natural grandeur, luxurious finish, and fine joinery of these cloisters were magnificent. They were adorned neither with paint nor engraving. The porticos on the outermost court where forty-five feet wide and stretched out over 1,000 yards, comprising the Antonia Tower. All these courts were open to the air and were paved with every imaginable sort of stone. When you went through these outer porticos into the temple's second court you saw a five-foot high stone wall surrounding it. It was elegantly constructed and featured pillars at its top set at equal distance from each other. Letters, some in Greek text and some in Roman, declared the Jewish law of purity, stating, "No foreigner should go within the sanctuary." [The temple's second court was called "The Sanctuary."]

Fourteen steps allowed one to ascend from the first court to the second. This second court was foursquare, had its own wall, and formed a level verandah. Facing the steps, the height of the buildings above seemed to reach sixty feet. But it was the steps that made it look like this. The actual height was a bit less then forty feet. For, as the floor of the second court existed on the higher level, some of what it contained was hidden by the hill. Moving up fourteen steps, one reached a flat area of fifteen feet between the steps and the wall. From here, other flights of five steps each rose to meet nine gates. Eight of these gates were on the north and south; four on each side, with two gates on the east. One of these gates on the east allowed entrance to a special place of worship for women. A wall separated the women's court, where they worshiped, and had a second gate cut out of the wall next to the first gate. On the court's other sides, a southern gate and a northern gate also existed, through which women could pass into their court. The women were not allowed to pass through the other gates, nor could they go beyond their own wall when they went through their own gate. This area was for Jewish women only, whether from inside the country or from other lands. The western part of the second court had no gate, for a wall encompassed that entire side. The cloisters between these gates extended from this wall inward in front of the temple rooms and were supported by very fine, large pillars. These porticos were of a single row, and except for their size, were in no way inferior to those of the lower court.

3. The gates on every side of the upper court numbered nine and were covered with silver and gold. Their jambs and lintels were also covered with these precious metals. But one gate had no such covering. It was the gate heading to the inner court of the holy house. It was made of Corinthian brass and was much larger than those gates covered with silver and gold. Each gate had two opposing doors, the height of each being forty-five feet, and the width of their opening almost twenty-three feet. On each side were large rooms, built like towers in breadth and depth, with each rising to sixty feet. Each of these rooms was supported by two pillars, which were each eighteen feet in circumference. All of the nine gates were of the same size, except the Corinthian gate, which opened on the east side, against the temple and its exterior gate. The Corinthian gate was much higher, at seventy-five feet with an opening sixty feet wide. It was decorated in a very costly way, bearing adornments of silver and gold much thicker than the other gates. The nine smaller gates had their silver

and gold poured on them by Alexander, Tiberius's father. Now, another fifteen steps led away from the women's court to this larger gate, whereas those of the other gates had only ten steps.

4. The temple building itself was situated in the middle of the innermost court. It was the most sacred part of the temple mount and was reached by ascending twelve steps. In front, its height and breadth with the same – each one hundred fifty feet. The rear of the holy house was sixty feet narrower, for its front fascia had thirty foot stylized shoulders on each side. The temple's entrance gateway was one hundred five feet high and almost thirty-eight feet wide. It had no door or metal gate, for it represented heaven's universal visibility—which heaven stood open to view. The front of the temple building was entirely covered in gold, as were all of its large inner walls. All of the surrounding parts of this entrance gateway gleamed brightly to those viewing them. The entire holy house was divided into two rooms, but we were only allowed in the first. Its ceiling was one hundred thirty five feet from the floor. The huge room was seventy-five feet long and thirty feet wide. The entrance gate, as well as this first room, as we've already observed, were covered with gold. The room also had golden vines on the ceiling, from which clusters of grapes fell that were as tall as a man's height.

Again, the house was divided into two rooms, the second of which appeared smaller than the first and had golden doors almost eighty feet tall and twenty-four feet wide. In front of the doors was a curtain of the same size. It was a Babylonian tapestry embroidered with blue, white linen, scarlet, and purple thread, and of a magnificent texture. The colors were mystical in their interpretation. They represented a sort of image of the universe. The scarlet thread seemed to signify fire; the white linen earth; the blue air; and the purple signified the sea. The scarlet and the blue thread might be identified by their color --fire and sky. The white linen and purple might be identified by their origin, as one comes from the earth and the other from the sea. And so a vista of the heavens was displayed on this curtain, but the signs of the Zodiac were omitted.

5. Entering the temple one is received on its ground floor. This part of the temple was ninety feet high and ninety feet long. But its width was only thirty feet. The first section, partitioned off at sixty feet, contained some of the most wonderful works of art ever created—the candlestick,

the table of showbread, and the altar of incense. The seven lamps rising out of the candlestick represent the seven planets. The twelve loaves on the table signify the months of the year and the Zodiac. On the altar of incense were thirteen kinds of sweet-smelling spices from both the sea and the land from which they were refilled. These spices signified that God owns all the earth, including both its inhabited and uninhabited areas, and that they are all dedicated to His use. Now the inmost room of the temple measured thirty feet. It was also separated from the outer room by a curtain or a veil. This inner room was completely empty. It was also inaccessible and sacrosanct, never to be seen by anyone. It was called "The Holy of Holies." Around the sides of the temple's lower reaches existed a great many small rooms with passageways between them. They were each three stories high and had entrances on each side from the temple gate. The upper part of the temple did not have these rooms because it was narrower and sixty feet higher — smaller in size than the lower part. The sixty feet, in addition to the height of the ground floor, brought the total height of the temple to one hundred fifty feet.

6. The front fascia of the temple was a sight to behold. It was completely covered with gold plates. When the sun arose in the morning, its light was reflected back in a fiery splendor, making those who tried to glance at it turn their eyes away. It was as bright as the sun itself. As strangers approached the city from a distance, the temple looked like it was covered with snow, for those areas not covered with gold plate were exceedingly white. The building had sharp spikes on its roof to keep the birds from polluting it. Some of the roof's enormous stone rafters were almost seventy feet long, eight feet high and nine feet wide. In front of the temple building stood the altar, twenty-two feet high and seventy-five feet wide. On its corners were horns. The priests walked up to its top on a sloping ramp. The altar had been constructed without the use of any iron tools. In fact, no iron instrument of any kind ever touched the altar. Additionally, a partitioning parapet about one and a half feet in height surrounded the altar and the sanctuary, separating the priests from the people. Anyone with leprosy or gonorrhea was always excluded from the city. Women were not allowed access to the temple during their monthly period, and as we've mentioned, and were never allowed beyond the women's court. Even the priests were not allowed there, if undergoing purification.

7. The priests who had some physical defect were not allowed to minister, but were allowed within the parapet wall, along with those without defect. They were also permitted to partake of the food allotted to them by birthright, but only the officiating priests donned the sacred garments. These priests without blemish went up to the altar dressed in fine linen. Those ministering abstained from drinking wine, for fear that they might trespass some sacred rules during their ministrations. The high priest went up to the altar with them, but only on the seventh day and during the New Moon festivals, as well as on any of the other festivals the Jews celebrate throughout the year. When on duty, the high priest wore linen shorts that reached down to cover his thighs. He wore an inner linen vestment, and over that a seamless blue gown with tassels that reached to his feet. Golden belts hung from these tassels, with pomegranates mixed between them. The bells represented thunder, and the pomegranates lightning. An embroidered girdle or sash bound his outer garment to his chest. It featured five rows of various colors: gold, purple, and scarlet. It was also made of fine blue linen. Its colors were the same as those on the temple curtains, which we've mentioned before.

The same embroidery work was also on the priests' ephod, but the ephod displayed much more gold. The ephod was like a short cape that hung over the shoulders by straps, and attached to the front of the priestly robe by two gold buttons shaped like small shields. On these buttons rested two very large and very excellent sardonyxes that had the names of Israel's twelve tribes engraved upon them. On the other part of the ephod hung twelve stones in three rows of four stones each. The rows were comprised of a sardius (sardonyx), a topaz, and an emerald; a carbuncle, a jasper, and a sapphire; an agate, an amethyst, and a ligure; an onyx, a beryl, and a chrysolite. Upon each of these stones was engraved the name of one of the aforementioned tribes. The high priest's head was covered by a fine linen mitre or headdress tied with a blue ribbon. Over that he wore a golden crown on which was engraved God's sacred name. It consisted of four vowels: YHWH. The high priest never wore these garments except when ministering in the temple. Outside the temple he wore a simpler garment. He wore the sacred garments when entering the temple's most sacred room, and then only once a year, on the day when every Jew keeps a feast to God: the Day of Atonement. I shall have a lot more to say later on about Jewish laws and customs relating to these things. But a great many aspects of the temple and the city need to be discussed now.

8. The Antonia Tower was situated at the northwestern corner of the two temple porticoes. The tower was built on a seventy-five foot high rock that stood on a formidable precipice. King Herod built the tower, and by doing so displayed his innate genius. The rock itself is covered from its base to its top with smooth pieces of flagstone, both for beauty and to cause anyone trying to climb up its face to slip backwards. On top of the rock, but in front of the tower, was a five-foot high wall. Within that wall the tower rose sixty feet from the rock. The tower's interior was very large and palatial. It had all kinds of rooms and apartments and other attributes, such as courts, places for bathing, and wide-open spaces used for troop encampments. By all these amenities it seem to be a city within itself, but at the same time it was a magnificent palace.

While the Antonia Tower was a tower in and of itself, it had four separate towers on each of its four corners. Three rose seventy-five feet, but the tower on the southeast corner was one hundred five feet tall. The entire temple area could be seen from that tower. On the corner adjoining the two cloisters were passageways through which troops could descend. A Roman garrison was always stationed in Antonia, and during festivals took up positions around the temple mount to watch the people and ensure that no rebellions arose. For the temple was a fortress guarding the city, while Antonia was guarding the temple. The garrison in the tower guarded all three: the city, the temple, and Antonia. Another peculiar fortress existed in the upper city—that of Herod's palace. The hill called Bezetha was separated from Antonia as we've said. It was the tallest of the city's hills and adjoined the new city. Bezetha was the only place that obstructed the view of the temple from the north. This description of the city and its walls shall have to suffice at present. I have planned to make a more comprehensive description of it elsewhere.

Chapter 6

Concerning the Tyrants, Simon and John. Also, as Titus Was Going around the City Wall, Nicanor Was Wounded by a Dart. The Incident Pressed Titus to Begin the Siege

1. The hostile men were still in the city. Simon's followers numbered 10,000, not counting the Idumeans with him. These 10,000 had fifty commanders over whom Simon ruled. The Idumeans paying Simon hom-

age were 5,000, and had eight commanders over them; the most famous of whom were Jacob ben Sosas and Simon ben Cathlas. John, when he had originally seized the temple, had 6,000 armed men under twenty officers. The Zealots who had defected to his ranks numbered 2,400. Eleazar was still their commander, together with Simon ben Arinus. While these factions fought one another, the people suffered under them both, as we've already said. Both sides pillaged those citizens who would not join in their wickedness. Simon's faction held the upper city and the great wall as far as the Cedron Valley. The old wall went from Siloam eastward, then westward down to the palace of Monobazus, who was king of the Adiabeni, northward above the Euphrates. Simon also held the fountain there as well as the Acra, which, of course, was the lower city. Simon also held that territory reaching to Queen Helena's palace. Helena was Monobazus's mother. John held the temple and all the wide area attached to it. He also held both Ophla and the Cedron Valley.

When the city between the factions had been burned, a no-mans-land was left between them, where most of the fighting occurred. The internal struggles of the Zealots and robbers did not cease with the Roman encampment near their very wall. They had grown wiser when the Romans first arrived, and that lasted for a little while. But then they returned to their former madness, separated from one another, and fought it out. The Romans could not have been happier. The internal warfare between the factions created more suffering for the Jews than the Roman siege ever could. The citizens of Jerusalem would not suffer any new calamity more horrible than that suffered under these men. Even before the Romans overthrew Jerusalem, it was not a happy place. Those who ultimately took the city did the citizens a kindness. I would venture to say that the sedition destroyed Jerusalem, and the Romans destroyed the sedition. That was much more difficult than destroying the city walls. We may justly ascribe all of our Jewish misfortunes upon our own people, and their just revenge done by Roman hands. But as to that matter, each must decide for himself.

2. Now the situation in the city was this: Titus rode around Jerusalem with some of his select cavalrymen looking for a weak spot in the walls. The Roman general was in doubt as to where he could make any possible attack, for it was impossible to take the city from the valleys. The walls on the other side were too strong for the Roman battering rams. So, Titus

thought it best to begin his assault on the monument to the high priest, John. The first fortification at that point was lower, and the second fortification was not joined to it. Later builders had neglected the strength of the walls when the new city wasn't very populated. The spot also provided an easy passage to the third wall, through which Titus believed he could take the upper city. Then, by going through the Antonia Tower itself, he could take the temple.

So while Titus was riding around Jerusalem, one of his friends named Nicanor was wounded in his left shoulder by a Jewish dart when he approached too close to the wall with Josephus. Nicanor had attempted to speak with those on the wall regarding peace terms, for he was known to them. Caesar, seeing the Jews' hostility, was provoked to press the siege. He understood that the people would not listen to anyone who approached them for their own benefit. He also permitted his troops to set the suburbs of Jerusalem on fire, ordering that they gather timber and raise banks against the walls. When he had divided his army into three divisions in order to begin the work, he placed the dart throwers and the archers in the middle of the banks to be raised. Before these, he set his engines that threw javelins, darts, and stones, in order to prevent the enemy from sallying out and assaulting their works. He also wanted to keep those on the wall from any ability to impede or obstruct the construction. So, the trees around the city were immediately cut down, and the entire area stripped bare. While the timber was being brought forward to raise the banks, and the Roman army was diligently engaged in their efforts, the Jews were not silent. The people of Jerusalem, who had suffered murder and theft before, now gathered their courage, hoping to gain a break while those that pressed them were now busy opposing the Romans. They hoped to have revenge upon the Zealots and robbers once the Romans took the city.

3. Fearing Simon, John stayed put in his own territory, even while his own men earnestly wanted to move beyond the walls to attack the foreigners. But Simon was very active, for the location the siege was to take place was near his territory. So, he brought his war engines forward and placed them at their positions on the wall. Some of these engines had been captured from Cestius, and the rest taken from the garrison that had been in the Antonia Tower. But even though Simon's men controlled these engines, they had little skill in their usage, and so the engines were

unusable. But, deserters had instructed a few of his men in their use, and so awkwardly, they were made ready for war. Simon's men also sallied out in groups and fell on the Romans. Those at work on the banks covered themselves with boards and wicker hurdles spread over the work, then they turned their engines upon those attacking. The Roman legions had admirably designed war engines ready to do their jobs. The tenth legion had the most extraordinary engines that threw darts and stones. They were larger and more forceful than any others in the Roman army. These engines not only repelled the Jewish sallies but also drove away those who occupied the walls. The stones these engines heaved weighed over seventy pounds and carried four hundred fifty yards, and even farther. The force of the blow when they landed took out everything in its path, both those who stood in the front lines and all those behind them, for a long distance. At first, the Jews watched the incoming stones. Not only could the gigantic stones be heard coming toward them, for they made an awesome noise, but they could also be seen from a distance, as they were bright white. Accordingly, the watchmen in the towers would shout a warning when a stone was released. They would shout in Hebrew, "The stone cometh!" Those in harm's way ran from its path or threw themselves on the ground. In this way, they were able to guard against the stones' fury and remain unharmed. But the Romans soon became wary of the Jewish defenses and blackened their stones. They were then able to aim them successfully because the Jews couldn't see the huge stones coming. So, with one shot, many of the Jews were demolished. But the Jews, even under this great distress, did not allow the Romans to raise their banks peacefully. With cunning and boldness they went to work repelling the Romans by day and night.

4. Upon completing the works, the Roman workmen measured the distance from their engines to the wall. They had to throw a lead weighted line from their embankments, for they would come in range of the Jewish missiles should they try to measure the distance bodily. When they had gauged what their engines' range was, the Romans brought them up to that distance. So when Titus got his engines near the wall where the Jews could not repulse them, he gave his men orders to commence firing. All of a sudden a tremendous noise echoed around the city from three places. The citizens within Jerusalem suddenly cried out as one, and terror struck the rebels as well. Understanding the common danger, the two factions

decided to join ranks against the Romans. So both John's and Simon's men cried out to one another that they were acting just like the Romans hoped they would. They ought instead—even if God did not allow them to establish a permanent peace—to lay aside their differences temporarily and unite against the Roman siege. Accordingly, Simon allowed John's men from the temple to go up on the wall. Even John himself went up, even though he did not believe Simon was being truthful. He also gave his men permission to ascend. So, both factions laid aside their hatreds and curious arguments and formed themselves into one body. They then proceeded to run around the walls with a large number of torches, which they threw at the Roman war engines. They also continually threw darts at the battering ram operators. The more courageous among them leapt out in scores upon the machines' coverings and tore them to pieces. Falling upon those Romans underneath, they beat them not so much by skillful tactics, but more by the mere boldness of their attack.

Meanwhile, Titus sent help to those who were being hit the hardest, placing cavalries and archers on all sides of the engines, beating off those rebels who tried to set them on fire. By this move, he also repelled those on the towers who hurled stones or darts. Titus then ordered the engines to get to work. But the wall would not yield from their blows, except at one place where the fifteenth legion's battering ram displaced a tower's corner. The wall below the tower was not in the same danger as the tower, for the tower rose far above the wall. Even if the tower was to fall it could not easily harm the wall underneath it.

5. The Jews stopped their sallies for a time, but when they saw the Roman troops widely distributed among the various works and several camps, they attacked. [The Romans believed the Jews to have stopped fighting from weariness and fear.] So the Jews made a sudden sally through an obscure gate at the Hippicus Tower, carrying their torches to burn the Roman works. They ran boldly up to the Roman machines and fortified camps, shouting loudly as they went. Those Roman troops near the machines and encampments quickly ran to assist their friends, while those farther away came running after them. This bold attack made the Romans forget their good discipline. The Jews overran those in their path and pressed hard upon those who were attempting to form up. The struggle around the machines was very intense as the Jews tried to set them on fire and the Romans tried to prevent it. Both sides were in confusion and

many in the thick of the fight were killed. The Jews had the better of the battle by their furious, wild assaults, and soon fire began to consume the Roman works. Both the works and the engines themselves would have been in danger of being destroyed had not a few of the select troops from Alexandria stood firm, displaying a courage beyond that of their recognized reputation. This was the state of affairs until Caesar took the most stalwart of his horsemen and charged the attacking Jews. Titus himself killed twelve of the first Jews he came near. The deaths of these men brought terror to those Jews nearby and sent them running. Titus gave chase and pursued his enemy into the city, saving his works from the fire. During the fight, a certain Jew was captured alive. Titus gave the order to crucify him in front of the wall, in hopes that, seeing his plight, the rest of the Jews might surrender. Then—following the Jewish retreat—John, the commander of the Idumeans, was speaking with one of the soldiers he knew while standing near the wall as an arrow from an Arabian's bow mortally wounded him. His death caused great grief in the Idumean camp and profound sorrow among the Jewish rebels, for John was revered both for his courage and for his good judgment.

Chapter 7

How One of the Roman Towers Fell of its own Accord; and how the Romans, Following a Great Slaughter, Got Possession of the First Wall. How Titus then Made Assaults upon the Second Wall. Also Concerning the Roman, Longinus, and the Jew, Castor.

1. The next night a surprising fracas came upon the Romans. Titus had given orders for his men to build three towers, each seventy-five feet high. By placing his soldiers upon the towers at every embankment, the Jews on the wall could be driven away. But one of the towers fell down around midnight, creating a loud racket. Hearing the noise, fear came upon the Roman army, which thought the enemy was attacking them. So, every soldier ran to pick up his weapons. All of the legions were in an uproar since no one knew what had happened. They spread out in search of the enemy, but finding none, they became afraid of one another. Every man demanded the password of every other man, thinking the Jews had infiltrated their camp. They were all in a panic until someone told Titus what had happened, and he gave orders to spread the word. And then, with some difficulty, everyone settled down.

2. The Roman towers became very troublesome for the Jews, who otherwise had fought their enemy with great determination. For the Romans shot missiles from their lighter engines from these towers, as did individual archers, dart throwers, and the men with slings throwing stones. Meanwhile, the Jewish missiles could not reach those high above them. Nor was it possible to overrun the towers, nor overturn them, as they were so heavy. Nor could the Jews set them on fire because of the towers' iron plate shielding. The Jews were forced to back away, out of the reach of the Roman projectiles, and no longer tried to stop the hammering rams, which, in continually beating against the wall, gradually began to make it give way. At length, the wall began to submit to the ram that the soldiers called, "Nico," as it was the largest of their siege engines and had always succeeded.

After the long days of fighting and standing guard, the rebels retired into their lodgings for the night some distance from the walls. They thought it would be superfluous to maintain a guard since two other walls remained standing. They were also indolent and weary from taking orders, so they lazily went off to bed. It was then that the Romans went through the breach that Nico had made in the wall. At this point the Jews remaining on guard retreated to the second wall. Those soldiers climbing through the breach quickly opened the gates and the Roman army streamed behind the first wall. And so the Romans got through this first wall on the fifteenth day of the siege, which was the seventh day of Artemisius [Jyar]. They then demolished a large portion of the wall, as they did the northern section of the city, which had formerly been ruined by Cestius.

3. So, Titus now pitched his camp within Jerusalem at a place called "The Camp of the Assyrians." He had taken all of the areas as far as the Cedron Valley. He then began his assaults upon the second wall. The Jews divided their forces into several divisions to begin their defense of that wall. John and his faction defended the Antonia Tower and the northern temple cloisters, fighting the Romans in front of the King Alexander monuments. Simon's men defended the ground in front of John's monument and fortified it all the way to the gate where the Hippicus Tower received its water. In their defensive plan, the Jews made frequent and violent sallies out of the gates from their positions to fight the Romans. Lacking

in the military skill of the Romans, they were beaten back to the walls. When the Jews fought on the walls the Romans also proved the strongest. Rome's soldiers were encouraged by their power along with their battle skills, while the Jews were emboldened and pressed on by the fear in their hearts and that chutzpah that our nation summons up when undergoing disasters. The Jews were also encouraged as they hoped for ultimate deliverance, while the Romans hoped to achieve a quick victory.

Neither side seemed to grow tired, but all day long there erupted assaults, battles on the wall, and continual Jewish sallies through the gates. Both sides practiced every sort of warfare. The fighting began at sunrise and hardly concluded at night. Neither side got much sleep, so the nights seemed more challenging than the days. One side was afraid the wall would be overrun, while the other side thought the Jews might launch a sally out of the walls at any time. Both sides slept in their armor at night so that they would be ready to go back to the battle at first light. The Jewish warriors each tried to be first into danger in order to satisfy their officers. Over them all, they feared and respected Simon to the degree that anyone serving under him would have killed himself with his own hands. The Romans, on the other hand, were emboldened by their custom of always being victorious and never having been defeated, because of their perpetual warfare and training exercises, and by the splendor of their vast empire. But what encouraged them most at present was that Titus was with them at every moment. It would have been unforgivable to grow tired while Caesar was there, standing with them. Titus fought as bravely beside his men, and could also see those who fought valiantly, and would reward them. It was thought to be a great advantage to have Caesar know every soldier's valor. On that account, many seem to have more enthusiasm than the strength to employ it.

The Jews at one time were standing in front of the wall en force while both sides threw their spears at one another. Longinus, a Roman of the equestrian order, suddenly leapt up out of his lines into the midst of the Jewish army, breaking their ranks and dispersing them. He killed two of their boldest soldiers. One of them he speared in the mouth as he ran to meet him, and the other he killed by the very spear withdrawn from the first. He ran the second man through as he tried to get away. When he had killed these two men, Longinus ran unscathed out of the Jewish ranks and back to the Roman side. So, this man showed off his gallantry and many Romans sought to imitate his valiant deed. At this point, the

Jews became unconcerned with their own welfare and only with what damage they could inflict upon the Romans. Death itself seemed like a trivial matter if they could at the same time destroy one of their enemies. But Titus took care to ensure his own soldiers' safety, all the while leading them to victory over their enemies. He declared that hotheaded violence was madness, and only courage teamed with right conduct was true bravery. He, therefore, commanded his men to be careful when they were in combat, and to prove their courage without taking unnecessary risks.

4. Titus now had one of his war engines moved to the middle tower of the north wall. In the tower, a certain Jew named Castor lay in ambush with ten other men. Roman archers had put the rest of the Jews to flight. These men lay still in their armor is if paralyzed with fear. But when the tower began to shake, Castor rose up, stretched out his hand in a petition and called for Caesar. Castor's voice moved with passion as he begged Titus to have mercy upon them. Titus, in his heart's innocence, thought Castor to be in earnest, and hoping the Jews would now repent, he stopped the battering ram and forbade his soldiers to fire on the petitioners. Titus then asked Castor to speak his mind.

Castor told Titus he would come down from the tower if he could have Titus's right hand of security. Titus replied that he was pleased to hear this and would be further pleased to give the same security to all Jerusalem. Now five of the ten men sustained the ploy by pretending to beg for mercy. The other five cried out that they would never submit to Roman slavery while it was in their power to die as free men. While all this quarreling was taking place, the Romans' delayed their attack. Castor sent a message to Simon telling him that they would save considerable time if they could keep the conversation going, and thereby delay Roman power for a while. At the same time he sent the message, Castor called openly for the five obstinate men to accept Titus's right hand of security. But the men seemed very angry with Castor. Brandishing their unsheathed swords on the tower breastworks, they struck themselves upon their chests and fell down as if dead. Seeing them fall, Titus and his men were amazed at the men's courage. Since they weren't able to see exactly what was done, they admired the Jews' fortitude and pitied their destruction.

During the lapse in fighting a certain archer shot an arrow at Castor, hitting him in the nose. Castor pulled the missile out of his nose and showed it to Titus complaining that he was being treated unfairly. So

Caesar reproached the man who shot the arrow and sent Josephus, who stood beside him to go and give his right hand to Castor. But Josephus refused because he knew this Jewish petitioner had a trick up his sleeve. He also restrained his friends who wanted to go up to Castor. But a Jewish deserter, Eneas, said he would meet Castor. Castor had also called out that someone should come up and take the money he had with him. That made the unarmed Eneas all the more willing to run to him. Upon Eneas's ascent, Castor picked up a large stone and hurled it at him. Eneas dodged the stone, but it wounded another man who was coming to help him.

When Caesar understood it was all a hoax, he realized that to show mercy in warfare is a destructive thing because such tricks would not take place under more sternness. So, in his anger Titus set his battering ram to work more vigorously than before. When the tower began to give way, Castor and his companions set it ablaze and leapt through the flames to a secret subterranean vault below. Seeing this move and believing Castor and the others had leapt into the fire, the Roman soldiers marveled at their audacity.

Chapter 8

How the Romans Took the Second Wall Twice, and Got Ready to Assault the Third Wall

1. Caesar took the second wall five days after he had overrun the first. When the Jews had fled before him, he entered with 1,000 armed choice troops at a marketplace where vendors sold wool, braziers, and cloth, and where the narrow streets ran obliquely from the wall. Here Titus erred. If he had immediately destroyed a wider part of the wall, or had acted in accordance with the rules of war and laid waste to what remained of it, his victory might have been achieved with no Roman losses. However, in hopes that he might make the Jews ashamed of their obstinacy by not being willing to strike them harder than necessary, he did not widen the wall's breach to allow for a safer retreat if needed. He did not think the Jews would set traps for him after he had treated them so generously.

When they breached the wall and the Romans poured in, Titus did not allow his soldiers to kill anyone they caught or set fire to their homes. He gave the rebels, if they so desired, a fight without harming the citizens, and promised to restore the citizens' personal property to them.

The Roman general was interested in preserving the city for his own sake, and the temple for the sake of the city. He knew the people were ready to comply with his peace proposals. The rebels saw Titus's humanity as a sign of weakness, and imagined he made the proposals because he saw victory over the entire city as impossible. The rebels also threatened the people with death if anyone dared to say a word about surrendering. They cut the throats of any citizen that even spoke of peace.

Then the rebels attacked the Romans that had come within the wall. Some of them they met in the narrow streets and others they attacked from the houses. The Jews launched a sudden sally out of the upper gate, assaulting the Romans on the wall, until in fear they leapt down from their towers and returned to their camps. The Romans remaining within the second wall shouted as enemy forces both inside and outside the wall were surrounding them. They became fearful of all the Jews left in the city. The Jews then gathered more in number and with their complete knowledge of the narrow streets, had a great advantage over the Romans. They wounded a large number of foreign troops, overwhelming them and driving them from the city.

The Romans resisted as best they could, but because the wall's breach was so small, only a few soldiers could pass through at a time. If Titus had not sent help, every soldier inside the wall would have probably been cut to pieces. But he ordered his archers to stand at the upper ends of the narrow streets. Titus himself also stood where the greatest number of Jews were attacking and stopped them with many arrows. Domitius Sabinus stood next to him, valiantly displaying his courage. Caesar did not stop shooting his arrows at the rebels, which detered them from killing his men until all the Roman soldiers had retreated from Jerusalem.

2. And so the Romans were driven out of the city after they had overrun the second wall. Their Jewish enemies were greatly encouraged, joyous at such a great victory. They began to think that the Romans would not attempt to come into the city again. The rebels and Zealots thought defeat was impossible as long as they remained in Jerusalem. For God had blinded their minds to the guilt of their sins, nor could they imagine any larger Roman forces than those they had just scattered. Neither could they see the hunger that was creeping up upon them. Before, they had fed themselves at the public's expense, sucking the lifeblood out of Jerusalem. But a famine had already come upon the city, and many citizens had already

died from lack of provisions. But the rebels supposed that the people's destruction was to their benefit, and hoped that no one would survive except those at war with Rome. They resolved to continue to oppose any who wanted peace. So, the rebels were pleased when those who thought differently believed death as a way of freeing them from a heavy load.

This is how the rebels thought about Jerusalem's citizens while they put on their armor to stop the Roman troops, who were trying to get back into the city. They made a wall out of their own bodies to fill the breach in the second wall. The rebels made a courageous stand for three days, but on the fourth day they were compelled to flee in front of the powerful onslaught of Roman forces, and return to their former hiding places. So, Titus quietly took back the second wall, completely destroying it. When he had placed a garrison on the towers facing the southern city, he began to plan how to assault the third wall.

Chapter 9

Titus Stops the Siege for a Season, but the Ceasefire Doesn't Appease the Jews. Titus Prepares to Begin Again, then Sends Josephus to Speak to His Countrymen about Peace

1. Titus now resolved to relax his siege for a while, hoping to allow the rebels time to reconsider things. Perhaps after seeing the second wall destroyed they would become more submissive and maybe even more afraid of starving to death. The food they had taken from Jerusalem's citizens meant their supplies would not last much longer. So, Titus made use of this lull in the battle to solidify his own plans. And so, as the time to distribute the soldiers pay had arrived, he gave his commanders orders to form up the army in full battle gear so the Jews could see them. He then gave each man his pay. So, the Roman soldiers, according to their custom, drew their swords from their sheaths and marched in their breastplates with the cavalry, leading their finely equipped horses. The entire area in front of Jerusalem gleamed with reflections from their armor.

None was as grateful as was Titus's army, nor so frightened at the sight than the Jews. The entire old wall and the north side of the temple were packed with spectators, as those observing from the city even stood upon their houses for a glimpse. Every part of the city was covered with crowds of Jews, and a great dismay seized the most courageous as they watched the Roman army gathered together with their weapons drawn and under such sharp discipline. I cannot but think that the rebels might have changed their minds at that point had not their crimes against the people been so horrendous. They knew the Romans would never forgive them. They understood that a tormenting death would be their punishment if they did not continue to defend the city. It would therefore be much better to die in battle. Their fate was settled, and the innocent would die with the guilty. Jerusalem was to be demolished with the rebels in it.

2. It took four days for all the Roman soldiers to receive their pay. But on the fifth day, after no signs of peace were forthcoming from the Jews, Titus divided his legions and began to raise embankments both at the Antonia Tower and at John's monument. His plan was to take the upper city at John's monument and the temple from Antonia. If he could not capture the temple, the entire city would be in danger. So, at each place he raised embankments with each legion raising one. The Romans who labored at John's monument received sallies from both the Idumeans and from Simon's men and had to pause their work. John's faction and the great many Zealots with him did likewise at the Antonia tower. The Jews became too tough for the Romans, not only in hand-to-hand combat—coming as they did from higher ground, but because they had now learned to employ their own war engines. By their continual usage, the operation of the war engines became familiar, and the Jewish skill increased. The rebels had three hundred engines used for throwing darts and another forty for throwing stones. By them, the Romans were greatly hampered in raising their banks.

But Titus, understanding that either the salvation or the destruction of Jerusalem was on his shoulders, pressed the siege more earnestly, all the while exhorting the Jews to repent. So he intermixed his good advice with the construction of his siege works. Knowing that expectations are frequently more effective than war weapons, Titus continued to persuade the Jews to surrender Jerusalem. For, in a way, the city was already taken so by surrendering they would save their lives. So, Titus sent Josephus to speak to them in Hebrew, for he thought the persuasiveness of a fellow countryman might bring success.

3. So Josephus went around the wall, trying to find a spot out of range of Jewish missiles and yet within their hearing. He coaxed the people to

spare them, their city and their temple, and not to be more obstinate than the Romans were. For the Romans, even though they were not Jews, had respect for Jewish sacred rites and places—even though they belonged to their enemies, and had until the present kept from damaging either one. Only the Jews, who alone had the benefit of these things, were now in a hurry to have them destroyed. They had obviously seen their strongest walls demolished and knew that the remaining wall is weaker than those already breached. They must now believe Roman power is invincible and that the Jews have long known what it was like to be their servants.

Josephus continued: "If they now believe it is just to fight for their freedom, why didn't they do that in the first place? But, having fallen under Roman power and submitting themselves to Rome for so many years, to now pretend to shake off their yoke was not to love liberty but to love death. Additionally, man may rightly resent the dishonor of having despicable masters over him, but not if those masters possess the entire inhabited earth. For what part of the entire world has escaped Roman control? Only useless areas of horrendous heat or extreme cold! It is now evident that the Romans have fortune in their hands and that God, who has all the world under his dominion, now lives in Italy. Moreover, even the wild animals are aware of this hard and fixed law; that men must bow down to those too strong for them and to allow those who are mighty in battle to have dominion over them.

"Consider your forefathers who were far superior in both soul and body to the Romans and who had many more advantages, yet they submitted to the Romans because they knew God was with Rome. And now that the Romans have already conquered the majority of Jerusalem, on what are you who oppose them going to depend? For although the city walls are still standing, all of you within them are suffering more now than if the city fell. For the Romans are not unaware that you have no food to eat. The citizens are already consumed with famine and those that fight soon shall be. For even if the Romans should stop the siege and not attack the city with their swords, the war inside the city wall continues to be insurmountable and grows worse every hour. You will die unless you are able to fight against famine or somehow control your own natural appetites."

Further, Josephus added that it was proper to alter their behavior before the catastrophe upon them was incurable, and to listen to the advice that might save them while the opportunity still existed. The Romans would not hold the Jews' previous actions against them unless they persevered in their insolence to the end. The Romans are naturally kind in their conquests, preferring what was profitable for all concerned rather than that which their more base passions required. There would be no profit for Rome in leaving Jerusalem destitute of inhabitants, nor Judea a desert. He said, "For these reasons, Caesar offers you his right hand of security. But if he is required to take the city by force, no one will be spared, especially since in the utmost peril you rejected his proposals. For these already penetrated walls should ensure you that the other wall will soon quickly fall. And even if the wall's foundation should prove too strong for the Roman war engines, famine within will join the Romans as an ally against you."

4. While Josephus was exhorting the Jews to surrender, many on the wall laughed at him and many more insulted him. Some even threw their darts at him. When he realized he could not persuade them by such openly good advice, he began to recite the history of the Jewish nation crying aloud, "Oh, miserable creatures! Have you forgotten the help you received in times past that you would now continue to take up arms against the Romans? When did the Jews ever conquer a nation by such means? And when did God, the Jewish people's Creator, not bring his revenge on those who had harmed you? Won't you turn and look back and think of the reason you fight with such violence and how great a Supporter you have so sacrilegiously abused? Won't you remember the amazing things God did for your forefathers and this holy place, and how He conquered your powerful enemies? Even I tremble telling you of God's works because you are unworthy to hear of them. However, listen to me! You fight not just against the Romans but God himself.

"There was a king of Egypt called Pharaoh Neco. He arrived with a vast army of soldiers and captured our nation's mother, Queen Sarah. What did our progenitor Abraham do? Did he defend himself from this harmful person by armed conflict, although he had three hundred eighteen officers with him, each with an immense army under him? He counted his army's numbers as nothing without God's help. He spread out his arms towards their holy place, which you have now polluted, and reckoned upon God as his individual helper and not upon his large army. And wasn't Queen Sarah sent back to her husband the next evening without being defiled? Meanwhile, King Neco ran away, leaving this very place that you have now defiled by shedding your own countrymen's blood. Neco trembled at the night vision he saw and gave the Hebrews both silver and gold as the people God loved.

"Shall I not recall the journey of our fathers into Egypt where they were treated terribly and fell under the power of foreign kings for four hundred years? They might have defended themselves by taking up arms and fighting, yet did nothing but commit themselves to God! Who among you does not know that Egypt was overrun by all sorts of wild creatures and consumed by all kinds of troubles? How their land would not bring forth fruit? How the Nile contained no drinking water? How the ten plagues upon Egypt followed one after the other? And how by all these events our fathers were sent out of Egypt under God's protection with neither bloodshed nor danger because God led them as his chosen servants?

"Moreover, when the Assyrians carried off our sacred ark, did not all Palestine suffer for their crime? Remember what happened to their false god, Dagon? Recall how those who took the ark were smitten with an internal disease? Their intestines came out filled with what they had eaten, until those who had stolen the ark brought it back with the sound of cymbals and timbrels and other oblations in order to appease God's anger for their violation of his holy ark. It was God who was our General then and who accomplished such great things for our fathers. He did it because the Hebrews did not fool around with war and fighting, but relied upon God to manage their affairs.

"When King Sennacherib of Assyria brought an army of all Asia and surrounded Jerusalem, was he beaten by men's hands? Were not instead our fathers' hands lifted up in prayer without touching their weapons, when the angel of God destroyed the entire extraordinary army in one night? The Assyrian king arose the next morning to find 185,000 dead bodies and then fled away from the Hebrews with the rest of his army though the Hebrews were unarmed, stayed in Jerusalem and did not chase after him.

"You also know of our slavery in Babylon where our people were in captivity for seventy years. They were not delivered into freedom again until God raised up Cyrus as his instrument in bringing it about. Cyrus set us free, and our people once again worshiped their Deliverer at His temple. In short, we can remember no example of our fathers achieving any success by force unless they first committed themselves to God.

When they remained quiet, they were victorious, as pleased their great Judge. But when they went out to fight they were always defeated. For example, when Babylon's king besieged this very city, and our King Zedekiah fought against him [contrary to the prophecies made against him by the prophet Jeremiah], he was taken prisoner and saw Jerusalem and the temple demolished. How much greater moderation did Zedekiah show than that of your present leaders, and the people of the city more than you now show! When Jeremiah proclaimed how exceedingly angry God was with them because of their sins, and told them they would be made prisoners unless they surrendered their city, neither the king nor the people put Jeremiah to death. But you abuse me and throw darts at me! I am only trying to exhort you to save yourselves. I am not speaking of what you have already done in the city, which crimes are so atrocious I cannot begin to describe them. I only hope to provoke you to remember your sins, and cannot bear to think of the crimes you continue to perpetuate each and every day.

"Here's another example. When Antiochus Epiphanes displayed his army before Jerusalem and was guilty of many indignities against God, our forefathers met him in battle and were killed. Our enemies plundered the city, and our sanctuary lay desolate for three years and six months. Why do you need more examples? Indeed, what could it be that has stirred up Roman anger against our nation? They did not come because of the seditious men among our forefathers when the insanity of Aristobulus and Hyrcanus in our internal quarrels brought Pompey upon the city, and when God placed those unworthy of the liberty they had enjoyed under Roman subjection? They were forced to surrender after a three-month siege even though they had not been guilty, like you, of offenses against our sanctuary and our laws. And they had much more capability to make war than do you. Don't we recall the end Antigonus, Aristobulus's son, came to? And don't we remember under whose reign God ordained that the city should again be taken due to the people sins? When Antipater's son, Herod, brought Socius upon us, and Socius brought the Roman army upon us, we were surrounded and besieged for six months, until as punishment for their sins they were overrun and the city plundered by its enemies. It seems as if weapons were never given to our nation unless we were always turned over to our foes to be overrun. I believe that all who live in this holy place should commit to God's sovereign control, resisting the help of men and resigning to their great Arbitrator who reigns on high.

"As for you people, what have you done to please our great Legislator? And what have you done of those things he has condemned? How much greater sinners are you than those who were overrun so quickly? You have not avoided mere secret sins. But you have stolen, devised treacherous plots against others, and committed adultery. You quarrel about rapine and murder and even invent new ways of wickedness. Now the temple itself has become a toilet — the Divine sanctuary polluted by the hands of fellow Jews. The Romans revered our sanctuary when they were at a great distance from it, when they gave up many of their own customs in favor of our law. After all you have done, do you expect God, whom you have so egregiously abused, to come to your support?

"Without a doubt, you have the right to pray to Him, to call upon God to help you because your hands are so clean! Did your King Hezekiah lift up such foul hands in prayer to God against Sennacherib when he destroyed his great army in one night? Have the Romans committed the same iniquity as the Assyrian king? If so, you may have reason to hope of like vengeance upon them. Did not the Assyrian king bribe Zedekiah and accept money from our king not to destroy Jerusalem? Yet contrary to the oath taken, still came to burn down the temple? The Romans do no more than demand the customary tribute, which our fathers paid to their fathers. If they get that, they neither want to destroy the city nor touch God's sanctuary. No! In fact, they allow you and your children to be free, to keep your private personal possessions, and allow you to live by your holy laws.

"It is pure madness to expect that God will not treat the wicked and the righteous alike, since He knows it is just to bring immediate punishment upon men for their sins. Accordingly, God broke the Assyrian power the very first night they pitched their camp. In the same way, if God had judged our nation worthy of liberty he would have immediately inflicted punishment upon these Romans as he did upon the Assyrians. Even when Pompey began to meddle in our national affairs, or, after him, when Socius rose up against us, or when Vespasian laid waste to Galilee, or finally when Titus first came upon Jerusalem. But not only did Magnus and Socius not suffer, they took the city by force. Vespasian went from his war against you to Rome to govern the empire. As for Titus, the springs that had almost run dry when they were under your control now run more plentifully than they ever did before he arrived. You recognize that Siloam failed, as well as did the other springs outside the city. Water was sold in precise measurements. But now, out of many of these springs flow copious quantities of water for your enemies, sufficient not only for them and their animals, but also to water their gardens. You experienced the same wonderful sign when the king of Babylon made war against us, when he captured the city and burned the temple. Yet I cannot believe that the Jews back then were as sinful as you are. So, I cannot but suppose that God has fled from his sanctuary and now stands on the side of those against whom you fight.

"Any man who is a good man will flee from a house full of sin and will hate those within it. In the same way, you have convinced yourselves that the God who sees all hidden secrets and hears all private conversations will now overlook your sins? But I ask you: What crime is kept secret or concealed among you? What is it you do that even your enemies don't know? For you pompously boast of your transgressions and contend with one another to see who can be the most wicked! You make a public demonstration of your injustices as if they were virtues.

"Yet there remains a chance to preserve your lives if you are willing to grasp it. And God is quickly reconciled to those who confess their sins and repent of them. Oh, such hard-hearted wretches you are! Throw down your arms and take pity on your country that already lies in ruins. Turn from your wicked ways and have regard for the superiority of the city you are betraying. Save that most excellent temple with the many worldwide offerings in it. Who could bear to be the first to set that temple on fire? Who is willing that the sacred articles should no longer exist? What in all the world deserves preservation more than they? O, you insensible creatures! Are you not more stupid than these stones themselves? If you cannot see all this with discerning eyes, have pity for your families. Look at each of your children, wives, and parents who are being gradually consumed by either famine or warfare. I know that the danger extends to my mother, my wife, and to that family of mine who have been by no means dishonorable. Indeed, my family line has been very imminent. Perhaps you're imagining that I give you my advice on their account. If that's how you think, kill them! Even shed my own blood if it may preserve your lives. For I am ready to die if you will only return to soundness of mind after I am dead."

Chapter 10

Many Citizens Attempt to Escape to the Romans. Also, the Intolerable Suffering of those Left Behind in the Famine, and its Sad Consequences

1. As Josephus was loudly proclaiming his message, the rebels would neither agree with him nor did they feel safe to behave differently. But the people wanted desperately to desert to the Romans. Accordingly, some of them sold their remaining possessions, even their most precious treasures – every trinket – and swallowed pieces of gold that they might not be found by the thieves. When they escaped to the Romans, they eliminated what they had swallowed and had enough funds to provide for themselves. Titus let a great many of those who escaped to go out into the country or wherever they pleased. The central reasons they were ready to desert were that they would be far away from the miseries endured in Jerusalem, and yet also be free from Roman slavery. However, John and Simon and their factions very carefully watched those wanting out of Jerusalem even more than they did the Romans who desired to get into the city. If any citizen showed the slightest suspicion of such intention, his throat was immediately cut.

2. For the more wealthy citizens, it made little difference whether they stayed in the city or sought to leave, for either way meant death. Every such person was executed on the mere pretense that they were going to desert, but in reality they were slain so the robbers might get at their riches. The thieves' insanity only increased with their hunger, and both of these miseries, madness and hunger, were inflamed more and more each day. No corn existed in public but that the robbers came quickly and searched private homes. If they found any, those in the household were tortured because they had denied the truth. If the bandits found none, the people were tortured even more severely because the rebels supposed the people had more carefully hidden their corn. The clue as to whether the miserable wretches had any food was the condition of their bodies. If they looked healthy they obviously did not need food, but if a person looked emaciated, the thieves went on their way without searching further. They didn't even bother to kill such as these, seeing they would die of famine anyway. Many sold everything they had for one measure, whether

of wheat if they were rich, or barley if they were poor. Once purchased, the buyers would shut themselves in their homes inmost rooms and eat the grain they bought. Some ate without grinding the grain because they were so hungry, while others baked bread with the grain depending upon what their needs and fears dictated. Nowhere were tables set for a proper meal. They snatched the bread out of the fire half-baked and quickly ate it.

3. The situation was so bleak as to make one shed tears. The stronger men had food aplenty while the weak were very sorrowful for want of it. All one's emotions are overcome by famine, but nothing does famine destroy like a person's shame. What was once admired and loved was now despised. Children took bits of food from their father's mouth, and what was worse—mothers did the same thing to their infants. When their most beloved were dying in their arms, they weren't ashamed to steal from them the little tidbits that might keep them alive. Neither was this done in secret, for the rebels would come upon them quickly and snatch from their mouths that which they had gotten from others. Or, when the people closed their doors it became a signal that those within had gotten some food. The rebels then broke the doors open and ran in, forcibly grabbing scraps of food the people were almost ready to swallow. The rebels beat old men who tried to hang onto their food. If a woman tried to hide food in her hands, her hair was ripped out. Nor was there any pity shown to the aged or to infants. The rebels lifted the children by their heels so that the morsels they had eaten were shaken on the floor. But the rebels' barbarity was even crueler with those who resisted those entering their house, taking food that had been swallowed as if it was their natural right.

They also invented horrible tortures in order to find food. They drove sharp stakes into the lower orifice of the miserable wretches, stopping up its passageway. Men were forced to bear things even more horrible in order to make them confess that they had a loaf of bread or a handful of barley meal. The tormentors did these things when they weren't even hungry. The tortures would have been much worse if the rebels actually needed food. But this was all done to exercise their madness and to provide themselves food for the coming days. Some citizens stole out of the city at night, as far as the Roman guards, to gather wild herbs and other plants. Thinking they had succeeded, the rebels then attacked them and snatched every morsel from them, even though their victims often called upon God's awesome name, and begged to be given back some of what they had brought into the city. Although the smallest crust would not be returned, the citizens considered themselves fortunate that they had only been robbed and not murdered.

4. In this way were the poor people treated at the hands of the rebel guards. Meanwhile, the more wealthy were taken before the tyrants, some of whom were falsely accused of planning treachery, and executed. Others were charged with betraying Jerusalem to the Romans. But the most common accusation was that a citizen was thinking about ways to escape. The person who had been robbed by Simon was sent to John while those robbed by John were taking again to Simon. They drank the people's blood together and divided the poor creatures' bodies between them. Yet, because of their common zeal to be in control, Simon and John opposed one another. The rebel leader who would not share his ill-gotten booty with the other was deemed a rotten criminal. He that did not share in what was stolen from the miserable populace was angry that he had been excluded, as if he had been defrauded of what was rightfully his.

5. It would have been impossible to list every instance of these men's crimes. I shall therefore just speak briefly of these things. Never before has any other city suffered such misery. And never before since the creation of the world has any age of mankind bred a generation more skilled in evil than this one. Ultimately, they brought the Hebrew nation into contempt while seeming less contemptible in comparison to their treatment of foreigners. They confessed what was true about themselves. They were the slaves, the scum of society, and the illegitimate and aborted children of our country. They authored our city's destruction by forcing the Romans, whether they had wanted to or not, to a sad, ultimate victory. By these men's continued defense of Jerusalem, it was as if they welcomed the fire that came upon the temple, hoping it would arrive even sooner. When they saw the temple burning from the upper city, they were neither troubled by it nor did they shed tears for the sacred sanctuary. Yet the Romans felt these intimate passions. We'll speak further of these matters in a more appropriate place when our context demands it.

Chapter 11

How the Jews Were Crucified Before the City Walls. Also Concerning Antiochus Epiphanes and how the Jews Overthrew the Roman Banks

1. Titus's siege embankment had by now advanced substantially, notwithstanding that his soldiers were under constant fire and extremely distressed by the enemy on the wall. He sent a cavalry unit out to lay ambushes for those venturing out of the city into the valley to gather food. Some of these were indeed rebel fighting men who were not content with what they got plundering Jerusalem's citizens. But the majority were poor people who didn't want to desert because of their concern for relatives left in the city. They couldn't hope of taking their wives and children with them without alarming the rebels. Nor could they consider leaving their family to be executed by the rebels because of their own escape. Only the famine's severity made them bold enough to venture out to find food. If the robbers did not catch them, perhaps the Romans would. If they were about to be captured, they had to defend themselves for fear of punishment. But if they fought the Romans, it would then be too late to beg for mercy. If caught, they would be first whipped and then tortured with all sorts of torments. Then, before they died, they were crucified in front of the city wall. This miserable procedure made Titus pity those captured, though the Romans caught at least five hundred Jews. Yet Titus didn't believe it was safe to merely release those taken by force. To try to guard the prisoners would have tied up too many of his men on guard duty. The central reason he didn't stop the crucifixions was the hope that the Jews inside the city, seeing the grizzly sight, might be induced to surrender. So, out of his soldiers' hatred for the Jews, Titus had those they caught nailed to the crosses, one after the other. The Romans joked that when the number of executions grew so great and no room existed for more crosses, neither would there be more crosses for Jewish bodies.

2. But the rebels were far from repenting at this sad sight. On the contrary, they made the rest of Jerusalem's citizens believe otherwise. They brought the families of those who had deserted to the walls along with those who were eager to desert to the Romans, and showed them what horrible pain was inflicted upon those who attempted to surrender. They told the peo-

ple that those crucified had sought Roman security, not those who had been taken prisoner. The sight kept many who were eager to desert inside the city until the truth came out. Yet some people ran right away into certain punishment, believing that death by Roman hands would be easier than starvation. So, Titus gave orders to cut off the hands of his prisoners so that they might not be mistaken for deserters. He sent them back into Jerusalem to John and Simon, exhorting the rebel leaders to pause a moment from their madness and not force him to destroy the city. By doing this, they would have the benefits of repentance, and in their utmost distress save their own lives, their city, and their sacred temple. Titus then went around to his embankments under construction and ordered the work to be hurried in order that his words might be backed up by actions. In response, the rebels shouted insults against Caesar and upon Vespasian, yelling that they despised death, greatly preferring it to slavery. While they still had breath within them, they would do everything possible to injure the Romans. As to their city, since Titus had already decided to destroy it, they didn't care. The world itself was a better temple than this one. They cried out that He who dwelt in the temple, the God whom they still had as their ally in the war, would spare the temple. So the rebels laughed at Titus's threats that they would come to nothing, because the entire outcome was in God's hands. They mixed their words with insults and made other loud noises as they shouted.

3. While this was all going on, Antiochus Epiphanes IV came to Jerusalem, bringing with him a large number of armed soldiers, together with a unit called "The Macedonian Band." They were all teenagers, tall and well armed and trained in the Macedonian way from which they took their name, though many of them did not come from Macedonia. Of all the kings established by Roman authority, Commagene, king of Macedonia, had enjoyed the greatest prosperity, until something changed. When he was well up in years, he proved that no man was considered happy until he died. But at this time, as his son Antiochus arrived at Jerusalem, the old king was still alive. His son criticized the Romans for taking such a long time in capturing Jerusalem's walls. He was a bold warrior and accustomed to exposing himself to danger. He was also very strong, and his boldness seldom failed to bring victory. Upon hearing Antiochus's boast, Titus smiled and invited the young man to share in the assault with him. So Antiochus took his men and his Macedonians and suddenly assaulted

the wall. He showed both personal skill and strength in protecting himself from the Jewish missiles, while he shot his own arrows back at them. But the young men with him had quite another experience. They soon realized that the boasts made of their courage and skill caused them to persevere in the fight. But at length many of them fell back wounded. They then understood that true Macedonians, if they were to be conquerors like their hero Alexander, needed his good fortune as well as his strength and skill.

4. The Romans had begun work on their embankments on the twelfth day of Artemisius [Jyar] – (May, 70 AD), and had labored hard to finish them in seventeen days. They had now raised four such banks. The fifth legion built one at the Antonia Tower near the pool called Struthius. The twelfth legion raised an adjoining bank at a distance of about thirty feet from the first. But the bank constructed by the tenth legion was farther away in the city's north quarter, at the pool called in Amygdalin. The fourth bank, erected by the fifteenth legion, was about forty-five feet from the third at the high priests' monument. Then Titus's war engines were brought up. But John had previously undermined all of the area near the Antonia Tower under the Roman banks. He supported his excavation with beams laid across one another, leaving the Roman works without a sure foundation. John then ordered flammable materials of pitch and bitumen brought in and set on fire. As the cross beams supporting the banks burned, the banks suddenly collapsed, falling into the excavated spaces and making a tremendous noise. Dust and smoke rose up from the pit as the fire was put out by the bank's soil. But the fallen materials were gradually set on fire and ultimately broke into brilliant flames. Seeing the flames, the Roman troops were bewildered, discouraged by the Jews' creativity. Further, coming right at the time when they thought victory was at hand, the incident diminished their hopes for the time being. They considered putting out the flames, but soon thought it useless since their banks had already become swallowed up, and were unserviceable.

5. Two days later, Simon and his men tried to destroy the other two Roman banks. The Romans had already brought up their engines and were beginning to make the wall shake. Suddenly, a group of rebels ran out of the city. Tephtheus of Garcis, a city in Galilee, and Megassarus, the offspring of one of Miriamne's servants, led them. With them was a man

named Chagiris ben Nabateus from Adiabene. His name Chagiris meant "The Lame Man," testifying to the poor luck he had suffered. These men grabbed some torches and ran out quickly to the Roman engines. Never before or after these Jews' bold move had any other Jews sallied out from the city who could surpass them, either in audaciousness or in the terror they caused the enemy. For they came upon the Romans not so much as enemies but as friends, without fear or caution. They rushed wildly through the Roman ranks and did not retreat until the siege engines were ablaze. And although darts came at them from every side, and on every side they saw the thrust of Roman swords, they did not seek refuge until the fire had taken hold of all the engines. But when the flames shot up, the Romans came running from their camp to save their engines. Then the Jews stalled them, fighting those who tried to quench the fire, without any regard to their safety. So, the Romans pulled the engines out of the fire while the covering hurdles were now also ablaze. But the Jews grabbed the flaming battering rams and held them fast, although their iron was becoming red-hot. The fire now spread from the engines to the Roman banks and stopped those coming to try to defend them. This entire time the Romans were enshrouded in fire. Finally, thinking they were unable to save their works, the Romans retreated to their camp.

The Jews, however, came in greater and greater numbers out of the city. Believing they were invulnerable in their violent assault, they advanced as far as the Romans' fortified encampments and engaged the guards. In front of the Roman camp stood an infantry unit in full battle array. These troops were governed by a Roman law that stated, "Any soldier who quits his post for any reason whatsoever is to die." So, that unit stood firm, preferring to die fighting courageously than to be punished for cowardice. Seeing these troopers standing firm, many of the soldiers who had retreated now turned back. When they had set up more war engines against their camp wall, they stopped the Jews who were pouring out of the city. The attacking Jews had not put on their armor nor made any provision for their personal safety. They impetuously fought in close combat against the Romans who met them, flinging themselves against Roman spears and attacking zealously. The Romans had to give way more because of the Jews' fearlessness than because of their own wounds.

6. Titus now came from the Antonia Tower looking for a place to raise new embankments. He furiously reprimanded his soldiers for allowing

their own fortifications to come into danger after they had overrun the city walls. They had allowed the Jews to turn the tables and now seemed to be penned up in a prison of their own making. Then, leading his chosen forces, Titus outflanked the Jews. Although they also fought against those in front of them, the Jews turned and continued to protect their flank. The two armies now were intermixed to such a degree in the noise and dust that they could neither see nor hear nor discern friend from foe. The Jews continued to do battle, though not as much with skill as from their fear of not surviving. Meanwhile, the Romans were driven on by the hope of glory and by retaining their honor, as Caesar himself fought alongside them. I would imagine that the Romans would have wiped out all the Jewish forces had not the enemy anticipated that the battle was lost, and retired to Jerusalem. However, the Romans had seen their works completely demolished, and were dejected by the loss. In the space of one hour, they lost what had cost them dearly. Many Romans thought they would never take the city with only their ordinary war engines.

Chapter 12

Titus Encompasses the City with His Own Wall; after which the Famine Consumed whole Houses and Families Together

1. Titus now held a war council with his officers. Those who had the hottest blood among them thought he should mass the entire army and storm the walls. Before this juncture, only a portion of the army had seen combat with the Jews. If the entire army were to attack in force, the Jews would not be able to defend against their overwhelming numbers, but would be slain by Roman darts. But those of a more cautious temperament voted to raise the banks once again. Others advised Titus to forget about the banks and merely lay outside the city, keeping the Jews walled in, unable to bring supplies into Jerusalem. In that way, they would leave the enemy to starve to death without suffering losses from direct combat. For the Jews were not afraid of being overrun, especially those who wanted to die by the sword, as long as a more miserable death awaited. However, Titus didn't think it was wise to allow so great an army to lie still, and yet neither was it wise to fight against men who wanted to destroy themselves. He also spoke of the impracticality of raising more embankments. The materials were simply not available. It was even more impractical to guard against the Jewish sallies. The city's circumference was too lengthy

to watch every exit, to say nothing of the dangers imposed by the Jewish attacks. For although they could guard the main gates, now that they were under great distress the Jews would certainly devise other passageways, since they were undoubtedly knowledgeable of such places. If any provisions were snuck into Jerusalem, the siege would only be delayed. Titus was also concerned that a long siege would diminish the glory of his victory. For while it is true that time solves all things, it adds to one's reputation to get the job done quickly. It was, therefore, his opinion to act with speed as well as security.

The Romans would completely encircle Jerusalem with a wall. It would be the only way to keep the Jews inside the city. Then the Jews would either completely give up any hope of saving Jerusalem and therefore surrender, or be more easily conquered when further weakened by starvation. In addition to this wall, he would begin raising new banks when the opposition had grown weaker from the famine. If anyone might think such a task too difficult to be completed without much effort, you should remember that Romans never undertake any small work. Only God Himself could accomplish such great tasks easily.

2. Titus's arguments convinced his commanders. So he gave orders to divide the task and each division share in the work ahead. By this time, a supernatural fury had come upon the Roman soldiers. Not only did they divide the tasks of wall building among themselves, each legion competed with another. Even smaller units entered into competition. Each soldier wanted to please his decurion, and each decurion his centurion. Each centurion desired to please his tribune and tribunes likewise their higher commanders. Caesar himself took notice and rewarded the same competitive spirit in his senior officers. He went around the work daily, inspecting the army's accomplishments. Titus began the wall at the camp of the Assyrians where his own camp existed. The wall went down to the lower parts of Cenopolis, then along the Cedron Valley to the Mount of Olives, and then curved southward, encircling the mountain as far as the rock called Peristerion and its adjoining hill over the valley that reaches Siloam. There, the wall angled around to the west, running down to the Fountain Valley, beyond which it went up against the monument to the high priest Ananus, and encircled that mountain where Pompey had pitched his encampment. The wall then returned to the north side of Jerusalem and went as far as a village called "The House of the Erebinthi,"

after which it went around Herod's monument, and there on the eastern side of the city joined Titus's camp again. The length of the entire wall was about five miles.

Additionally, thirteen forts were built around the wall to garrison soldiers. These forts added more than an additional mile to the construction. The army completed the wall in three days. What would have normally required several months was finished in an incredibly short time. When Titus had, therefore, encircled the entire city with his wall and had placed garrisons in the forts, he rode around the entire wall on the night's first watch and inspected the guard. He sent Alexander to inspect the second watch, and the third was given to the legion commanders. The commanders cast lots between themselves to determine which of the night watches each would take, and who would go around throughout the night to watch the spaces in the walls between the fortified garrisons.

3. The Jews were now cut off from all hope of escape, as well as any freedom to move around outside of the city. Famine got progressively worse and devoured the people in entire families. The upper rooms were full of women and children dying of starvation and the city streets filled with the corpses of the aged. The young men and older children wandered about the marketplaces like shadows; their stomachs swelled by hunger. They fell down and died wherever their misery took them. Those who were ill were not able to bury the dead, while the great number of bodies and the uncertainty of their own approaching deaths stymied those who felt better. Many died as they were burying someone else while many others had lain down in their coffins before they were dead. Nor was any mourning heard under these distressing calamities. Nor did one hear sighs of complaint. The famine suppressed all natural feelings. Those about to die looked upon those already dead with dry eyes and open mouths. The city was also deathly quiet while the robbers continued exceed these miseries by their evil conduct. They broke into those homes that were no more than the graves of dead bodies, and plundered even the corpses they found of their meager possessions, taking even the clothes they wore and then would leave the house laughing. They tested their sword points against the corpses, and in order to prove the strength of the metal ran some of them through those who were still alive and prone on the ground. Anyone who begged the robbers to lend them their right hand and sword to finish them off heard sneers returning the request, and was then left

to be consumed by starvation. Every one of these people died with their eyes fixed on the temple, blinded to the rebels who were still alive. At the onset, the rebels gave orders that, due to the stench of decaying bodies, the corpses should be buried at public expense. But later, when that was impossible, they threw the corpses down from the city walls into the valley below.

4. When Titus made his rounds inspecting these valleys, he groaned as he saw them fill with corpses, with a thick fog of putrefaction hanging over them. He spread out his hands to heaven and asked God to witness that this was not his doing, but was the sad plight of Jerusalem itself. But the Roman soldiers were filled with joy, since the rebels could no longer sally out of the city to fight. The rebels were now suffering as the famine had also touched them. The Romans had plenty of corn and other supplies from Syria and the neighboring provinces. Many soldiers would stand near the city wall in plain sight of the people and show them what large amounts of provisions they had, making the enemy feel their hunger more tangibly by the great storehouses the Romans had. But when the rebels showed no sign of surrendering, Titus, out of his sympathy for the remaining citizens and from his earnest desire to rescue those still alive, began to raise his banks again. Materials were hard to come by, as almost all of the local trees had been cut down to build his former embankments. But the soldiers prowled the countryside and brought in materials, some from as far away as eleven miles. The Romans began to raise even greater banks in four places, all in the vicinity of the Antonia Tower. So, Caesar made his rounds through the legions, hastening the work and showing the rebels that they were now virtually in his hands. But these men—and only these men—were incapable of remorse for the terrible wickedness of which they were guilty. Like wild animals, the rebels didn't care for either body or soul, and treated each of their own souls and bodies as if it belonged to someone else. No human affections arose from their souls nor could physical pain register in their brains. They still ripped open the dead as if they were dogs and filled the prisons with the sick.

Chapter 13

A Great Slaughter and Sacrilege in Jerusalem

1. The rebel Simon would not allow the high priest Matthias, who had first

welcomed him into Jerusalem, to die without being tortured. Matthias ben Boethius had been very faithful to the people, and they had held him in high regard. When the Zealots, of whom John was one, were distressing the citizens, Matthias had persuaded the people to admit Simon into the city to help them. But when Simon and his henchmen had come in, Simon took control of the city. Simon saw Matthias as an enemy, just as he did everyone else, and looked upon his advice to allow him entrance as foolishness. So Simon had Matthias brought before him and condemned him to death for taking the Roman side. Matthias was not allowed to speak in his own defense. Simon also condemned Matthias's three sons to die with him. The fourth son had previously escaped to Titus. Matthias begged the favor of being executed prior to his sons on the basis that he had opened the city gates to Simon. But Simon proclaimed that Matthias would die last of all and would live to see all of his sons executed before his eyes and in full view of the Romans. Simon then charged Artanus ben Bamadus, whose barbarity exceeded the rest of Simon's guards, to do the work. Simon laughed at the condemned man, telling him he would now see whether or not the Romans would come to help him. Simon then forbade burial of their corpses. After these executions, Ananias ben Masambalus, an eminent priest, and Aristens, the Sanhedrin's scribe from Emmaus, were executed along with another fifteen prominent citizens. They also kept Josephus's father in prison and ordered that no one speak to him, nor even enter his company with others, fearing he would betray them. They also executed anyone who mourned for these men who died without a fair trial.

2. When Judas ben Judas, one of Simon's subordinates and the person charged with guarding one of the towers, saw what Simon had done, he gathered ten of his faithful men and said to them: "How long shall we bear this misery?" [He spoke perhaps partly from pity for Matthias and his sons but mainly to provide for his own security.] "What hope do we have of saving our lives if we continue to follow these wicked wretches? Are we not already caught up in the famine? Are not the Romans ready to enter the city? Is not Simon unfaithful to his friends? Shouldn't we fear that he might soon place the same punishment on us while the Romans offer us certain security? Come on, let's surrender this wall and save our city and ourselves. Simon won't be harmed very much, since even now he has lost all hope of deliverance. It may be that we bring him to justice a

little sooner than he thinks."

    Judas's ten underlings were swayed by his arguments, so he disbursed the rest of his men in various directions in order that the plot might not be discovered. Then, in the third hour, Judas called to the Romans from the tower. Some of the soldiers made prideful sneers at what he said. Others didn't believe him to be telling the truth. But most of the soldiers just shrugged their shoulders. They believed that they would soon overrun the city without any danger. But when Titus was approaching their position with his armed guard, Simon, already hearing of Judas's treachery, took over the tower before it could be surrendered to the Romans, grabbed Judas and his men and executed them in full view of the Romans. When he had mangled their corpses, he threw them down from the city wall.

3. Meanwhile, as Josephus was on his way around the city, he was hit in the head by a stone and fell stunned to the ground. As he lay there, the Jews ran out to capture him. Had not Caesar immediately sent soldiers to protect him, Josephus would have been carried into the city. As the fight ensued over his capture, Josephus got back on his feet but was unaware of the action around him. So the rebels thought they had killed their most wanted man and began loudly to rejoice. News of the incident soon spread throughout the city, and upon hearing it Jerusalem's citizens became very gloomy, thinking Josephus was dead. Their only hope of ever deserting to the Romans was with Josephus's help. When Josephus's mother heard her son was dead, she said to those with her that since the fall of Jotapata, she knew he would die and that she would never again enjoy her son's company. She also mourned privately with her maidservants, asking what advantage remained for one who brought such an extraordinary son into the world? She would not be able to bury her son, whom she expected to bury her. However, the false report of Josephus's death did not cause his mother pain much longer, or the rebels joy. For Josephus soon recovered from the blow to his head and came out to the wall shouting that it would not be long before the rebels were punished for the wound they had given him. He also renewed his reassurance to the people of the security Rome had promised to them. Seeing Josephus alive encouraged the people tremendously, but brought great anxiety upon the rebels.

4. Hearing Josephus, some men immediately leapt down from the wall

to desert, having no other choice. Others went out of the city carrying stones as if to fight the Romans, but instead fled to them. But a worse fate befell those men than that which they would have met within the city. Those fleeing in great numbers were dispatched quicker among the Romans than they would have been from starvation among the Jews. When they arrived in the Roman camp, their stomachs were bloated by starvation. Choking down the Roman food, their stomachs quickly became overfilled and burst. Those who knew enough to restrain their appetites were unharmed, and then little by little ate their fill until their stomachs grew accustomed to it.

Yet another curse struck those who had eaten but little. One of the Syrian deserters was found collecting pieces of gold from the excrement of dead Jews. As we said before, the deserters would swallow such gold pieces before they came out of the city. The rebels had made a thorough search of their possessions, and there was a vast quantity of gold in the city. In fact, so much gold existed in the Roman camp that it now sold for less than half what it brought before. But when the Romans discovered the Syrian deserter, the news that the runaways came laden with gold traveled quickly throughout the camps. So, a crowd of Arabians and Syrians cut up those who sought security and searched their stomachs. It seems to me that no other misery that the Jews endured exceeded this one. At least 2,000 of these deserters were dissected in the space of one night.

5. When Titus heard of this travesty, he wanted his cavalry to surround those guilty and shoot them all dead. And he would have done exactly that had not their numbers been so large, and the innocent most likely killed with the guilty. However, he called together the commanders of the auxiliary troops as well as the commanding officers of the Roman legions, and with great indignation said to them: "What? Have any of my own soldiers done this thing hoping for gold, without remembering that their own weapons are made of silver and gold? Moreover, have the Arabians and Syrians begun to govern themselves as they please in this foreign war, indulging their appetites? And so, barbarously murdering men out of their hatred for the Jews, yet blaming the Romans? [For the abominable tragedy was said to have been instigated by some of Titus's own soldiers.] Titus then threatened to execute any man who would be so insolent as to repeat the deed. Additionally, he ordered the legions to make a thorough search of those suspected of involvement and brought to him. It seemed

that the love of money was greater than their fear of punishment. A passionate desire for gain is natural to man, and no passion is so audacious as that of covetousness. Other passions have limits and are tempered by fear. In reality, it was God who condemned the whole nation of Israel, and turned every path that seemed to lead to preservation into one that ended in destruction. Thereafter, Caesar's threats forbidding the practice moved it underground. The barbarians would still go out secretly to meet the deserters before anyone saw them. Making sure no Roman eyes were watching, they dissected those leaving Jerusalem and pulled the little polluted money out of their bowels. A great many died in the mere hope of getting by the Arabians and Syrians. But the wretched treatment caused many deserters to return to the city.

6. When John found he could no longer steal from the citizens, he turned to sacrilegiously melting down the sacred utensils in the temple. Many of these holy vessels, such as cauldrons, dishes, and tables, had been given to the sanctuary and were used in the priests' ministry. John took pitchers sent to the temple by Caesar Augustus and his wife, for the Roman emperors both honored and adorned the temple. So this man John, also a Jew, seized these sacred objects given by aliens and told all those around him to fear not, that this was a proper use of Divine things. After all, they were fighting for the Divinity and so the temple should pay for their efforts. So, on that basis John poured out all the sacred wine and oil and distributed these elements to the Zealots around him. The wine and oil found in the temple's inner court were intended to be poured on the burnt offerings by the priests. The Zealots anointed themselves with the oil and drank the wine, each of them using over eight quarts of the sacred ointments. At this point, I must pause and speak of my concerns. I suppose that if the Romans had delayed any longer in coming against these villains, Jerusalem would have either been swallowed up by the ground opening underneath her, or destroyed by a great flood, or burnt to cinders by the fire and brimstone that struck Sodom and Gomorrah. For the city had brought forth a generation of men who were more ungodly than those burned up by God's earlier punishments. It was because of their madness that the people of Jerusalem were devastated.

7. Indeed, why is it that I'm telling you of all these disasters? Manneus ben Lazarus ran to Titus at this very time and told him that no fewer

than 115,880 dead bodies had been carried out of the city gate entrusted to his care. [This count was conducted between the fourteenth day of Xanthicus [Nisan], when the Romans pitched their camp next to the city, and the first day of Panemus [Tamuz]. Though this body count in itself was a huge number, it was by no means the total. Manneus was not the administrator of that gate. It was his duty to pay the public stipend for each body carried out and so he had to count them one-by-one. Families and relatives buried the rest of the dead. But a burial only meant that the corpse was carried out of the city and thrown away. Following Manneus, many of Jerusalem's prominent citizens told Titus that no fewer than 600,000 corpses had been thrown out of the gates, though those bodies remaining inside could not be counted. They further said that when they were no longer able to carry the corpses of the poor outside the city, they stacked them in mounds in large houses and then shut the doors. Titus was also advised that a medimnus (about twelve gallons) of wheat sold for a talent of gold (seventy-five pounds), and that when the Roman wall made gathering of herbs impossible, some people were given to searching the common sewers and old cattle dunghills and eating the dung they found. What earlier generations had not even wanted to have before their eyes was now used for food. When the Romans heard of these miseries they had pity on the people, but when the rebels saw these conditions, not only did they not repent, but permitted the same fate to become their own. They were blind to the fatal extremity that was coming not only upon Jerusalem, but upon them as well.

The Jewish Wars
Book VI

Containing the Interval of about One Month.
From the Great Extremity to which the Jews Were Reduced, to the Taking of Jerusalem by Titus

Chapter 1

The Miseries of the Jews Grow Worse; and About the Roman Attack upon the Antonia Tower

1. And so Jerusalem's misery grew greater and greater with each passing day. The rebels grew angrier and angrier at their own misfortunes as the famine began to prey upon them; after it had it had already consumed so many citizens. A vast multitude of corpses was stacked one upon another throughout the city. They made not only a horrible sight but gave off a powerful stench. The piles of rotting corpses hindered those rebels who made sallies out of the city to fight the Romans. As they prepared to go out to do battle, these rebels who had already murdered tens of thousands of victims weren't frightened as they marched on top of the dead bodies. Nor did they feel any pity for those on whose backs they marched, nor think of this treatment of the dead as a severe portent of what would come upon them. But, as they had their right hands already polluted with their own countrymen's blood, they ran out to fight against foreigners. It seemed to me that in this condition they insulted God Himself, as if He were too slow in heaping punishment upon them. For they had no hope of winning the war. They found glory even in their deep hopelessness of either victory or escape. Meanwhile, the Romans were still having diffi-

culty in finding materials for their banks. But they did succeed in raising them one hundred twenty days after they had cut down all the trees eleven miles into the surrounding countryside, as I've already said. The countryside was a depressing sight to behold, for everywhere where trees and gardens used to grow, now was desolate. The troops had felled every tree. Every visitor who remembered the lush Judean landscape and the beautiful suburbs of Jerusalem now saw only wilderness and mourned, shedding tears over such an abrupt change. For the war had laid waste to any sign of beauty. If anyone who had previously known the area were to now come upon it suddenly, he would not recognize it. Though standing right at the city gates, he would not know where he was.

2. With the Roman embankments completed, they formed the underpinnings of fear both for the Jews and the Romans. The Jewish rebels expected that the city would be quickly overrun, unless they could burn the Roman banks again. The Romans feared that if the Jews burned the banks again as they expected, Jerusalem would never be taken, for there were no more materials. Besides, the extremely difficult labor had sapped the soldiers' strength. The soldiers' emotions also suffered from their many failures. And the tragedies within the city proved more discouraging to the Romans than to the rebels. For the Romans had found the rebels were not held back by their many appalling afflictions, even with less and less hope of victory. The Roman banks had been undermined and burned by the enemy's shrewdness. The walls of Jerusalem also had proved firm, and the fighting very difficult, as the Jews sallied out boldly. But the Roman soldiers' greatest discouragement was that the Jewish passions were stronger than all the calamities they had suffered in their rebellion—their want of food, and in the war itself. The Romans began to imagine that the Jewish violent attacks were invincible and that the Jews' eagerness and boldness would not be depressed by any defeat. If the Jews were encouraged by their defeats, what would it be like if favored with victory? So, the Romans guarded their banks more closely under these conditions.

3. Meanwhile, John and his men made sure that they would be secure even if the wall were demolished. So, they began to work even before the battery rams came forward. Yet success eluded them. Running out with their torches, they came back discouraged at not being able to reach the banks. The reasons they failed were these: first, the Jews seemed to be

disorganized, timidly running out and in small groups, at separate times. In a word, they lacked the Jewish audacity. They had none of that chutzpah that only our nation possesses. They lacked boldness and violence in coordinating their attacks and in following through to their goal. When they first failed, they sallied out even more dispirited than before, finding the Romans in formation and more audacious than usual. The Romans guarded their banks so fervently on all sides with their bodies and their armor that they left no room for the Jewish torches to get through. Every Roman soldier was so passionately committed; he would rather die than break ranks. Furthermore, the very thought that they should lose these banks made them even more committed, as all hope of taking the city would be lost. The soldiers knew what shame would be theirs if Jewish treachery should defeat Roman courage; or Jewish madness defeat their armor; or the many defeat their skill; or the Jews ultimately defeat the Romans. But the Romans had another advantage. Their siege engines were able to throw darts and stones at the Jews as they sallied out of the city. The man who fell became a stumbling stone to the man behind him and the danger of attempting to go further made the Jews lose confidence. Those who had gotten under the barrage of darts overhead were now terrified by the good order and the solidity of the Roman ranks even before they came into hand-to-hand combat. Other Jews, cut by Roman spears, turned back without going farther. This attack was carried out on the first of Panemus [Tamuz].

So, when the Jews had retreated, the Romans, under a shower of stones from the Antonia Tower, also received Jewish fire and sword and many darts. Although the Jews relied heavily on their wall and held contempt for the Roman engines, they did everything they could to keep them off the embankments. Meanwhile, the Romans struggled hard to bring up their engines, understanding that the Jewish fervor was because they knew of the wall's weakness at Antonia and that its foundations were insubstantial. But the wall still wouldn't yield to the battery rams' many blows. The enemy darts came raining down on the Romans, yet they did not give way, as they fought with their artillery. But, fighting below the Jews' superior position, many Romans were grievously wounded by the stones raining down upon their heads. Some Romans covered themselves with their shields, and, working with crowbars, undermined the wall. With great difficulty, they removed four large stones. Then night fell, putting a temporary end to the struggle. That night, however, the battering rams had so distressed the wall in the same place where the tyrant John had undermined the Roman banks, the ground gave way again, and the wall suddenly collapsed.

4. Both sides felt the consequences of this unexpected event. While one would expect the Jews to be disheartened because they held such high confidence in the wall's strength, and had made no plans in the event of its collapse. Yet they summoned up their courage as Antonia still stood. On the other hand, the Romans were dismayed at the wall's failure when they saw another wall inside the one that John and his men had built. However, an assault upon the second wall looked easier than the first, since it appeared weaker than Antonia and the outer wall, and more easily assaulted. The Romans also imagined it had been hurriedly constructed, and so they should be able to overrun it quickly. But none of them ventured to approach this inner wall, knowing that anyone who did would certainly die.

5. Titus understood that a soldier's willingness to fight is chiefly animated by encouraging words of hope and that expectations and promises often make men forget the dangers even to the point of despising death itself. He, therefore, assembled the most courageous of his army and tried these methods on them saying, "My fellow soldiers, to spur men on to accomplish that which entails no peril is an inglorious waste of time to those hearing such a speech. It is the same for the speaker, even an indication of his own cowardice. I believe, therefore, that such encouragements should only be made in dangerous circumstances. Men are able to encourage themselves in other situations. And so I am in full agreement with you that it will be very dangerous to attack this inner wall. But it is right that those who want to be remembered for their valor undertake to accomplish hazardous missions. It is a particularly brave thing to die with glory. The courage necessary to empower those who will go first will not go unrewarded.

"My first reason to encourage you to attack seems like an argument that might be used to the contrary — to persuade you not to go. I refer to the firmness and persistence of the Jews even under their difficult reverses. You are all Romans and my troops. You have been trained to fight in peacetime. It would therefore be a shame to you to now suffer defeat by these Jews who are our inferiors, both in skill and courage. This is especially true when victory is at hand, and God Himself has helped you. We have suffered setbacks, mainly due to Jewish madness. But their sufferings have been caused by your valor and God's assistance. The Jews have been in rebellion against one another and are now undergoing famine. They must now endure our siege engines and the destruction of their walls. What can this be but demonstrations of God's anger at them and his help for us? It is, therefore, not right to show yourselves as inferior to those who are your inferiors or betray God's divine providence which is your ally.

"Even more disgraceful would be to not conquer these Jews who are expecting defeat. They rebel against being slaves, and in order to run from slavery, they prefer to die. So they continually sally out against our forces, not expecting victory, but only to show off their fearlessness. Meanwhile, we Romans have taken possession of all the earth's land and seas. What a great shame it would be if we could not conquer such as these because we were afraid to go where danger lies, but would rather while away the hours in our armor while our enemies die of starvation. All this, when we have them in our power, with only a little danger standing between us and victory!

"If we can hold the Antonia Tower the city will be ours. Once there, I think the fighting in the city will end, since we will have the high ground and will pounce upon our enemies before they know what happened. These advantages will afford us a quick and certain victory. As for me, I will not now speak of commending those who die in warfare. Nor will I speak of the immortality of those who are slain fighting bravely in battle. On the contrary, I cannot cease to speak against those who want to die in peacetime by some disease or another, whose bodies and souls are forever condemned to the grave. For what virtuous man among you doesn't know that those souls severed from their fleshly bodies by an enemy sword are received into the ether, the purest of all elements, and joined to those whose home is amidst the stars? Who among you does not know that they become angels and famous heroes, showing themselves as such to all who come after them? But upon those souls that succumb in their diseased bodies comes a subterranean darkness into which they dissolve into nothingness, never to be remembered even though they are innocent of this world's defilement. In this death not only do the soul and body die, but any memory of them as well.

"But since God has determined that death will come upon every man, a sword is a much better instrument for that purpose than any disease could be. Why then shouldn't we yield to death for the greater good, rather than yielding death up to fate? I have made my speech supposing that those who make the first attempt to overrun the second wall will be killed in the attempt, though men of real courage have a chance to escape even the most menacing endeavors. In the first place, that demolished portion of the first wall can be easily ascended, and the new wall easily destroyed. Will you all now screw up your courage, reassuring and assisting one another to set about this work? Your bravery will soon break the enemy's hearts and just possibly, your glorious mission may be accomplished without bloodshed. For it may be rightly presumed that the Jews will try and stop you at the onset, but if only a few of you get over the wall, the Jews will not be able to keep the rest of you out. The person who gets over the wall first I will bestow upon him so many honors that he will be the envy of all the army. Shame on me if I do not! If that man escapes alive, he shall command those who are now his equals, although the greatest rewards will accrue to those who die in the fight."

6. Hearing Titus's speech, the whole gathering was afraid of the great danger. But one man, named Sabinus, stepped forward. He was born in Assyria and served among the cohorts. Sabinus had great courage that he had already displayed in battle. But Sabinus didn't look like a hero. He had a very weak-looking body; not seeming fit for a soldier. He was also black in color, and all skin and bones. But inside that small body lay a heroic soul, in a house too slender for his expansive courage. Accordingly, Sabinus was the first to speak up. He said, "O Caesar, I readily surrender to you. I will be the first to climb the wall, and I heartily hope that good fortune will accompany my courage and my resolve. If bad luck takes away my success, know that my death will not be unexpected. I voluntarily choose to die for you." When Sabinus had said this, and had lifted his shield over his head with his left hand, and had drawn his sword with his right hand, he marched up to the wall at about the sixth hour. Only eleven others, determined to imitate Sabinus's bravery, followed him. But Sabinus was first among them, and went forward as if empowered by a divine fury. Those Jews on the wall shot arrows from every side at these few attacking Romans. They rolled huge boulders down upon them that ran over some of the eleven. But Sabinus met the missiles thrown and shot at him, and though they overwhelmed him, he did not stop his violent

charge until he was at the top of the wall and had made the enemy flee.

The Jews were astonished at Sabinus's great strength and bravery. They imagined that more Romans had climbed to the top than had, and they turned and ran. At times like these, one can understand how ill fortune is jealous of valor and is always getting in the way of glorious achievements. This was true in Sabinus's case. When he had just reached his goal, he stumbled over a large rock and fell down upon it with a loud crash. Turning back, the Jews saw that he was but one prostrate soldier, and threw missiles at him from every direction. Sabinus managed to rise on one knee and protect himself with his shield. He defended himself as best he could and wounded many who rushed at him. But his right hand soon grew tired because of the many lesions he had suffered until finally, pierced by many darts and arrows, he gave up the ghost. Sabinus and his bravery deserved a better fate, but as might be expected, he tried to do more than one could expect of a single soldier. Of the remaining eleven soldiers, the Jews crushed three of them with their boulders or killed them as they reach the top of the wall. The eight others were wounded and carried back to camp. This all happened on the third day of Panemus [Tamuz], (July, 70 AD).

7. Two days later, twelve soldiers guarding the Roman banks near the wall got together and called over the fifth legion's standard bearer. With two other cavalrymen and a trumpeter with them, they all quietly crept through the ruins in the darkness of the ninth hour and climbed the Antonia Tower. They cut the sleeping guards' throats, got possession of the wall and ordered the trumpeter to sound off. Hearing the blast, the rest of the Jewish guards suddenly awoke and ran away before anyone could tell how many Romans had climbed up. From their fear and the trumpet's sound, they imagined it to be a great number. When Caesar heard the trumpet, he ordered the army to don their armor immediately and then he rushed to the scene with his commanders. He ascended the wall first with his chosen men. As the Jews were running away to the temple, the Romans entered the tunnel leading to the sanctuary that John had dug when he undermined the Roman banks.

Both rebel factions, those with John and those with Simon, teamed up to face the Romans. They took on the Romans with every ounce of their strength and ferocity, knowing that if the Romans took the temple they would be ruined. The Romans saw it in the same way, as the beginning of the end of Jerusalem's conquest. So, a horrendous battle was waged at the temple entrance. The Romans were forcing their way in to take over the sacred sanctuary, and the Jews were intent on driving the Romans back to the Antonia Tower. They fought in such close quarters that arrows, darts, and javelins were useless. Both sides fought with their swords in hand-to-hand combat. While the battle raged, Romans and Jews intermixed with one another and fought randomly, indistinguishable in the confusion. The place where they fought was narrow, and the violent noise added to the confusion. Each side took heavy casualties as the combatants walked over the bodies and armor of the dead, mangling them. When one side gained the advantage they exhorted one another to press on, while those driven back uttered lamentations.

But the space allowed no room to flee nor pursue. For those in the front lines, it was kill or be killed with no possibility of escape. Those behind them pressed forward, so no clearance existed between the armies. At length, the Jews' violence was too much for the Romans' skill. The fight lasted from the evening's ninth hour until the morning's seventh hour. While the Jews joined the battle by the hundreds—for the temple's preservation was their motivation—the Romans had only a portion of their army in the fight. The legions on which the Roman soldiers depended had not joined them. So, the Romans retired to the Antonia Tower, thinking the possession of that crucial place to be sufficient.

8. A centurion named Julian was from Bithynia. He had a well-known reputation – famous for his martial skills, his strength, and his courage. Seeing the Romans giving ground and in sad disorder [for he stood by Titus at the Antonia Tower], Julian alone leaped out and put the already victorious Jews to flight, making them retreat to the far corner of the temple's inner court. The Jews fled from him in droves, believing his strength and ferocity to be that of no mere human being. He rushed through the Jews as they ran away in every direction, killing those he caught. No sight could have cheered Caesar more nor been more terrible to the Jews. But Julian was also subject to the fate from which he, and any mortal man, cannot escape. Like all the other Roman soldiers, Julian's shoes had soles full of thick, sharp nails. When he ran onto the temple's pavement, he slipped and fell onto his back, his armor making a loud noise. The sound made those Jews running away turn back. Fearing for Julian, the Romans watching the action from Antonia shouted encouragement. But a crowd

of Jews had already surrounded Julian, striking out at him with their spears and swords. Julian caught many thrusts with the shield. He tried time after time to stand up but was thrown back down by those striking at him. Yet while on his back he was able to stab many Jews with his sword. It took a long time for Julian to die, covered as he was by his helmet, breastplate, and other armor in all those areas of his body that might receive a mortal wound. He also pulled his chin close to his body until all his limbs were shattered. When no one arrived to help him, Julian yielded to his fate.

Caesar was deeply moved by Julian's great courage, and particularly that he met death in full view of so many. He wanted to go to help him, but his position as commander would not allow it. Everyone else was too afraid to attempt a rescue. Julian fought against death for a long time. He had wounded many Jews in his struggle. But finally his throat was cut, though not without some difficulty. He left behind his great fame, not only among the Romans and with Caesar himself, but also among his enemies. Then the Jews took up his corpse and put the Romans to flight again, shutting them up in the Antonia Tower. Of the many Jews in the battle, these men distinguished themselves, fighting zealously for their side: of John's faction—Alexis and Glyphtheus; of Simon's faction—Malachias and Judas ben Merto, as well as James ben Sosa, the Idumean commander; and of the Zealots—Simon and Judas ben Justus, two brothers.

Chapter 2

How Titus Gave Orders to Demolish the Antonia Tower, and then Persuaded Josephus to Once Again Try to Persuade the Jews to Surrender

1. Now Titus gave his soldiers orders to dig through the Antonia Tower foundation and open a passage for his army to get into Jerusalem. Titus then had Josephus brought to him, for he had been informed that on that day, the seventeenth of Panemus [Tamuz], the daily sacrifice had not been offered to God because no priest remained to offer it. This omission terribly grieved the people. Titus commanded Josephus to say to John precisely what he had said previously: that if he intended to fight the Romans, he should bring as many men as he wanted out of the city to fight, and so relieve Jerusalem and the temple from the threat of destruction. Titus did not want John to defile the temple and thereby offend God.

356   *The Jewish Wars*

John might, if he so pleased, offer the discontinued sacrifices by any of the Jews he might select. So, Josephus stood in the spot where John and the many Jews with him could hear him, and declared in Hebrew what Caesar had ordered him to say.

He earnestly pleaded with the rebels to spare their city and to prevent the fire that was about to consume the temple, and to instead offer the sacrifices pleasing to God. Hearing these words, a great and desolate silence fell upon the people. But John himself threw out many insults at Josephus, along with many threats. Finally, John added this: that he had never feared the Romans taking Jerusalem because it belonged to God alone. Answering John, Josephus cried out in a loud voice, "To be certain, you've kept the entire city marvelously pure for God's sake, and the temple completely unpolluted, as well! Nor have you been guilty of any sin against Him in whom you hope of help! He still receives His daily sacrifices! You are a vile wretch! If anyone took your daily food away, you would consider him an enemy, but you hope to have God's support in this war, while taking away His eternal worship. And then you blame the Romans for your own sins, while the Romans have taken great care to allow us to observe our holy laws, and almost compel these sacrifices to continue to be offered to God! These sacrifices that you have stopped! Who remains that does not join in moaning and lamenting at what Jerusalem has become? Even Gentiles and the enemy now want to right the wrongs that you have committed, while you, a Jew and educated in our traditions, have become a greater enemy of our laws than have the Romans!

"But John, it is never dishonorable to repent and seek to make right that which as been wrongfully done, even at this late hour. If you have any desire to save the city, remember the actions of the Jewish king, Jeconiah. When the king of Babylon assaulted Jerusalem, the king by his own volition had left the city before it was overrun. He and his family voluntarily went into captivity, that the sanctuary might not wind up in enemy hands and that he might not witness God's house set afire. His actions are still celebrated by all the Jews at their sacred memorials, and his memory will be remembered down through the ages and immortalized. His example is appropriate in dangerous times, and I dare promise that the Romans will forgive you. Take notice that I am a Jew, one of your own nation; I make this promise to you. And you might also consider who it is that gives you this counsel, and where my home is. For while I am alive I will never be subjected to slavery, that I would betray my own kinsman and forgot my forefathers' laws.

"Again, you shout that you hate me, and you shoot at me and insult me. Indeed, I cannot deny that I am worthy of an even greater reproach, for opposing providence and instead making this kind of merciful invitation to you—hoping to bring deliverance to those whom God has already condemned. And who does not know what the prophets of old have written? Particularly, that oracle which is now about to be fulfilled against this miserable city. They prophesied that Jerusalem would fall when a Jew began the slaughter of his own countrymen. Are not both the city and the temple filled with the corpses of your fellow countrymen? It is therefore God Himself who is bringing on the fire to purge both Jerusalem and its temple, using Roman hands. He is going to destroy this city that is so full of your abominations."

2. As Josephus spoke these words in a choking voice and with tears in his eyes, he was interrupted by his own sobs. Hearing him, the Romans could not help but pity his afflictions and marvel at his bearing. But John and his rebels became even more exasperated at the Romans on his account. They wanted to take Josephus prisoner, and yet his speech directly influenced a great many of the people. Some of them were so afraid of the rebel guards that they didn't move, understanding that both they and the city were doomed to destruction. But there were also some who looked for a good opportunity to quietly escape and flee to the Romans. Of these were two high priests, Joseph and Jesus, along with three sons of the high priest Ishmael, who had been beheaded in Cyrene. Also, another four men, all sons of Matthias, and one son of the other Matthias—who ran following his father's death at the hands of Simon ben Gioras, as I have already related. Three of his sons went with him. Many others of the Jewish nobility also went over to the Romans, as well as other high priests. Caesar not only received these men hospitably, but, knowing they would not willingly live by other nations' customs, sent them to Gophna and asked them to remain there for the present. He told them that when the war was over he would restore all their possessions to them. So, they cheerfully went to that small city he had given them without fearing any more danger. But when they were found missing, the rebels put out the word that the Romans had executed them in order to deter the rest of the citizens from running away. Their subterfuge succeeded for a time, as did the rebels' previous chicanery. For a time, the fear of Roman execution

stopped all attempts at escape.

3. When Titus had recalled the men he had sent to Gophna, he ordered them to parade around the wall with Josephus, showing themselves to the people. Seeing the men, another crowd of citizens fled to the Romans. These escapees also gathered in great numbers and stood in front of the Romans, begging the rebels with loud groans and tears to allow the Romans into Jerusalem to save their homes. But if they could not agree to do this, they would at least get out of the temple and save the holy sanctuary for the citizens' use. The Romans would not set the temple on fire unless under duress. But the rebels yelled back their contradictions, and voiced loud and bitter insults against those who had deserted the city. The rebels also set their war engines on the temple gates, preparing to throw darts, javelins, and stones. They were placed at set distances from one another, so that the area around the courts resembled a cemetery, as so many corpses littered the floor. The temple looked like a fortress. Accordingly, the rebels rushed out upon the unapproachable holy places in their armor, while their neighbors' blood was still warm upon their hands. The rebels proceeded to show the same indignation and suffered the same offenses against the sanctuary as they had naturally shown against their Roman enemies. Their impiety regarding their own religious customs was as if the temple had abused them in the same way the Romans had abused the Jews. Every one of the Roman soldiers who saw the scene was horrified. They adored the Jews' holy house and hoped the rebels would repent, before there was no turning back.

4. Titus was deeply affected by the state of things, and reproached John and his henchmen saying, "Have not you vile wretches put up this wall before your sanctuary with our permission? And haven't you put up its pillars at equal distances engraved in Greek letters the prohibition that no alien or Gentile should go beyond this wall? Have we not allowed you to execute anyone that goes beyond it, even a Roman? So what are you doing now, you malicious criminals? Why do you trample over the dead bodies on the temple floor? Why do you pollute the holy sanctuary with the blood of both Jews and Gentiles? I appeal to the gods of my own country and to all gods who have held this place in high esteem. [For I suppose none of them now hold it in such high honor.] I also testify to my own army, to the Jews who are now standing with me, and even to you rebels,

that I am not forcing you to defile your holy sanctuary. If you, therefore, choose another battlefield, no Roman will come near your temple nor cause any offence against it. On the contrary, I will make every effort to preserve your holy house whether you will or not."

5. As Josephus translated these words from Caesar's mouth, both the rebels and the tyrant John thought Titus's exhortation to come from fear and not out of any goodwill toward them. So, the tyrants grew even more insolent. But when Titus realized the rebels were neither moved to have pity upon themselves, nor to have any concern over the temple's welfare, he unwillingly proceeded to press the war against them. He could not deploy his entire army, as the place was too small. Instead, he chose thirty of the bravest soldiers out of each hundred men. He gave 1,000 to each tribune and made Cerealis their commanding officer. Titus then ordered that they should attack the night watch at the ninth hour that very evening. As Titus put on his armor in preparation for joining them, his friends would not let him go, as the danger was too great. The commanders suggested that he would serve the cause better if he remained in the Antonia Tower, and would dispense rewards to those soldiers who proved exemplary in the battle, rather than join them in the fight and come into danger alongside them. On the other hand, the soldiers would fight boldly with Caesar watching their heroism from Antonia. Caesar complied with their advice, but for one reason only—that he might be able to judge the courage of their combat and that no action of any brave soldier might be hidden, and, therefore, go unrewarded. So he sent the soldiers forward at the ninth hour, while he went up to the Antonia Tower where he could look down upon the battle. Once there, Titus waited impatiently for the fighting to commence.

6. But when the soldiers began their assault, they did not find the temple guards asleep as they had hoped. Instead, they fought them in close combat right from the onset, as the rebels charged and fell upon them with a loud shout. As soon as the rest of the rebels in the temple heard the shout of the night watch, they ran out in force to meet the Romans. So, the Romans received the first wave of rebels, but those rebels following fell upon their own troops. They treated many of their own men as if they were enemies. The great tumult that arose from both sides was so loud and the night so dark that confusion reigned and the Romans became

indistinguishable from the Jews. Additionally, a natural blindness comes over those inflamed with both passion and fear so that one soldier looks like any other. The confusion did more damage to the Jews than to the Romans, because the latter held their shields high and were more disciplined in their movements than the Jews. Each Roman remembered his password, while the Jews were widely dispersed, attacking or retreating erratically. So, the Jews frequently seemed to be their own enemy. As the Jews retreated, running back toward the temple in the darkness, those in the rear thought them to be Romans and cut them off. More Jews were wounded by their own men than by the Romans until the dawn gave them the ability to discern friend from foe. Then the Jews stood in distinct battle formations, throwing their darts and defending themselves more appropriately. Neither side yielded nor grew weary.

The Romans tried to be braver and stronger than the other, both individuals and whole regiments, for they knew Titus was watching. Every soldier believed that if he fought bravely, this day would bring his promotion. Meanwhile, the Jews were encouraged to fight forcefully by their fear, both for themselves and the temple, and because John kept exhorting them on while beating and threatening others to fight more fearlessly. The fight, as it turned out, became stationary. One side would gain some ground then lose it again. Suddenly, there remained no room to move forward or backward because of the corpses. But still, an uproar of loud yells kept coming from Antonia, as those in the tower cried out for their men to press on with audacity. The shouts of encouragement kept coming when they forced the Jews backward or even when the troops had to retreat a bit.

It was like an entire theatre of war where every move was seen by Titus and those around him. At length, it appeared to be a draw, as no one army could claim victory. It had begun at the ninth hour of the night and ended at the fifth hour of daylight. Both armies were uncertain as to which had won. A great many Romans had distinguished themselves. On the Jewish side were Simon, Judas ben Merto, and Simon ben Josas. Of the Idumaens were James and Simon ben Cathlas, and James ben Sosas. John's men who stood out were John, Glyphtheus, and Alexas; and of the Zealots, Simon ben Jairus.

7. In the interim, the rest of the Roman army had opened some of Antonia's walls in the space of seven days and prepared a broad avenue to the

temple. So, the legions came near the temple's outer court and began to raise their siege works. One bank was up against the northwest corner of the inner temple. They raised another at the northern building between the two gates. Of the remaining two, one was located at the western cloister of the temple's outer court, and the other against the northern cloisters. The Romans accomplished the work with much travail and difficulty, particularly since they had to import their materials from as far away as almost thirteen miles. Other problems abounded. They always had to be prepared for the traps the Jews laid for them. The Jews were even bolder since they had now almost lost every hope of survival.

Some of the Roman cavalry had let their horses run free without bridles while they foraged for wood or hay. The Jews then sallied out of the city in large groups and seized the horses. When this thievery had been going on for a while, Caesar began to believe that the horses were being stolen more by his own men's negligence than by the enemy's bravery. He determined to use more severe measures to ensure that every horseman took care of his horse. He ordered the execution one of the men who lost his horse. This terrified the rest of the troops. Their horses became safe for a time, as they were no longer allowed to feed unguarded. But, as if they were physically attached to their backs, the men always rode horses wherever they went. So the Romans continued to wage war against the temple and to raise their banks against it.

8. One day had passed since the Romans ascended through the breach in the Antonia wall. Many of the rebels were so pressed by the famine and the failure of their other wild attacks they assembled together and assaulted the Roman guards on the Mount of Olives. They planned the attack at the eleventh hour of the morning, thinking the Romans would not expect it, for at that time the Romans would be eating lunch. They thought that with a surprise attack they would win easily. But the Romans had been informed of the assault. Suddenly running together from their neighboring encampments, they prevented the Jews from climbing over their fortifications or from penetrating the human wall that now surrounded the Jews. A sharp fight ensued with many heroic deeds performed on both sides. The Romans, of course, displayed both their courage and martial skills, while the Jews came upon them with wanton violence and extreme passion. One side was urged on by shame and the other by necessity. The Romans thought it would have been shameful to allow the Jews to escape

now that they had them in a kind of net. The Jews, on the other hand, had but one hope of saving themselves, and that was to push violently against the Romans surrounding them and break the encirclement.

One cavalryman, named Pedanius, saw the Jews already beaten and being forced down into the valley. He vigorously spurred his horse onto their flank and caught a certain young Jew by his ankle as he was trying to get away. Pedanius was strong, particularly in his hands and arms. Wearing his armor, and as he was galloping, he bent far down over his horse and seized the young man as if he were a precious treasure. Holding him upside down by his ankle, he took his captive to Caesar. Titus admired Pedanius's great strength and ordered the Jew executed for his attack against the Roman fortifications. Titus then left to oversee the temple siege and the raising of the Roman embankments.

9. Meanwhile, the Jews became very distressed by the war that now surrounded them. The Romans had advanced higher and higher, creeping up upon the temple itself. It was as if their infected arms and legs were being amputated to keep the gangrene from spreading further. They set fire to the northwest tower, which adjoined Antonia. After that, they knocked down about thirty feet of that cloister and began to burn the sanctuary. Two days later, on the twenty-fourth day of Panemus [Tamuz], the Romans set fire to the cloister adjoining that one and it burned another twenty-five feet. In the same way, the Jews cut off the cloister's roof and did not stop until Antonia was separated from the temple. This was done even when it was within their power to put the fire out. They did not move when the temple was first ablaze and thought the fire's spread was to their advantage. But the armies still fought one another around the temple, each side sallying out against the other.

10. At the same time, a man arose among the Jews named Jonathan. He was short in stature and appalling in appearance. He had nothing to recommend him, nor his family. Jonathan went out to the high priest John's monument and slandered the Romans, challenging their best fighter to personal combat. Many in the army who stood listening laughed out loud, but others were afraid of him, as they should have been. Some soldiers justly reasoned that it wasn't appropriate to find a man who wanted to die. Those who had lost all hope of survival had in addition to other passions a violence in attacking men that defied any opposition. Further-

more, such a one may have the Deity's pity. To expose oneself to danger with such a man would not be considered courage, but rashness. If you win, you haven't done much, and if you lose, there's a chance of winding up as a prisoner.

So, no Roman came out to accept Jonathan's challenge. The Jew continued to insult them as cowards, for he was an arrogant man who hated the Romans. Then, one of the Roman cavalrymen, named Pudens, enraged by the Jew's words, insolence, and arrogance—and perhaps because Jonathan was so small in stature, ran out to fight him. He would have beaten the Jew soundly, but as he was running, Pudens was betrayed by bad luck. He fell down. As he was lying there, Jonathan ran up to Pudens and cut his throat. Then, standing on his corpse and brandishing his bloody sword while shaking a shield with his left hand, he screamed at the Romans while gloating over Pudens's corpse. While he laughed at the Roman, one of the centurions named Priscus shot an arrow at Jonathan that pierced him through while he was leaping and cavorting like a fool. Shouts went out from both Jews and Romans although for different reasons. Jonathan twisted in anguish and fell down dead on Pudens's body. This is just one instance of how suddenly fortunes can turn on one who achieves an unlikely victory.

Chapter 3

Concerning a Jewish Strategy by which they Burned many Romans; and Another Description of the Famine in Jerusalem

1. The rebels within the temple continually tried to beat back the Roman soldiers who mounted their embankments. Then, on the 27th of Panemus, the Jews came up with another strategy. They filled a portion of the western cloister between the rafters and the roof with dry materials, and bitumen and pitch. They then withdrew from the cloister, as if wearied from their wounds. Seeing their retreat, several of the more aggressive Romans got carried away by their furies and ran after them. Placing their ladders on the cloisters, they quickly went and stood on top. Other, more prudent Romans, stayed where they were when they saw the unaccountable Jewish retirement. However, as the cloister was packed with the Romans who had climbed the ladders, the Jews set the cloister on fire. Flames suddenly burst forth. The Romans who were out of danger were suddenly in great fear, while those in the fire were in great anguish. When

they saw themselves surrounded by flames some of the Romans hurled themselves off the temple court, down into the city. Others fell among the Jews in the temple. Some leapt back down among their own men and suffered broken legs and arms. But even more who would have leapt off the cloister were prevented from doing so by the raging fire. Others killed themselves with their own swords, preventing the fire from destroying them. Meanwhile, the fire quickly spread and surrounded many who would have perished by some other means.

Caesar, meanwhile, could only pity those who had perished in the flames, even though they had gone up the ladders without awaiting orders. There was no way to come to their rescue. Knowing that Titus grieved would have been a comfort to those who died, for it was for his sake they perished. The Roman commander shouted to them, leaping up and exhorting those around him to do something to help. So every last man died cheerfully, carrying with him Titus's words and commitment as a sacred tribute. Some of the Romans had escaped to the cloister's wide wall and avoided the fire, but then were surrounded by Jews. They resisted for a long time but were punctured with wounds and ultimately fell dead.

2. One young man stood out among those who died. His name was Longus. He became a symbol of the sad event. While all who died were worthy of commendation, this man deserved applause more than all the rest. The Jews admired Longus's courage and wanted him dead. So, they persuaded him to jump down from the burning cloister to them, giving him security for his life. But Longus's brother, Cornelius, persuaded him not to do this, as he would only tarnish his own glory and that of the Roman army, as well. Longus complied with his brother's request, and then lifting up his sword in full view of both armies, killed himself.

One Roman soldier surrounded by the fire escaped using his head. His name was Artorius. He shouted to his friend Lucius, a tent mate, saying, "I will leave you all my possessions if you will come and save me!" Hearing him, Lucius came running over to catch his friend. Artorius then jumped down into Lucius's arms and saved his own life. But Lucius struck the stone pavement so violently he died immediately. This sad incident depressed the Romans for a while, but it made them more cautious in the future of the Jewish trickeries. For the Romans were at a disadvantage not knowing the city architecture or the nature of its inhabitants. Now this cloister was destroyed as far as John's tower, which he had built over the

Xystus gates during the war with Simon. After they had eliminated the Romans who had burned after they climbed up to the porticoes, the Jews also tore down the rest of the cloister where it joined the temple. But the next day, the Romans burned down the entire northern cloister that adjoined the Cedron Valley and extended out over it at a very great height. So, this was the state of the temple at the time.

3. The number of those dying from starvation within the city grew enormously. The people's misery was unspeakable. If even as much as a shadow of something to eat appeared, a battle over it immediately began. The closest of friends fought each other over food, snatching from each other's mouths any visible means of life. The rebels would not believe those who were dying of famine. They searched them even in their death throes, lest anyone had hidden some morsel in his clothing, and now was merely pretending to die. The robbers were bent over with hunger, running about staggering like mad dogs and reeling against house doors like drunken men. They were so swollen with hunger they would sometimes enter the same house two or three times in the same day. Their hunger became so intolerable that they chewed anything, even that which most wild animals would not touch. They tried to eat anything, even belts and shoes. The leather in their shields was pulled off and gnawed. Little wisps of straw became food for some. Some gathered up grass and sold small amounts of it for four Attic drachmas. But why do I tell about the shameless impertinence of famine that caused people to eat all sorts of inanimate objects? I'm going to tell of much worse, the likes of which no historical documents tell, either of the Greeks or Barbarians. It is horrible to speak of it and unbelievable when hearing of it. I was going to omit this awful calamity completely, as I didn't want to disturb posterity with what might become a portent of things to come. But since I have so many witnesses to it in my own time, my country would have little reason to thank me for suppressing the sufferings that went on at this time.

4. A certain woman named Mary lived beyond the Jordan River. Her father was Eleazar of the village called Bethezob, which means, "the House of Hyssop." She was prominent both for her heritage and her wealth. She had fled to Jerusalem with so many others and was now besieged with no means of escape. The possessions she brought with her from Perea had already been stolen. Other things she had been able to treasure up as well

as the food she had saved, also had been carried off by the ravenous guards who came running to her house daily to see what more they could steal. All this had put this poor woman into a great passion, and by her frequent reprimands and maledictions against at the rebels, she had incited their anger against her. None of them, however, either from their anger against her or out of pity for her, would take her life. If she did find food, she thought she had searched for others and not for herself.

But it was now impossible to find more food, and the pangs of starvation pierced through her bowels and her bones. When her hunger had swollen beyond mere hunger, and in her great need had lost any soundness of mind, she did a most unnatural thing. She snatched up her young son, a mere infant sucking at her breast, and said, "O, you miserable infant! For whom in this war shall I allow you to live—the famine or the rebellion? If the Romans prevail against us, we will be their slaves. But the famine will destroy us even before we can become slaves. Yet are not these rebels more terrible than either the famine or the Romans? Come on now. Be my food for the rage of vengeance upon these rebellious varlets, and a proverb to the world, to tell them how deep the suffering of the Jews truly was." As soon as she had said this, Mary killed her infant son and then roasted him. She ate half of the boy and then hid the other half. The rebels soon arrived and smelling the scent of the food, threatened to kill her immediately if she did not show them where she hid it. She replied that she had saved a choice cut just for them. Then she uncovered the remains of her son. Seeing the half-eaten body, the rebels were horrified and amazed, standing astonished at the site. She said to them, "This is my own son, and what you see is my own doing. Come, eat of this food! For I have eaten of it myself. Do not pretend to be more tenderhearted than a woman or more compassionate than a mother. But if you are too scrupulous and think my sacrifice to be an abomination, as I have already eaten half, let me eat the other half as well." The men left the house trembling. Never had they been so frightened by anything such as this. With some difficulty, they left the rest of the meat to the mother. News of the cannibalism spread quickly throughout the city. Hearing it, people shook with fear, as if they had done this unspeakable thing themselves. So, those in great distress from starvation hoped to die, as those already dead were happy because they had not lived long enough to either see or hear of such dreadful tragedies.

5. The word of this sad occurrence soon reached the Romans. Some would not believe it while others' pity grew for the Jews' distress. But for many Romans, their hatred of our nation grew more bitter than usual. As for Caesar, he excused himself before God in the matter, saying that he had proposed peace and freedom for the Jews, as well as pardon for all their previous audacious offenses. But instead of agreeing with him, the Jews had chosen rebellion. Instead of peace they had chosen war. Instead of plenty and pleasure, the Jews chose famine. They had started the fire to burn down the temple, the temple that he had so carefully preserved. They, therefore, deserved to eat this kind of horrifying food. The atrocious act of eating one's own child ought to be buried under the rubble of this country itself. No one should leave such a city like this standing upon the earth, where mothers are fed in such a way. Such food would be more appropriate for fathers than mothers, since it is the fathers who continue to wage war against us, even after they have undergone such misery. At the same time Titus said this, he reflected upon the desperate condition of his enemy. He could never expect men to be able to think properly after undergoing such suffering. Only by avoiding such catastrophe might they have repented.

Chapter 4

When the Banks Were Finished and the Battering Rams Powerless, Titus Gave Orders to Set Fire to the Temple Gates. Shortly Thereafter the Holy Temple Itself Was Burned Down, even against Titus's Wishes

1. The two Roman legions completed the work on their banks on the eighth day of Lous [Ab]. Titus then gave the order for the battering rams to be brought forward and set up against the western wall of the inner temple. Before the Romans brought these forward, the largest and most powerful of all the other engines had struck the wall continually for six days without making the slightest impression on it. The gigantic stones and the tight joints between them were no match for these engines or to the other battering rams. Other Romans had succeeded in undermining the foundations of the northern gate, and after much hard labor had removed the outermost stones. And yet the gate still endured on the inner stones.

The gate stood unharmed until the workmen, losing all hope

in their engines and crowbars, brought their ladders up to the porticos. When doing this, the Jews stood back and did not interrupt the work. But when they brought up the ladders, they fell upon the workmen and fought them. Some of the Jews threw the Roman soldiers down from the heights, casting them headlong from the cloisters. Others they met and killed. The Jews also beat many who tried to escape down the ladders, killing them with their swords before they could grab their shields for protection. Some of the ladders, full of armed men, toppled over backward. But many of the Jews were also slaughtered during the melee, as those who bore the Roman ensigns fought hard to keep them, realizing their loss would bring shame. At length, the Jews did capture the Roman engines and killed all the soldiers that had gone up the ladders. The rest of the soldiers, intimidated by the suffering of those who died, withdrew. None of the Romans died without having fought bravely. Of the rebels, those had fought bravely in previous actions did the same in this one, as did one Eleazar, the tyrant Simon's nephew. When Titus realized that his endeavors to preserve a foreign temple were now working to injure and kill his soldiers, he ordered to set the gates on fire.

2. Meanwhile, two rebels deserted to Titus, hoping for forgiveness. One was an Ananus from Emmaus, the bloodiest of all Simon's guards, and the other was named Archelaus ben Magadatus. Their hope stemmed from leaving the rebels while the Jews still had the upper hand. Titus saw the desertions as one more of the Jews' scheming ruses. He was informed of the barbarity these two men had displayed against their fellow countrymen, and was in a hurry to have them executed. He told them he believed that they were driven to desert because of their suffering and didn't come out of the city due to any natural inclination. He said that no one who had set their own city on fire deserved to live. It was out of that very fire that the two had hurried to leave Jerusalem. Ultimately, however, Titus's promise of security for deserters won over his anger, so he dismissed the men accordingly, although he withheld the same privileges he had given to others.

The soldiers had now set fire to the gates. Their silver casing quickly transferred the fire to the wood underneath. It spread rapidly, and soon all the porticos were ablaze. When the Jews saw fire surrounding them their spirits sunk, as did their strength. They were frozen in astonishment. They could neither try to defend themselves nor put out the fire, but only stood gazing at the blaze. But the Jews were not grieving because of the loss of what was burning. As though the temple itself was on fire, their hatred of the Romans only increased. The fire burned all that day and all the next, for the soldiers were not able to burn all the cloisters at once, but only little by little.

3. The next day, Titus ordered that the fire be quenched and the road for the legions constructed, in order to make the march into the temple and city easier. He then met with his commanders. Among the assembled officers were six principal men: Tiberius Alexander, the second in command of the entire army; Sextus Cerealis, commander of the fifth legion; Larcius Lepidus, commander of the tenth Legion; and Titus Frigius, commander of the fifteenth legion. Also among them was Externius, the leader of the two Alexandrian legions, and Marcus Antonius Julianus, procurator of Judea. After these six came the rest of the procurators and tribunes. Titus asked their advice about what should be done regarding the holy temple. Some thought to demolish it would be in accordance with the rules of warfare, because the Jews would never quit their rebellion while the house still stood. Others were of the opinion that should the Jews vacate the temple, and no arms remained in it, Titus might want to save it. But in case the Jews continued their resistance from its courts, he should burn it, because then it should not be considered a holy sanctuary but a fortress. The sinfulness of seeing it burn would then rest on those who forced it done and not on the Romans. But Titus responded, "If the Jews continue to occupy that holy house and keep fighting us from there, we ought not to avenge ourselves upon an inanimate object instead of the men themselves." He was not in favor of burning down so vast a building anyway, since its destruction would require so much effort on the Romans' part. Also, preserved, the temple's presence would accrue glory to Rome as an adornment to their sensible authority. Then Frontus, Alexander, and Cerealis boldly announced their agreement with Titus. The assembly then dissolved after Titus had ordered the commanders to have most of their forces lie still, but that he would use the best and bravest men among them to mount his attack. So, he commanded that men should be chosen out of the cohorts to make their way through the ruins and extinguish the fires.

4. The Jews were so weary on that day and under such great bewilderment

that they ceased their attacks. But the following day they assembled all their forces and boldly assaulted the Romans who were guarding the temple's outer courts. They came through the east gate at about the second hour of the morning. The guards received the attack bravely. They covered themselves with their shields like a wall and drew close to one another. But it was evident that they could not hold out for long but would soon be overrun by the violence, as a great number of Jews sallied out upon them. But Titus was looking on from the Antonia Tower and saw the guards' squadron was in trouble, so he sent some select horsemen in support. The Jews could not sustain their attack as some in the forefront of the battle were slaughtered, and most of the rest were forced to flee. But as the Romans were leaving the battle scene, the Jews turned back to attack once more. But the Romans did an about-face and sent the Jews running again. This continued for about three hours, until the Jews finally had enough and withdrew to the temple's inner court.

5. So Titus retired into the Antonia Tower, resolving to storm the temple the following morning with his entire army. He pitched the Roman camp around the holy house. God had doomed the temple to be burned long ago, and now that fateful day was at hand – a day that the ages had determined. For it was on the tenth of Lous [Ab] that the Babylonian king (Nebuchadnezzar) burned the temple (some six hundred years before). The current flames, however, were lit by the Jews themselves. When Titus retired to the tower, the Jews lay still for a while but then attacked the Romans again when the guards extinguished the fire burning in the temple's inner courtyard.

But the Romans again put the Jews to flight, and then proceeded into the holy house itself. It was then that one of the soldiers set fire to a golden window without hearing any orders to do so, or without any fear and trembling at undertaking such an action. As if he was hurried along by a supernatural fury, he snatched some burning wood out of the Jewish fire, was boosted up by another soldier, and set it ablaze. Through that window, one could enter a passageway that led to all the rooms on the north side of the temple building. As the flames shot skyward, the Jews let out a great shout as such a horror required, and ran to stop the fire—caring less for their lives and not allowing anything to stop them as their holy temple was burning. It was for the temple's survival that they had guarded it so carefully.

6. A certain soldier came running to Titus and told him of the fire as the Roman commander was resting in his tent following the previous battle. Titus immediately jumped up and ran to the temple to try to stop the fire. His officers and the several legions followed him, all greatly astonished. A great uproar and furor arose among them, as is only natural on the disorderly advance of so many troops. Then Caesar called loudly to the soldiers fighting the Jews, and gave them a signal with his right hand to quench the fire. But they could not hear his shouts over the uproar, nor did they pay attention to his hand signals. Some were distracted by fighting the Jews, while fury deafened and blinded the others. The legions pressed quickly forward; no persuasions or threats could restrain their violence. Each man's vehemence completely controlled him. As they crowded into the inner courts, many of them were trampled by others coming from behind, while many also fell among the cloister's ruins that were still hot and smoky. They were destroyed in the same miserable fashion as those whom they overcame. Also, the soldiers who had come near the temple not only disobeyed Caesar's orders, they acted contrary to them, encouraging those in front to set it all on fire. The rebels, meanwhile, were in too great a despair to help put the fire out. The bodies of rebels and the corpses of citizens lay everywhere. If caught, the people had their throats cut in their unarmed weakness. Dead bodies were stacked one upon another near the altar, and the steps leading up to it were covered in blood so thick that the cadavers near the top slipped down to the floor below.

7. Caesar, soon realizing that he could not restrain the enthusiastic fury of his men, and with fire consuming more and more of the temple, went into the holy place and saw the sacred rooms and their furnishings. He found them to be far superior to the reports he had heard from other Gentiles and not inferior to what the Jews believed and boasted about the rooms. As the flames had not yet reached the temple's inner rooms but were still consuming its outer rooms, Titus thought it still possible to save the temple. So, he hurriedly tried to persuade the soldiers to extinguish the fire. He ordered the centurion Liberalius and one of the spearmen with him to restrain those who would not listen – to beat them with their poles. But the soldiers' passions outweighed the respect for and fear of Caesar. Their hatred of the Jews and their vehement desire to slaughter them added to their refusal. Additionally, the hope of plunder pressed

many soldiers onward, seeing money all around them, and everything made of gold. When Titus had to run out to restrain his men, one soldier brought fire upon the hinges of the inner gate, and sudden flames burst out within the holy place itself. With this, Titus and his officers left the scene and no longer ordered anyone to extinguish the fire. And so, even against Caesar's wishes, the sacred temple burned to the ground.

8. Since the temple was the most marvelous building the world had ever seen, anyone would justly mourn its destruction. It was enormous and beautiful, as well as extremely costly, and had a worldwide reputation for its sacred uses. Yet we might take comfort at this though—that fate certainly decreed its destruction in the same way as fate determines the future of all living creatures, their works and their dwellings. One cannot but marvel at the specificity of fate's timing. Herod's temple was destroyed on the same month and day that the former temple was destroyed by the Babylonians. King Solomon laid the first foundations 1,130 years, seven months, and fifteen days before its destruction in the second year of Vespasian's reign. The second temple, constructed by Haggai in the second year of King Cyrus, lasted until its obliteration under Vespasian – a period of six hundred thirty nine years and forty-five days.

Chapter 5

The Incineration of the Temple Causes Great Distress Among the Jews. Also Concerning a False Prophet, and the Signs that Preceded the Temple's Destruction

1. While the holy house burned, everything that could be stolen was stolen. Those caught pilfering were executed, without consideration of their age or position. Children, old men, normal citizens, and even priests died in the same way. Executions affected all sorts of people, even those who begged for clemency and those who tried to save themselves by fighting back. The temple flames were seen from a great distance and echoed the groaning of the people being killed. Because the temple structure was so large and sat on a high hill, one would have thought the whole city to be ablaze. And no one could imagine anything more horrible than the noise, for the Roman legions shouted as they marched together, while the rebels, surrounded by fire and sword, howled desolately. The people left in the temple's vicinity faced their enemy with shrieks as they lay dying. Those

in the city below echoed their moans. When many of those nearly dead from starvation saw the temple flames, even though their mouths were almost permanently closed, they used all of their strength to groan and wail. Perea returned an echo, as did the mountains surrounding the city, swelling the force of the noise. Thus was the misery more terrible than the chaos. One would have thought that the hill on which the temple stood was white-hot, as every inch of it was ablaze. But Jewish blood loomed even larger than the fire, and those killed more in number than their executioners. Dead bodies so completely covered the ground it became invisible. The soldiers ran on the corpses' backs as they chased those who tried to get away.

By this time, the Romans had forced out the last of the rebels from the temple's inner court. They had much difficulty getting from there to the outer court and then into the city, as the rest of the people attempted to rush into the outer court. Some of the priests ripped the spikes from the temple roof and with their heavy lead bases shot them at the Romans like darts. But the priests didn't gain a thing, as the fire broke forth upon them, forcing them to retreat to the broad wall, where they found a temporary respite. Two of the most prominent among them, who might have saved themselves by deserting to the Romans—or who might have had more courage and suffered with the others—threw themselves into the flames, and were burned along with their holy sanctuary. Their names were Meirus ben Belgas and Joseph ben Daleus.

2. When the Romans realized that it was useless to try to save all the structures surrounding the holy house, they burned all of those constructions, including the remaining cloisters and gates, except for two—those on the east and south side. But afterward they burned those also. They also torched the treasury chambers, in which a vast amount of money resided, as well as an immense number of priestly garments and other precious goods. In a few words, it was the entire wealth of the Jewish nation all stacked in one place, along with furniture the rich citizens had stored there from their decimated homes. The soldiers then went to the outer temple court where a large mixed crowd of about 6,000 Jews had fled, including women and children. The soldiers who came upon the multitude were in such a rage, that before Caesar had determined the fate of these people or their commanders had issued any orders regarding them, they set the area on fire. Some Jews tried to escape the flames by throwing

themselves from the heights, but most died in the fires. Not one of these people escaped alive.

It so happened that a false prophet was responsible for these citizens' deaths. He had that day made a public proclamation that God had commanded the people to go up to the temple, and there receive miraculous signs of their deliverance. A great many of these false prophets had been hired by the rebels to tell the people to wait on God's deliverance. Their lies were a ruse to keep the people from deserting the city, buoyed up by such hope. A person in adverse circumstances will naturally comply with such false promises, for being promised deliverance from his oppression he will patiently wait in hope of rescue.

3. And so, these men deluded the wretched people of Jerusalem, claiming to hear messages from God. People would not believe the signs that stated so clearly the city's eventual devastation. But like men in love, with blinded eyes and befuddled minds, they would not see the desolation that God was bringing upon them. One of these signs was a star resembling a sword that hung over Jerusalem, and a comet that remained visible for an entire year. Even before the Jews rebelled and before all the confrontational incidents that preceded the war, these signs were in evidence. When multitudes gathered in Jerusalem for the Feast of Unleavened Bread on the eighth of Xanthicus [Nisan], at about nine at night a great light shone around the altar at the holy house, illuminating the temple as if it were broad daylight. The sign lasted for half an hour. The sight seemed good to the laity, but it was accurately interpreted by the sacred scribes to portend those gloomy events which soon followed. At that same festival, as a heifer was being led to be sacrificed by the high priest, she gave birth to a lamb in the middle of the temple court.

Further, the temple's eastern gate to the inner court was made of very heavy brass. It took twenty men to close the gate, and then not without difficulty. The gate rested on an iron base that had bolts fastened deeply into the floor, which was comprised of one gigantic stone. The gate was seen opening of its own accord at about the sixth hour of the night. Those that watched over the temple went running to their captain and told him about it. He then went to the gate and without any trouble was able to shut it by himself. This also appeared to the unlearned to be a good sign as if God himself was opening the gate of happiness. But the wise and learned men among them understood that their holy sanctuary's security was now removed and that the gate had been opened to allow entry to their enemies. So, they declared publicly that the sign foreshadowed the desolation soon to come upon them.

Besides all of these signs, a few days after the Festival of Unleavened Bread on the 21st day of Artemisius [Jyar], an incredible and immense phenomenon occurred. I suppose one might judge its telling to be a fable had not there been so many eyewitnesses to it, and if the events that followed were not so significant as to deserve such a sign. Just before sunset on that day, chariots and troops of soldiers in their armor were seen running about in the clouds surrounding the city. Moreover, at the festival we call Pentecost, as the priests were customarily going by night into the temple's inner court to perform their sacred duties, they felt a quaking and heard a loud noise. Following that, they heard a sound as of a great multitude saying, "Let us leave this place."

But even more terrible were the premonitions of Jesus ben Ananias. Jesus was a peasant farmer who for years before the war, at a time when Jerusalem enjoyed great peace and prosperity, came to the Feast of Tabernacles and began to shout, "A voice from the east, a voice from the west, a voice from the four winds! A voice against Jerusalem and the holy house, a voice against the bridegrooms and the brides, and the voice against this entire people!" This is what he cried out as he walked the city lanes both day and night. But, some of the more prominent citizens, who were greatly indignant at this dire message, grabbed the man and beat him severely. But Jesus did not try to defend himself nor say anything against those who hit him. He merely went on crying out the same words as before. Then our rulers began to think that this was a supernatural fury in the man [as proved to be the case]. They took him to the Roman procurator, who whipped him until his bones were exposed. Yet he still didn't ask for pity nor did he shed a tear. But, lowering his voice to the most mournful tenor possible, Jesus shouted, "Woe, woe to Jerusalem!" at every stroke of the whip. When the procurator, Albinus, asked him who he was, where he came from, and why he uttered such words, Jesus made no reply. He just kept uttering the same refrain. Finally, Albinus figured Jesus was a lunatic and dismissed him. Then, during the span of time that passed before the war began, the man did not speak to any of the citizens, nor was he seen anywhere while not saying these words. But every day he uttered the same sad words, "Woe, woe to Jerusalem." He never said a bad word to those who frequently beat him, nor did he speak any kind words to those who gave him food. This sad refrain was his reply to everyone. Indeed, the people heard no other prophecy of Jerusalem's fate but this unhappy one. He cried out the loudest at the festivals, and he continued to cry out for seven years and five months without ever growing hoarse or getting weary, until the moment he saw his prophecy fulfilled when the siege ended. For, when he was circling the wall he cried out with all the energy his frail body could muster, "Woe, woe to the city again! And woe to the people and to the holy house!" And just as he had finished, he added, "Woe, woe to me also!" At that moment, a rock flew out of one of the Roman engines and killed him instantly. As he was uttering the last of his dire prophecies, Jesus gave up the ghost.

4. If anyone will but think on these things, he will discover that God takes care of mankind and helps us to understand what is best for our protection. But mankind perishes by the misery that we voluntarily and insanely bring upon ourselves. For the Jews, in demolishing the Antonia Tower, had made their temple four-square. At the same time, it had been written in their sacred oracles that, "Whenever their temple becomes four-square, then would their city be taken as well as their holy house." But, what had done more to push them into the war was an ambiguous prophecy also written in their holy scriptures, that, "About that time, one from their own country will become governor of the habitable earth." Many wise men were fooled in its interpretation as the Jews took the prediction too mean one of their own race. The prophecy clearly indicated Vespasian's rise to power in Rome, he who had been emperor in Judea. However, it is not possible for men to avoid Fate, even when seeing it in advance. But men interpret signs according to their own pleasure. Some prophecies were utterly despised until their accuracy was demonstrated, both by the overthrow of Jerusalem and their own destruction.

Chapter 6

How the Romans Carried their Ensigns to the Temple, and Made Joyous Acclamations to Titus. Also, the Speech Titus Made to the Jews when they Asked for Mercy. The Jewish Reply, and how their Reply Made Titus Indignant

1. When the rebels had fled into the city, and the holy temple and its surroundings had been burned to the ground, the Romans brought their

ensigns into the temple and set them against the eastern gate and began offering sacrifices to them. They acclaimed Titus as imperator with joyous exclamations. All of the soldiers held vast accumulations of the spoils of war that they plundered from the temple. In Syria, a pound of gold sold for half its former value. There was a boy who had hidden himself on top of the broad wall on which the priests had also found refuge. He was in great thirst. While confessing his thirst, he asked some of the Roman guards to give him their right hands of security for his life. The guards pitied him for his young age and his distress, and so gave him their right hands. So, the boy came down and drank some water, and also filled the vessel he had brought with him. Then he fled to join his friends. None of the guards could catch him, but shouted reproaches after the boy for his perfidiousness. The lad called back, "I have not broken the agreement. The security you gave me did not require me to stay with you but only that I come down safely and get some water. I got down and took the water, so I think I've been faithful to our agreement." Hearing this, the guards admired the boy's shrewdness, particularly given his young age.

On the fifth day, the famine finally brought the priests down from their hiding places and before Titus. There, they begged for their lives. But Titus replied that the time for a pardon had ended with the end of their holy house, on which account alone they might be spared, though now destroyed. It was right that they should now perish with the house to which they belonged. So, he ordered the priests executed.

2. The tyrants and their henchmen now found themselves surrounded on every side, without any hope of escape. They, therefore, wanted to meet personally with Titus. And so, as it was Titus's nature that he desired the city saved from destruction, he took his friends' advice—who believed that the tyrants had now become rational—, and sat down on the west side of the temple's outer court to meet the rebels. There were gates on that side above the Xystus where a bridge connected the temple to the upper city. Titus sat on one side of the bridge while the rebels set on the other. A great crowd gathered on each side. The Jews sat near Simon and John with great hopes of pardon, and the Romans gathered around Titus with great interest as to how Caesar might receive their requests. Titus ordered his troops to restrain their rage and drop their darts. He then appointed an interpreter to serve between the parties, which move signaled that he was the conqueror and the Jews the conquered.

Titus began the discussion saying, "I hope you gentlemen are satisfied with the miseries of your country. You have harbored no defensible opinions, either of our great power or your own weakness. But like madmen, in violence and selfishness, have taken such actions as to bring your people, your city, and your holy house to destruction. You men have not stopped your rebellious ways since Pompey first conquered you. Since that early time, you have not stopped making war on the Romans. Have you depended on your great numbers, when even a small cadre of Roman soldiers has been strong enough to quash you? Have you relied upon the fidelity of those who fought alongside you? What nations are there outside Roman rule that would choose to come to the Jews' assistance against Rome? Are your bodies stronger than ours? No. Of course, you realize that even the strong Germanic people are our servants. Have you stronger walls than we have? Answer me! What wall is greater than that of the ocean which surrounds the Britons, and yet they live peaceably under Roman care. Do you exceed us in your courage or the wisdom of your leaders? No. Indeed you obviously know that even the Carthaginians have been conquered by Rome. It, therefore, can be nothing but our kindness that has caused you to rise against us.

"In the first place, we have given you this land to possess. Next, we have set kings of your own nation over you. Third, we have honored the laws your forefathers handed down to you and have allowed you to live as you please, either by yourselves or among others. And what is the central favor of all we have given you? To gather up that tribute you pay to God and all the other gifts dedicated to him. We have not admonished those who collected these gifts nor did we stop them. In the end, you became wealthier than us, even when you had become our enemies. You prepared for war using our money! In fact, when you enjoyed all these advantages, you turned your vast riches against the very people that gave them to you. Like merciless serpents, you spewed out poison against those who treated you with kindness.

"Perhaps you might have despised Nero's slothfulness, and like a broken or dislocated arm or leg, you lay still waiting for another time, still harboring malicious intentions. But you have now shown your true character and allowed your ambitions to grow to the extent that your ample and impudent hopes would take them. At the time my father arrived in your country, he did not come to punish you for what you had done under Cestius, but to admonish you. Had he come to overthrow your nation he could have immediately gone to your fountainhead, Jerusalem, and laid waste to the city. But he chose to ravage Galilee, and its surroundings instead to give you time to repent. But you saw his humanity as a sign of weakness and quickened your impudence on the milk of our compassion. When Nero died, you acted as the most wretched of wicked men would've acted by encouraging each other to rise against us in civil dissensions, and used the time when my father and I had gone to Egypt to prepare for war. Neither were you ashamed to raise troubles against us when we were advanced to be emperors. All this time you had experienced how mild we had been as mere generals of the army. But when the Roman government was placed upon our shoulders, and all the rest of the world's people were at peace, and foreign nations sent ambassadors to congratulate us upon our ascension to power, you Jews still showed yourselves to be our enemies. You sent emissaries to the countries beyond the Euphrates to come and help you foster disruptions. You built new walls around your city. Seditions arose with one tyrant pitted against another. A civil war broke out in your midst that only served to illustrate what wicked people you truly are.

"I then came to Jerusalem, unwilling to come but sent by my father, and received depressing orders from him. But I rejoiced when I heard the people here wanted peace. I exhorted you to quit your warring factions before I began the siege. I even would have spared you after you had fought against me for a long time. I gave my right hand in security to deserters. What I promised I did. I had pity on those who fled to me, and on those we captured. I didn't torture those who wanted war except to restrain them. I unwillingly brought my war engines against your walls. I always stopped my soldiers from treating you severely, even when they were anxious to slaughter you. After every victory, I sought peace as though I had been conquered. When I approached your temple, I set aside the rules of warfare and begged you to spare your own sanctuary, to preserve the holy house for yourselves. I allowed you safe passage out of the temple and gave you security for your lives. I even said that if you had a mind to do so, I would allow you to move the battle to another location.

"Yet you have despised every one of my proposals and set your holy house on fire with your own hands. And now, you vile wretches, do you expect me to negotiate with you using mere words? For what purpose would you have saved such a holy house that is now destroyed? What can I now preserve for you since your holy temple has now been leveled? Yet you still stand in your armor! You can't even pretend to be supplicants, even in this, your ultimate angst. O, you miserable creatures! On what do you depend? Are not your co-conspirators dead? Is not your holy house gone? Are your very lives not in my hands? Do you still esteem death as part of fearlessness? However, I will not imitate your insanity. If you throw down your arms and surrender yourselves to me, I will grant you your lives. I will act like the kind father of a family. What I cannot heal, I will punish. The rest I will do with as I see fit."

3. The rebels then made this reply to Titus's offer: that they could not accept it because they had sworn never to do so. But they wanted to be allowed to go through the wall thrown up around them with their wives and children. They would then go into the desert and leave the city to him. Titus was more than indignant. Here they were, men who had already been captured, trying to make their own terms of surrender as if they were the victors. So he ordered this proclamation be made to them: that no others should come out to him as deserters or hope for any further security. He would henceforth spare no one but fight them all with his entire force. They should therefore save themselves as best they could, for from that point on he would treat them all according to the rules of war. So, Titus gave orders to the Army to plunder and burn the city.

The Romans did nothing that day, but on the following day they set fire to the Jewish archives, to Acra, to the council-house, and to the place called Ophlas. The fire proceeded as far as Queen Helena's palace in the center of Acra. The streets and houses packed with corpses from the famine also burned to cinders.

4. On that same day, the sons and relatives of King Izates assembled with many other prominent men of the city and asked Caesar to give them his right hand of security. Hearing their pleas, Titus received the men with his earlier moderation even though he was still angry with all those remaining in the city. At the time, he kept them all in custody but bound the king's sons and relatives and took them with him to Rome as hostages, due to their country's allegiance to Rome.

Chapter 7

What Happened to the Seditious After the Roman Victory when they Had Caused so much Trouble; and how Caesar Became Vic-

torious Over the Upper City

1. The rebels now all rushed to their stronghold in the royal palace, where many of them had stored their personal belongings, and drove the Romans away. They also killed all the citizens that had crowded into the palace, 8,400 of them, and stole what they had on them. They took two Romans alive: one a cavalrymen and the other an infantryman. They cut the infantryman's throat and immediately had him drawn through the entire city as if to avenge themselves on the entire Roman army by this one event. But the horseman claimed that he had some suggestions as to how they all might save themselves. He was, therefore, brought before Simon, but having nothing to say he was delivered for punishment to Ardalas, one of Simon's commanders. Ardalas tied the man's arms behind him and put a blindfold over his eyes, then paraded the horsemen out before the Romans as if to cut his head off. But while the Jewish executioner was drawing his sword, the man jumped up and ran away to the Romans. When he had escaped, Titus could not think of executing the horseman. But, because he was no longer deemed worthy of being a Roman soldier because the enemy had captured him alive, Titus took away his weapons and armor and discharged him from his legion. The punishment and the sense of shame it carried was far worse than death itself.

2. The following day, the Romans drove the robbers out of the lower city and set it on fire as far as Siloam. The soldiers were happy to see the city's destruction. But they got no plunder, as the rebels had already taken all that remained and had withdrawn to the upper city. They also remained unrepentant for the damage they had done. On the contrary, they were insolent as if they had done well. The rebels were cheerful as they watched the city burned, and smiled from ear to ear. They saw the burning as an omen that would soon bring them death and put an end to their misery. And so, as the people were now all dead, and the holy house burnt down, and the city itself ablaze, nothing was left for the enemy to do. Yet even to the end, Josephus did not grow weary of begging the rebels to spare what remained of Jerusalem. He mainly spoke to them of their sinfulness and barbarity and gave them advice as to how to escape. All he got in return were sneers and laughter. They could not imagine surrendering because of the oath they had taken, nor were they strong enough to continue an equal fight against the Romans. They were already surrounded on all

sides, and were like prisoners. Yet, they had grown so accustomed to killing people they could not do otherwise.

So, the rebels dispersed around the city and laid ambushes amidst the ruins to catch any citizens who might attempt to desert to the Romans. They were successful in this, and slew many trying to run away. Their prey was too weak from starvation to flee from them. So, the weak citizens' dead bodies were thrown to the dogs. Every kind of death was believed better than death by starvation, so although the Jews had now lost all hope of mercy, they tried to flee to the Romans, and even of their own accord fell victim to the rebels as well. No place in the city was without its corpses. The corpses of those who had died by either famine or in the rebellion littered the ground. Everywhere one looked, the dead bodies from famine and sword covered the pavement as far as the eye could see.

3. The only hope the tyrants and the crews of robbers with them lay in the caves and caverns under the city. They did not expect the Romans to search there, if they could only take refuge underground. Once the city had been ultimately demolished and the Romans had left, they would then emerge and escape. This proved no better than wishful thinking, for in reality they could hide from neither God nor the Romans. Nevertheless, they depended on the subterranean refuges, and set fire to more buildings than did the Romans. Those running out of their blazing homes into the ditches were killed without mercy and robbed of any valuables. If they discovered anyone with food, the rebels swallowed it down, along with the blood. They began to fight one another over their plunder. I cannot think but had not their destruction prevented it; the rebels would have even tasted the corpses themselves.

Chapter 8

How Caesar Raised Banks Around the Upper City [Mt. Zion]. When Completed, He Ordered the War Engines to Be Brought, and Soon Possessed the Entire City

1. On the 20th of Lous, when Caesar realized that the upper city could not be overwhelmed without raising embankments against it, he divided the work among the several legions of his army. Carrying materials to the site was difficult work, since all the trees within thirteen miles of Jerusalem had their branches already shorn off to build the former banks.

The works belonging to the four legions were begun on the city's west side, against the royal palace. Meanwhile, the auxiliary troops and the multitude with them erected their banks on the Xystus. They reached the bridge and the tower that Simon had built for himself against John when they were at war with one another.

2. The commanders of the Idumeans got together privately at this time and spoke about surrendering to the Romans. They sent five men to Titus asking for his right-hand of security. So, thinking the tyrants would yield if the Idumeans—who had played a large part in the war—were taken out, Titus, after some thought and delay, agreed to spare their lives and sent the five men back. But as the Idumeans were preparing to march out, Simon caught wind of their treason and quickly killed the five men who had met with Titus. He then threw their officers into prison, the most prominent among them being Jacob ben Sosas. The rest of the Idumeans were at a loss as to what to do without their commanders. Simon had them guarded and secured their walls by numerous of his men. But his garrison could not stop the deserters, for although many died, many more Idumeans escaped. The Romans received them all, as Titus neglected to fulfill his earlier orders to execute them, and because the ordinary Roman soldiers grew tired of all the executions. The soldiers also hoped to extort money from the Idumeans.

So, only the people of Jerusalem remained in the city. The women and children were sold for a piddling sum, since buyers were few. Even though Titus had proclaimed that no single deserter would be welcomed alone, that they should bring all their families with them, even single persons were also allowed to come out. However, he set men over them who were able to distinguish friend from foe, to determine if anyone deserved punishment. The number of people sold into slavery was enormous, but over 40,000 of the populace were spared. These Titus allowed to go wherever they pleased.

3. At this time, one of the priests, Jesus been Thebuthus, had Caesar's oath of security given him on the condition that he deliver up certain precious articles that had been taken from the temple. Jesus took him two candlesticks like those that had resided in the holy sanctuary, as well as tables, containers, and vials, all very heavy and made of solid gold. He also gave Titus the veils, and the garments with the precious stones,

and a great number of other precious vessels that accompanied Jewish sacred worship. The man who had been treasurer of the temple, Phineas, was also captured. He showed Titus the priestly coats and belts, along with a large quantity of purple and scarlet cloth for use in the veil, and a huge amount of cinnamon and cassias and other sweet spices which were formerly mixed together and offered each day as incense to God. Many other treasures were also delivered to Titus each day, along with many of the temple's sacred ornaments. When all these objects had been brought to Titus, Jesus ben Thebuthus received the same pardon as those who had deserted privately.

4. The Roman embankments against the upper city were finished on the seventh of Gorpieus [Elul], within the space of eighteen days. The Romans then brought their engines against the wall. Meanwhile, some of the rebels, losing hope of defending the wall, left it and withdrew to their stronghold—their citadel. Others went down into the subterranean vaults. But many others fought those Romans who attempted to bring their battering rams up to the wall, yet were soon overwhelmed by Roman strength and numbers. The Romans prevailed in the main by going about their work cheerfully, while the Jews were dejected and weak.

As soon as the Romans battered down a breach in the wall and a few of its towers yielded to the hammering rams, the defenders scattered, and terror fell upon the tyrants, probably to a greater extent than the occasion required. Even before the Roman troops got through the breach, the tyrants were stunned and immediately fled. To see these men who had been so arrogant and brazen in the past to now be trembling in fear would make one's heart pity the change in their demeanor. They began to run violently toward the wall that surrounded them to break through those guarding it in an attempt to get away. But when they realized that the men who had been faithful to them in the past had fled in their great distress, and also when they learned that the entire western wall had been overrun and the first wave of Romans were entering the upper city with others close behind, they fell on their faces and loudly lamented their own insane behavior. Their nerves were so shattered they couldn't even run away.

Here one may appropriately reflect on the power that God exercised on these wicked wretches, and on Rome's good fortune. For, the tyrants completely deprived themselves of the security they had achieved and came running from their strongholds of their own accord. Otherwise, they could have never been taken by force—or by any other means, other than famine. And so when the Romans had taken such pains to weaken the walls, they got what they could have never gotten by their engines. For three of these towers were stronger than any battering ram in existence, a subject we have formerly discussed.

5. So, the Romans bypassed the towers, their occupants having been ejected by God himself. After recovering from the trance they were under, the rebel leaders and their men fled immediately to the valley below Siloam. They ran violently against the Roman wall that existed on that side, but their courage was so depleted they could not attack with sufficient power. Their clout, broken by fear and affliction; they were consequently repulsed by the guards. Dispersing themselves in every direction, the rebels made their way into the subterranean caverns. The Romans now held all the walls and had placed their engines on the towers, joyfully cheering their victory. Winding down the war was a much easier task than beginning it. When they had overrun the last wall without bloodshed they could scarcely believe what they found to be true. But, seeing no one opposing them, they wondered what this unusual isolation could mean.

But then, going in companies into the city streets with drawn swords, the Romans killed those they found and set fire to the houses into which the Jews had fled, incinerating every person in them. They destroyed a great many homes. Entering each to find valuables, they found entire families of the dead, the upper rooms filled with corpses of those who had died from starvation. The troops were horrified at the sights and left without touching anything. But, although they had pity on those who had died in this way, they didn't feel the same for those who still lived. Every person they found they ran through, filling the lanes with corpses. Indeed, the entire city ran with blood. The fires begun in some homes even seemed to be quenched with the blood these men shed. As the killers went back to their camps in the evening, the fires blazed more vehemently that night in Jerusalem, on this eighth day of Gropius [Elul]. The city that had enjoyed such happiness since its founding and had been the envy of the world, now saw such joy overcome by the terrible misery of the siege. Jerusalem deserved such awful misfortunes solely because it had produced a generation of men who were the authors of its downfall.

Chapter 9

What Injunctions Caesar Gave to the Captives on His Arrival in Jerusalem. The Number of Captives and Dead in the Siege. And also Regarding those who Escaped into the Caverns, among whom Were the Tyrants John and Simon.

1. When Titus arrived in the upper city, he admired its strong fortresses, in particular those towers that the tyrants had relinquished in their madness. When Titus saw their height, their gigantic stones, their joints' precision, and their length and breadth, he expressed himself this way: "God has certainly been our Ally in this war. It was none other than His hand that expelled the Jews from these fortifications. For what men or machines would be up to the task of overthrowing these towers?" He had many such conversations with his friends. Then he let those who had been bound by the tyrants and those imprisoned go free. To conclude, when Titus had completely demolished the rest of the city and destroyed all its walls, he left these three towers as a witness to his good fortune. Good fortune had proven to be his ally and had enabled him to take that which could never have been taken without it.

2. Since his soldiers had grown weary of the executions and since the vast number of people still remained alive, Caesar ordered no one killed except those who carried arms. The rest were to be taken alive. But in addition to those they were ordered to kill, the soldiers also slew the aged and the infirm. But those who were young and who might be useful, they drove into the temple and imprisoned them within the court of women. Caesar sent one of his freed-men and one of his friends, Fronto, to oversee them. Fronto was to determine every captive's fate according to his or her usefulness. He executed all those whom others had accused of being robbers or rebels. But of the young men, Fronto selected the tallest and most handsome and set them aside for the triumphal march into Rome. All the rest over seventeen years of age he put in chains and sent to the Egyptian mines. Titus had also sent many as presents into the provinces to be destroyed in the games by the sword and wild beasts. But those under seventeen were sold into slavery. While Fronto was making these decisions, 11,000 of those in his care starved to death, due to the hatred the guards had for the Jews. The others would not eat even the food pre-

pared for them. There were so many prisoners, not enough corn existed for their survival.

3. The total number of those taken captive during the war was estimated to be 97,000. The number who died during the entire siege was believed to be 1.1 million. Most of the dead were from the same country but were not citizens of Jerusalem. For most Jews had come into the city for the Feast of Unleavened Bread and had been suddenly shut inside by the Roman army. At first, the overcrowded conditions brought on disease, but soon after the famine hit and ended many of their lives. The fact that the city could contain so many people was established by the census that Cestius had taken earlier. Cestius just wanted to let Nero know how formidable the city was, as Nero had always been contemptuous of the nation. So, he asked the high priests to count the people. When the festival call Passover arrived—when the priests slay their sacrifices from the ninth to the eleventh hour, and a group of no less than ten and as many as twenty Jews attend each sacrifice [for it is unlawful for individuals to feast alone]—they counted the number of sacrifices to be 256,500. Calculating no more than ten Jews accompanying each sacrifice placed the number of persons made pure and holy at 2,700,200. This number did not count those with leprosy or gonorrhea, or women in their menses, or anyone otherwise polluted. It is not lawful to have them partake of the sacrifices, nor any Gentiles who had come to Jerusalem to worship.

4. The vast multitude of worshipers came out of the remote country, so the entire nation had been shut up by fate, as if in prison. As the people crowded into Jerusalem, the Roman army surrounded the city. Accordingly, the number of people that perished in Jerusalem had exceeded all other destructions that either man or God had wrought upon the known world. Of course, we only speak of what we know publicly. The Romans killed some, and some of them they carried off as captives, and some of them they killed underground and left there. Over 2,000 persons were found dead in the catacombs, partly by their own hands and partly by someone else's. Most died due to the famine. The stench produced by the dead bodies was most offensive to those discovering them, inasmuch as they were obliged to get away as fast as they could. Others were so greedy, they would go walking on stacks of corpses, for much treasure was to be found in these caverns, and the lust for wealth made any way of getting it

to be thought lawful. Many of those imprisoned by the tyrants were now brought out. The tyrants did not cease their barbarous cruelty, even to the very end. But God avenged Himself on them both, according to his perfect justice. As John and his men lay in the caverns, they had nothing to eat and begged the Romans to give them the right hands of security, which John had so proudly rejected earlier. Until forced to surrender, Simon struggled mightily in his distress. We will take up his story later. Ultimately, he was reserved for the triumphal march into Rome and then executed. John was condemned to a life term of imprisonment. At this juncture, the Romans set fire to all the rest of the city and completely tore down the remaining walls.

Chapter 10

A Brief Account of Jerusalem's History: How it had Been Taken Five Times Before, and Destroyed Twice

1. And so Jerusalem was taken in the second year of Vespasian's reign on the eighth of Gropius [Elul]. The city was overrun five times before in its history, though this was only the second instance of its destruction. Sishak, the king of Egypt, was first, and after him came Antiochus Epiphanies. After him was Pompey, and then Socius and Herod took the city but preserved it. But before all these conquerors came Nebuchadnezzar, the king of Babylon, who not only conquered Jerusalem, but left it desolate, 1,468 years after its founding. The man who first built the city was a Canaanite. In Hebrew, he is called Melchizedek, the righteous king. For he was a righteous man. He was the first of God's priests and the first to build a temple. He called his city Jerusalem. Salem was the city's original name.

However, The Jewish king David expelled the Canaanites and settled his own people in the city. It was demolished by the Babylonians 477 years and six months after David. From King David, the first Jew to reign there, to this destruction under Titus, 1,179 years passed But from its earlier inception until this last destruction; 2,177 years had elapsed. Nothing prevented it from being destroyed, not it's great antiquity, nor it's vast wealth, nor the dispersal of its people over all the known world, nor the greatness of the religious veneration being paid to it. Nothing was sufficient to keep it from its ultimate destruction. And so ended the siege of Jerusalem.

The Jewish Wars
Book VII

Containing the Interval of about Three Years.
From the Taking of Jerusalem by Titus, to the Sedition of the Jews at Cyrene

Chapter 1

How the Entire City of Jerusalem was Destroyed except Three Towers; and how Titus Commended His Soldiers in a Speech, and Distributed Rewards to them, then Dismissed many of them

1. As soon as the army could find no more people to either rob or kill, Caesar gave orders that they would now destroy the entire city. [No one remained as the object of their fury, nor would any have been spared had not Titus ordered more work to be done.] The army was to save the most eminent of the towers—Phasaelus, Hippicus, and Miriamne—and that part of the wall that enclosed the city on the west side. The wall and the towers were spared in order to provide an area for the garrisons' camp, and to show posterity what kind of a city and how well fortified Jerusalem once was. This would prove that Roman courage had not conquered an easy target. But the rest of the walls were demolished, down to the ground level. Nothing was left of the city to cause those who came afterward to think that it was ever inhabited. And so Jerusalem came to its end, prompted by the insanity of those who wanted to rebel against Rome. Jerusalem had been the most magnificent city known throughout the inhabited earth.

2. Caesar decided to leave a garrison in the city, comprised of the tenth legion with certain cavalry troops and companies of infantry. So, having successfully brought the war to a conclusion, Titus now wanted to com-

mend his entire army, because of the great achievements they had accomplished. He wanted to hand out appropriate rewards to those who had distinguished themselves in action. Titus, therefore, had a large tribunal constructed in the center of his former camp and stood on it, surrounded by his commanders. Speaking in order to be heard by the whole army, Titus said that he thanked them profusely for the goodwill they had shown to him, and commended them for the active obedience they had shown over the course of the war. Their obedience was exhibited in the great and many dangers they had so courageously faced. He also remarked how the army's courage had augmented Rome's power. They had made it evident to all men that neither the size of any enemy army, nor the strength of enemy fortresses, nor the large size of enemy cities, nor the rash boldness nor brute rage of their antagonists could ever match Roman courage, although some enemies may have Fortune on their side, at one time or another. He further said that it was now reasonable to conclude the war that had lasted so long. Victory and peace are what they had all hoped for when the war began.

But even of greater glory for them was that the men whom they had elected to be their empirical governors and administrators had been sent off to Rome for that purpose. They were welcomed there and were being treated with unanimous approval. Accordingly, Titus expressed his admiration and affection for every man, because he realized that they all had gone cheerfully about their work as their capabilities and opportunities allowed. But, he continued, he would immediately bestow rewards and honors upon those who proved the bravest and strongest and had distinguished themselves in the most glorious fashion. These had made his army more famous because of their noble heroism, and no man who had given more of himself should not gain just payment for it. Titus said he had been very careful with the rewards. He had given special attention to these honors because he would much rather reward the virtues of his fellow soldiers than to punish those who had committed offenses.

3. At that point, Titus ordered the officers to begin reading the list of all who had performed heroically in the war. Each recipient was called out by name to meet Titus, and commended before those present. Titus rejoiced in the same way a man might have rejoiced in his own heroic actions. He put golden crowns on their heads and gold chains around their necks, and gave them the long golden spears and silver ensigns. He then promoted

every hero to a higher rank. Besides all of this, Titus distributed among them copious amounts of gold and silver, and garments that the army had taken as spoils of war.

So, when the men had all these honors bestowed upon them as each had earned, and Titus had wished the entire army every happiness, he came down from the tribunal amid loud shouts and applause, and proceeded to offer thank offerings to the gods. He sacrificed a vast number of oxen that stood ready at the altars, and distributed the meat to the army for their feast. When he had stayed three days feasting with his principal officers, Titus sent his army to the diverse places where they would be stationed. He permitted the tenth legion to remain at Jerusalem as a guard, and did not send them beyond the Euphrates River, where they had been before. As Titus remembered the twelfth legion, which under Cestius had fallen to the Jews, he removed them completely out of Syria [for they had previously been at Raphanea], and sent them to a place called Meletine near the Euphrates—between Armenia and Cappadocia. He also thought it best to have two legions stay with him until he went to Egypt. Titus then led his army to Caesarea on the sea, and there stored the vast quantity of spoils. He ordered that the captives be held there, as the winter season now hindered him from sailing to Italy.

Chapter 2

Titus Put on all sorts of Shows at Caesarea Philippi. Concerning how Simon the Tyrant was Captured and Reserved for the Triumphal Entry into Rome

1. While Titus Caesar was besieging Jerusalem, Vespasian took passage on a merchant ship and sailed from Alexandria to Rhodes. He sailed on ships bearing three rows of oars. As he landed at several cities along the way, the citizens all joyfully received him, so he crossed over from Ionia to Greece. From there, he sailed from Corcyra to the Iapyx isthmus, where he continued his journey overland.

But now, Titus marched from Caesarea, that was by the sea, to the city named Caesarea Philippi, and remained there quite a while, putting on shows and spectacles. A large number of captives met death in the circuses. Some were thrown to wild beasts, while many others were forced to kill each other. While he was in Caesarea Philippi, Titus was informed of the capture of Simon ben Giora. It happened like this: Simon was in the upper city during Jerusalem's siege. But when the Roman army got within the walls, and were demolishing the city, Simon took his most faithful friends and went underground. Among those he took were stonecutters, who took along their iron occupational tools. Simon also took a large supply of provisions, that were to last them a long time. Simon led them all down into a certain cavern, the entrance of which wasn't visible from above ground.

Where they had dug previously, and the earth was softened, things went well. But when they met solid earth, and began to dig a tunnel that they hoped would provide them with a safe place to rise above ground and escape, the going soon got difficult, and they found their hopes dashed. They made only little progress, and that with great adversity. Their provisions, which had been distributed carefully, began to run out. So Simon, thinking he would be able to surprise and hence escape the Romans, put on a white robe and buttoned a purple cloak over it. He arose out of the ground at the exact location where the temple formerly stood. Those who saw him were astonished at first sight, and did not move. But as they approached him they asked him who he was. Simon wouldn't say, but asked them to call their captain, which they did. The man left in charge of the Jerusalem garrison, a man named to Terentius Rufus, went over to Simon and heard the whole truth. Simon was placed in chains while Rufus advised Titus of the robber's capture.

So, God brought this man to be punished who had exercised such savage tyranny against his own countrymen. Those who had been his worst enemies would now chastise him. The Romans did not seize Simon violently. He delivered himself up voluntarily for punishment. Ironically, he had made false accusations against many Jews on that same charge. He had barbarously executed them, as if they were trying to desert to the Romans. Wicked actions do not escape Divine anger, nor is Justice too weak to punish offenders—but will in time overtake all who transgress its laws. Justice inflicts her punishments in a way that seems even more severe, because the culprits don't expect it after so long a time has elapsed since their crime. Simon realized this after falling under Roman anger. His arising from underground led to the arrest of many other rebels who had hidden there. Then Simon was taken in chains to Caesar, who ordered that he be kept alive until the army's triumphant entry into Rome, which he was soon to celebrate.

Chapter 3

Titus has many of the Captive Jews Executed in Celebration of His Father's and Brother's Birthdays. Also Concerning the Danger the Jews were in at Antioch Because of the Transgression and Impiety of a Jew named Antiochus

1. While Titus was in Caesarea, he celebrated his brother Domitian's birthday in high style. He had earmarked much of the Jews' punishment to be in his brother's honor. Over 2,500 Jews were slaughtered fighting wild beasts, by fire, and through fighting one another. Even though the Jews suffered in ten thousand different ways, the Romans believed that their punishments to be far less than they deserved. Following his stay in Caesarea, Caesar went to the Roman colony Berytus in Phoenicia and stayed there a considerable time. In Berytus, he exceeded his other celebrations to honor his father's birthday. Magnificent, and at great expense, a great many captive Jews were executed in the same way.

2. After this same time, the Jews in Antioch were under pressure for their lives from the many Gentile accusations against them. Many false allegations had reached ears in Antioch, aggravated by certain incidents that had recently taken place. I will describe these incidents briefly, so as to make my coming narrative more understandable.

3. The Jewish nation has been widely distributed among all the earth's peoples. Jews are particularly present in Syria, due to its proximity to Judea. The largest Jewish populations were in Antioch, which is a very large city and a place where the Syrian kings following Antiochus had given the Jews safe refuge. Although Antiochus Epiphanes had leveled Jerusalem and desecrated its temple, those succeeding him had restored the brass articles of worship to Jewish synagogues and given the Jews equal citizenship with the Greeks. As each succeeding king treated the Jews kindly, they grew in number and magnificently adorned their synagogues with many gifts. Many Greeks were being continually converted to Judaism, then assimilated into Jewish congregations.

But about the time the Jewish war began and Vespasian had arrived in Syria, a great hatred of the Jews arose among the people. It was then that a certain Jew arose named Antiochus. He was greatly respected on account of his father, who had been the governor of the entire Jewish population in Antioch. This Antiochus arrived at an assembly of citizens at a certain theater, and proceeded to accuse both his father and others of plotting to burn the city down. Antiochus also turned over to the crowd certain foreign Jews, who he said were partners in the plot. When the people heard this, they could not restrain their anger and vowed to burn those who had been brought before them, which they immediately did. The people then turned violently upon the local Jewish population, thinking they would save their city by punishing the Jews. Antiochus continued to aggravate the citizens' rage by demonstrating his own hatred of Jewish laws. He sacrificed to pagan deities. Antiochus also persuaded the Greeks to compel the rest of the Jews to do the same, and in that way determine those who had plotted against the city by their unwillingness to commit apostasy. So, the people of Antioch tested his theory. A few Jews complied, but they slaughtered those who refused. Antiochus then acquired some soldiers from the local Roman commander and became a strict despot over his own people. He prohibited any observance of the Sabbath day by forcing them to work as they did on other days. He caused such massive distress among them that the Sabbath was not only suspended in Antioch, but soon in other cities as well.

4. Following these Jewish difficulties in Antioch, a second calamity came upon them, which was linked to what I have just related. It seems that the four-square marketplace burned down. The royal palaces were also destroyed, as well as the archives where the public records in Antioch were held. [If they had not quenched the fire, with great difficulty, it would've consumed the entire city.] Antiochus quickly accused the Jews of arson. In their confusion, the citizens of Antioch immediately believed the fabrication. They probably would have believed Antiochus, even if they had not previously been enraged against the Jews because of his former slanders. The citizens violently attacked the accused like enraged madmen, as if they had actually seen the Jews in the act of setting fire to the city. Finally, and with great difficulty, Cneius Collegas, the legate, persuaded the rioters to set the matter before Caesar. [Vespasian had sent Syrian president Cneius Petus on a mission and he had not yet returned.] But when Collegas had gathered the facts, he found not one of the Jews who had been accused by Antiochus guilty. Some enterprising debtors had started the fire; they figured that if they could burn the public records their debts

would also disappear. So, the Jews were under great confusion and fear, due to the ultimate outcome of these false accusations against them.

Chapter 4

How Vespasian was Joyfully Received in Rome; also how the Germans Revolted against the Romans but were Subdued. The Samaritans Overran Mysia, but were Compelled to Return to their own Country Again

1. Titus Caesar received the news joyfully that his father Vespasian had been jubilantly received with great eagerness and splendor by all Italy, particularly Rome. Vespasian, now free from his immediate responsibility of command, had rejoiced with pleasure upon his enthusiastic reception. All Italy had anticipated his arrival, excited by its prospect as if he was already there, as they so greatly desired to see him. They held Vespasian freely and unrestrainedly in the highest regard. The Roman senate, remembering the troubles undergone by the recent changes of emperors, were thrilled to receive an emperor adorned with the wisdom of old age and the utmost ability in prosecuting warfare. They believed that Vespasian's promotion would accrue to their abundant benefit. Additionally, the citizens had been so harassed by the civil strife, they were all the more earnest that Vespasian should come immediately to deliver them from their misfortunes. They believed he could recover peace and bring prosperity back to the empire. The soldiers also held Vespasian in high regard, and knew of his great exploits in war. Other commanders had lacked both skill and courage, causing the army great shame. They wanted a leader who would give them security, and one of whom they could be proud.

The support for Vespasian was universal. The people of means couldn't wait for his arrival in Rome, but traveled great distances to meet him. Indeed, no one could rest until they saw him, but poured out of the city en masse to get a glimpse of their new emperor. They all thought it was easier and better to go out to see him rather than to remain in the city. Indeed, the city was almost devoid of its citizens, for those remaining behind were fewer than those who hurried out. As soon as the news came that he was close to the city and that he had received all who came to him graciously, everyone in Rome—with wives and children in tow, went out to the roadway and waited for him there. As Vespasian passed by, the crowd applauded wildly in their joy to see him and at how pleasantly he smiled down upon them. They called him their Benefactor and Savior, the only person worthy to be emperor in Rome.

  Meanwhile, the city was decorated like a temple, full of flowers and sweet spices. It wasn't easy for Vespasian to reach his royal palace, due to the press of the adoring crowds. On his arrival, he at last offered sacrifices to his household gods for his safe return to Rome. The crowds then took to feasting—celebrating in their tribes, families, and neighborhoods, and asking God to allow a dynasty of Vespasians—all his sons and their posterity, to rule Rome for a long time. They prayed that his reign would be preserved from any opposition. This tribute was how Rome joyfully received Vespasian, and immediately began to thrive.

2. Prior to Vespasian's arrival, while he was still in Alexandria and Titus was besieging Jerusalem, a multitude of Germans arose in rebellion against Rome. The Gauls living nearby joined them in their conspiracy, in hopes of casting off Roman rule. The German motivations were these: first, the nature of the German people was to go against reason. They were ready to throw themselves rashly into danger even with little hope of success. Second, they bore great hatred against the Romans. Their nation had never been subjugated by any other nation nor forced to be under submission. In addition to these motives, the opportunity now arose which was the strongest motive of all. They believed the Roman government to be in great confusion from the continual changes in its emperors. They also noted the unsettled nature of the entire empire. It seemed to totter on the verge of destruction. So, the Germans thought the time was right to rise, as Rome was on its heels.

  Classicus and Vitellius, two German commanders, puffed the Germans up with such hopes. They had wanted such a rebellion for a long time, and were propelled by the current situation to make their sentiments known. The populace was also ready, so when they heard their commanders' intentions, they gladly received the news. So, as the majority of Germans decided to rebel, and the remainder did not object, the matter was decided. Vespasian, unaware of the rebellion but guided by Divine Providence, sent a letter to Petilius Cerealis, who formerly had governed Germany. The letter ordered Cerealis to assume the governance of Britain with full consular authority. Cerealis had heard news of the German revolt while was on his way to Britain. He then had come upon the Germans, just as they were uniting. Putting his army in full battle array, he attacked, killing a great number of Germans and forcing the rest to retire from their insanity and to return home.

Had Cerealis not attacked at that moment, it still would not have been long before the uprising was smothered. As soon as Rome heard of the revolt, and Caesar Domitian became aware of it, he did not hesitate to act, even at his young age. Domitian had his courageous father's mind and was wiser than his youthful age would admit. He immediately marched against the barbarians. Hearing rumors of the Roman approach, the Germans' hearts failed them, and they fearfully submitted to Domitian, considering it a good thing that they were brought back under the Roman yoke without further suffering. When the mission had restored order to Gaul, and ensured that affairs there would remain so, Domitian returned with honor and glory to Rome. He had performed exploits beyond what his age would normally permit, and was found worthy of his great father.

3. Concurrent with the German revolt, The Scythians also rose against Rome. The Scythians, also called Sarmatians, are very numerous. They marched over the Danube into Mysia without being detected. They then proceeded to attack the Roman frontier guard, and in a surprise and violent assault, slew a great many of them. The Roman consular legate, Fonteius Agrippa, was killed in the battle after fighting courageously against them. The Scythians then overran Agrippa's entire region, ripping apart all that stood in their way. But, when Vespasian was informed of what had happened and how Mysia was laid waste, he sent Rubrius Gallus to punish the Sarmatians. Many Scythians perished in the battles with Gallus, while those who escaped returned to their own country. So, when Gallus had ended the war, he increased the security of the region by placing many more garrisons there. It was soon impossible for the Barbarians to cross the Danube again into Mysia. The Mysian war was over almost before it began.

Chapter 5

A Description of the Sabbatic River that Titus saw as he was Journeying through Syria; and how the People of Syria Came to him with a Petition against the Jews, which he Rejected. Also concerning Titus's and Vespasian's Triumphs

1. Titus Caesar remained at Berytus for a time, as we said earlier. He then left there and brought some magnificent spectacles through all the Syrian cities he visited. He publicly exhibited his Jewish captives as those that had brought destruction on their nation. As he journeyed, Titus saw a river that is of such a nature that it warrants recording for posterity. It runs between Arcea, in Agrippa's kingdom, and Raphanea. It is peculiar in as much as when it flows, its water-filled current is strong. After that, its springs fail for six straight days leaving the channel dry. Then on the seventh day, it runs as it did before, seemingly undergoing no change. It has been observed to keep this exact pattern perpetually. For this reason, they call it Sabbatic River, the name of course derived from the sacred seventh day of Judaism.

2. When the people of Antioch heard that Titus was approaching, their joy caused them to leave the city and go out to meet him at a distance of almost four miles. The crowd was not only men, but included women and children as well. When they saw Titus drawing near, they stood on both sides of the road, stretched out their right arms in salute, and loudly proclaimed their joy in seeing him. They then fell behind and marched with his procession. All the way back, the multitude asked Titus to eject the Jews from Antioch. Titus bore their petitions in silence and did not yield to them. However, the Jews feared terribly under the uncertainty of his opinion and what he might do to them. But Titus did not stay in Antioch, but proceeded immediately to Zeugma on the Euphrates. There, messengers came from Vologeses, king of Parthia, bringing Titus a gold crown for his victory over the Jews. Titus accepted the crown and feasted with the king's messengers, and then returned to Antioch.

The town senate and Antioch's citizens begged Titus to join them in their theater, where a large multitude awaited him. He didn't join the throng, but when they earnestly and continually pressed him to eject the Jews out of the city, he gave them this pertinent answer: "How can this be done, since their country where they would have to go lies in ruins, and no other place will allow them to come in?" Hearing his denial, the people of Antioch made two requests. They asked Titus to remove the brass plaques on which the Jews' privileges were engraved. Titus would not grant this request either, but allowed the Jews of Antioch to enjoy the same privileges they previously enjoyed.

Titus then left for Egypt, and on his march went by Jerusalem. He compared its ruins with the ancient glory of the city. He could not but pity the great city, once so arrayed in splendor and now totally destroyed. He was far from boasting about taking the once great city by force. No, he frequently cursed those that had started the revolt and had been responsible for the city's punishment. Titus never seemed to want to enhance his own reputation by annihilating the rebels. Yet a large quantity of the city's riches still lay under its ruins, much of which the Romans dug out. But those who were captives found most of the wealth, and so they carried it all away. The riches were not only silver and gold, but also the Jews' precious furnishings and that which the owners had hidden underground against the uncertain fortunes of war.

3. Titus then resumed his march to Egypt and quickly passed over the desert, reaching Alexandria, where he prepared to go to Rome by sea. As he was accompanied by two legions, he sent each of them back to the place from which they had come. He sent the fifth legion to Mysia and the fifteenth to Pannonia. He gave orders that the Jewish leaders, Simon and John, along with seven hundred Jews that he had selected as the tallest and strongest, to be transported to Italy to take part in his triumphant march into Rome. So, following a successful voyage to his homeland, the people of Rome received Titus much as they had his father, meeting him at a distance. The finest moment in Titus's eyes was the appearance and reception of his father. But, the crowd of Romans received their greatest joy when they saw the three Caesars standing together.

Only a few days went by before it was decided that they should all share the glory of triumph, because of the glorious exploits each had achieved. [The Senate had previously decreed each Caesar a separate triumph.] So, when notice was given of the appointed day when the victory celebrations should take place, every Roman left the city and went out as far as necessary to find a place to stand and watch the procession. Only a narrow passageway was left for those who would be seen going through the procession.

4. Now, all the military units had already marched by night out of the city, in parade order, under their several commanders. They formed near the gates, not those of the upper palaces, but those near the Temple of Isis. It was there that the emperors had rested the previous night. As soon as the sun rose, Vespasian and Titus appeared wearing laurel crowns and

clothed in royal purple—proper for their family, and went as far as Octavian Walks. Waiting for them there were members of the senate and high officials and those of the equestrian order. A tribunal had been constructed in front of the cloisters with ivory chairs set upon it. The Caesars came and sat down.

The soldiers immediately began a loud and joyful ovation, giving testimony to the valor of both father and son. The latter were clothed in silk garments, unarmed, and crowned with laurel. Vespasian received their shouts, but seeing that they were disposed to continue incessantly, he signaled the troops to be silent. When everyone had become quiet, he stood up, covered his head with his cloak and began to pray. Titus also prayed, and then Vespasian made a short speech to all the people and then sent the soldiers off to a breakfast the emperors had had prepared for them. Then Vespasian and Titus left for the Pomp Gate, through which all triumphal processions passed. There, they enjoyed some food, put on their celebratory garments and offered sacrifices to the gods. They then moved forward, marching triumphantly through the theaters so they might be easily seen by the multitudes.

5. It is impossible to describe these incredible spectacles, as they deserve. Their magnificence was such that one could not think of them as being the work of mere men, nor of wealth, nor yet of nature's rarities. Every enjoyable sight afforded to men in their lifetimes was all heaped up one upon another. Everything admirable and costly brought together that day demonstrated the vast Roman authority. A massive quantity of silver, gold, and ivory, shaped and fashioned into all sorts of things, did not seem like it was carried along the procession, but flowed out like a river. The rarest purple draperies were carried along, some displaying scenes from life embroidered by Babylonian artists. Transparent precious stones passed by in the procession. Some were set in golden crowns, and the some in other modes such as pleased their craftsman. These came forth in such abundance it was impossible to think that any of them were rare!

Images of the gods were also conveyed along, being both large and beautifully crafted of costly materials by skilled artisans. Many species of animals also came forth, all characteristically ornamented. The great multitude of men who paraded before us was clothed in purple interwoven with gold. Each man so chosen also displayed magnificent and extraordinary ornamentation. Additionally, the large crowd of captives did not lack adornment. Their garments, made of fine textured cloth in various colors, hid their bodily deformities and wounds from view.

But the greatest surprise was the structure of the banners that passed in the march. They were so huge that one worried that the men that bore them might fall under their weight. Many of them were three, even four stories tall! These magnificent banners brought both pleasure and surprise, for many boasted golden carpets. Wrought gold and ivory pieces bearing resemblance to the war were fastened all about them. There were scenes of a happy country reduced to rubble, entire companies of dead enemies. Other enemies were shown running away, while some were shown being carried away into captivity. Walls of great height and thickness were shown being ruined by war machines. Strong fortifications were shown being overrun. They showed a mountaintop city's walls being overthrown, and armies pouring into the city. Every place was shown full of slaughter and the enemy begging for mercy, too exhausted to raise their heads in opposition. Temples blazed with fire, and houses were overthrown and collapsed on their owners. Rivers flowed out of vast arid deserts and ran into land where it was used neither for cultivation nor as drink for men and cattle; the land was ablaze on every side. This was a picture of how the Jews viewed what they had undergone during the war. The workmanship exhibited by these representations was so magnificent and realistic that it seemed to have been created by those who were present in Israel. The top of each gigantic banner held the name of the governor of each city taken and the way in which the city was taken.

Following this pageantry came a large number of ships and other spoils of war, carried in great quantities. But the objects taken from Jerusalem's temple were the most significant. The very weighty Golden Table, and the Golden Candlestick passed by the spectators. The latter was constructed more like a trident. Its long shaft was attached to the base, with small branches that were elongated, each having a socket for a lamp at the top. The lamps were seven in number; representing the honor the Jews hold the number seven. Last carried of all the spoils was the Jewish Law. Following that, a great many men carried images of Victory, constructed out of ivory and gold. Then marched Vespasian in the first place, followed by Titus. Domitian also rode with them, making a glorious appearance, on a magnificent horse.

6. The last part of the triumphant procession was at the temple of Jupiter

Capitolinus, where it ended. There, everyone waited. It was the Roman custom to wait until someone brought the news that the enemy general was dead. The general in this case was Simon ben Gioras, who had been among the captives in the procession. A rope had been looped around his neck, and he was then pulled into his proper place on the forum, tormented as he went by those who pulled him along. Roman law required that evildoers condemned to die were to be executed on that particular spot. Accordingly, when the news came that Simon was dead, the multitude shouted for joy and began to offer sacrifices that had been consecrated, uttering the prayers used in such solemn moments. When all this had been completed, the Caesars went into the palace. The emperors entertained some of the spectators at their own feast, and the rest had made preparations for feasting in their homes. But this was a Roman day of Festival, celebrating the victory of their army over their enemies. The end of their civil strife had arrived, and their hopes for future prosperity and happiness had begun.

7. After the end of the triumphant parade, with Roman affairs set on a solid foundation, Vespasian resolved to build a temple dedicated to Peace. The temple was finished in a short time, and was so glorious in its design as to be beyond human expectation. For Vespasian now had, by the hand of Providence, a vast quantity of wealth in addition to that gained in his former exploits. He had this temple adorned with paintings and statuary, and had placed inside all the collections of rare items such as men might wander the known world to see. Here they were laid out one after another. He also placed there the golden vessels and instruments from the Jewish temple, as symbols of his glory. But, the Jewish Law and the purple veil of the holy place—those were kept in the royal palace.

Chapter 6

Concerning the City Called Machaerus, and how Lucilius Bassus took the Citadel and other Places

1. Meanwhile, Lucilius Bassus had been sent as legate into Judea, where he took command of the army from Cerealis Vitellianus. He then subdued the fortress of Herodium along with its garrison. Then he concentrated the Roman troops from their several locations with the tenth legion, and resolved to attack Machaerus. The destruction of the Machaerus fortress

was a high priority, since its presence might be an inducement to further rebellion. It was very strong, and by its nature able to assure the hopes of safety for anyone possessing it, at the same time bringing fear and time-consuming trouble for anyone attacking it. The fortress was walled and on a very rocky hill, which rose to a great height. This circumstance alone made its assault difficult. Machaerus was also protected naturally, inasmuch as its ascent could not be easily made surrounded as it was by deep valleys, the bottoms of which seemed almost beyond view. Not only could these valleys not be easily crossed, but would be impossible to fill up with earth. The valley that crosses on the west extends over nine miles and does not end until it comes to Lake Asphaltites. On that same side, the peak of the Machaerus hill was elevated above all the others. The valleys on the north and south were not as large as that on the west, yet like that valley, are extremely impractical to cross. The valley on the east extended downward for one hundred fifty feet. It extends as far as a mountain that faces Machaerus.

2. When the Jewish king Alexander (Jannaeus) initially saw the nature of this area he built a fortress there—the first to do so. Gabinius later destroyed that stronghold when he battled against Aristobulus. But when Herod became king, he thought the location to be worthy of his highest praise and therefore of being fortified to the utmost, particularly because of its location so near Arabia. For, it was very convenient that it faced that country. He therefore encompassed a large area with walls and towers and built a city there. From the city there was a path that led up to the citadel on top of the mountain. Furthermore, Herod built a wall around the top of a hill with two-hundred-forty-foot high towers at its corners. In the middle of the fortress, he built a magnificent palace with large and beautiful buildings. He also created many cisterns to collect rainwater that there might be plenty of it ready for all uses. And so, Herod added to the natural strength of the place in order to make it virtually impregnable by those fortifications made by men's hands. Additionally, Herod placed a large quantity of darts and other trappings of war into it, and worked to put anything in it that would contribute to its inhabitants' security, even under the longest possible siege.

3. Within Machaerus grew a type of rue that deserves to be mentioned due to its immense size. (Rue is an ornamental herb used in ancient times

for its medicinal qualities.) Normally a small plant, this rue plant was higher and thicker than any fig tree. It is reported to have lasted since Herod's time, and it would have lasted much longer had it not been cut it down later by the Jewish inhabitants. But in the valley that surrounds the city on its north side is a certain place called Baaras. It produces a root of the same name, the color of which is like fire. In the evening, it sends out a certain ray like lightning. Harvesting the root is not easy for it recedes from an intruder's hands. Nor will it be taken quietly. Only a woman's urine or her menstrual blood poured upon the root will subdue it. But even then, it is certain death to those who touch it, unless it is hung down from his hand and taken in that way. Another method also affords a way of taking the Baaras root without danger. One digs a trench around it until the hidden part of the root is very small. Then a dog is tied to the root. When the dog tries hard to follow his master, the root is plucked up easily, though the dog immediately dies—as if taking the place of the man who is the primary cause of the root's demise. After its uprooting, no one need be afraid of handling the Baaras root. Yet, even with all this trouble, the root has only one value: that if it is brought to a sick person, it drives away the demons that caused his illness. These demons are the spirits of the wicked that enter into living men and kill them unless they can find help.

Also here are hot water springs that flow, each having a different taste. Some of these springs are bitter, and others are plainly sweet. But also here are many eruptions of cold water that are close by each other, and not only in the valleys. Even more wonderful is a rather shallow nearby cave. A prominent rock covers the cave. Above this rock stand two small breast-like hills, a short distance from each other. One hillock sends out a stream of cold water and the other issues forth water that is scalding. If you mix the two waters together, they compose a very pleasant bath. They also have medicinal qualities for treating certain maladies, but are especially good for strengthening the nerves. This place also has sulfur and alum mines.

4. When Bassus had fully reconnoitered the stronghold, he decided to lay siege by filling up the valley that lay on the east side. So the Roman set his men to work, taking great pains to raise the banks quickly and so make assault easier. The Jews now separated themselves from the strangers there. The latter were forced into the lower regions of the city, where the principal danger lay, while the Jews held the upper citadel, the strength of

which might ensure their safety. The Jews also thought they might receive pardon from the Romans in the case that, ultimately, they may be forced to surrender the stronghold. But first, they sought to see if any hope existed for avoiding the siege altogether. Accordingly, they made sallies out each day and fought with the Romans. Many Jews lost their lives but took many Romans with them. Each side gained the upper hand when its opportunities were advantageous. The Jews gained victories when they caught the Romans off guard, while the Romans were victorious when they foresaw the Jewish sallies and were ready for them.

But the siege did not reach its conclusion as a result of these minor battles. On the contrary, a surprising accident relating to the siege ultimately forced the Jews to surrender the citadel. Eleazar, a certain bold and aggressive Jewish youth who had distinguished himself in the sallies, encouraged the Jews to go out in great numbers to hinder the Romans from raising their banks. They did much damage to the Romans in their attacks. Eleazar so directed their sallies that the Jews suffered few casualties. Eleazar himself served as rear guard.

One day, when a sally had ended and the Jews had retired safely, separated from the enemy, Eleazar stayed outside the city gates. Thinking that the battle had ended, and out of his contempt for Rome, he casually spoke with the Jews manning the wall, fully focused on his conversation. Just then, an Egyptian-born man in the Roman camp named Rufus ran toward Eleazar unexpectedly and seized him, carrying him off in his armor. Those on the wall were so shocked they couldn't move. So Rufus carried Eleazar off to the Romans.

Bassus ordered that Eleazar be stripped naked, stand before the city, and then be sorely whipped before their eyes. Seeing this tragedy, the people of the city were bewildered and cried out with one sorrowful voice. This mourning produced more of a result than one might expect from a single man's misfortune. When Bassus saw the citizens' reaction, he began to concoct a plan that would exacerbate their grief in order to get them to surrender the city in hopes of saving this one man. His hopes were soon realized. He commanded his men to set up a cross, as if he was going to hang Eleazar on it immediately. The sight of the Roman cross brought another expression of grief from those in the citadel—and they groaned intensely, crying out that they could not bear to see Eleazar crucified. Eleazar asked the citizens for help, knowing he was about to suffer a most miserable death. He exhorted the people to surrender to Roman power and fortune, and save themselves. The citizens were greatly moved by Eleazar's cries, and many were interceding for him, as he was from a large and eminent family. So, contrary to their custom, the Jews yielded to their sympathies. They sent out an envoy of messengers who met with the Romans, and surrendered the citadel to them on the condition that they may be permitted to leave and take Eleazar with them. Bassus and the Romans accepted their terms.

When the many strangers in the lower part of the city heard that the Jews had made this agreement for themselves only and not with them, they decided to flee the city secretly that night. But, no sooner had they opened the gates and prepared to run than the Jews told Bassus of their plan. Whether the Jews envied the strangers' deliverance or whether they feared being held accountable by the Romans for their departure is uncertain. But, the more courageous of the strangers got past the guards and fled for their lives, while the Romans slaughtered those men remaining in the city. The slain numbered 1,700, and their women and children were subjected to slavery. But Bassus, realizing he should honor the agreement he made with the Jews who surrendered the citadel, released them and restored Eleazar to their number.

5. When Bassus concluded the events at Herodium, he marched to the Jordan forest. He had heard that many of those who had fled from Jerusalem and Macherus had gathered there. When he arrived, Bassus understood that what he heard was no rumor. He first surrounded the entire area with his cavalry so that any Jews who wanted to escape could not. He then ordered his infantry to cut down the trees hiding the Jews in the forest. So, the Jews realized that the only way out was to engage the Romans, and by some miracle to escape. So the Jews attacked, bringing a loud shout from those surrounding them. The Romans fought bravely, while the Jews fought desperately. The Romans would not yield, so the battle went on. The Jewish expectation of escape was not to be. Only twelve Romans fell with a few more wounded, yet not one Jew escaped. Their dead numbered more than 3,000, along with their commanding general, Judas ben Jairus. We spoke of him earlier, as having been the captain of a certain Jerusalem band. When there, he had gone underground to make his escape in secret.

6. At the same time, Caesar had sent a letter to Bassus and to Liberius

Maximus, procurator of Judea. In it, he ordered that the entire country—including its cities, be given over to his ownership. He did exclude property in Emmaus, which he gave to eight hundred veterans as a place to make their homes. Emmaus is about seven and a half miles from Jerusalem. Caesar also levied a tax upon all Jews, requiring that everyone bring two drachmae annually into the capital. It was the same as the temple tax had been in Jerusalem. This was the Jewish state of affairs at the time.

Chapter 7

Concerning the Calamity that Befell Antiochus, King of Commagene, and also Concerning the Alans and what Great Mischief they did to the Medes and Armenians

1. In Vespasian's fourth year as emperor, it so happened that king Antiochus of Commagene and his family suffered some great tragedies. Cesennius Petus, the president of Syria at the time, sent a letter to Caesar informing him that Antiochus's son, Epiphanes, had decided to rebel against the Romans. [It could've been the truth, or it could've been done out of Petus's hatred for Antiochus. At any rate, the real motive for the letter was never discovered.] Petus said that Epiphanes, in his rebellion, had formed an alliance with the king of Parthia. Therefore, Caesar ought to do something, as the uprising might cause a general upheaval within the Roman empire.

Hearing the news, Caesar was very concerned, as the proximity of these kingdoms to one another was strategically important. Samosata, the capital of Commagene, sits on the Euphrates, and such an alliance would allow the Parthians easily to cross over it and be received cordially. Caesar believed Petus and gave him the authority to do whatever he deemed best. So, Petus immediately attacked Commagene when Antiochus and his people least expected any trouble. He took the tenth legion as well as other cohorts and troops of cavalry. Coming to Petus's assistance were Aristobulus, king of Chalcidene, and Sohemus, king of Emesa.

The Romans entered Commagene finding no one lifting up his hand in opposition. Antiochus was completely surprised. He had given no thought to making war against Rome. Conversely, Antiochus had determined to live out his retirement years in peace with his wife and children, and had thought he would demonstrate this to the Romans to avoid any accusation against him. So, he left the city and went out fifteen miles to a level place and there pitched his tents.

2. Petus then sent some of his troops to capture Samosata, while he and the rest of his army went to attack Antiochus and his troops. Antiochus, meanwhile, wanted nothing to do with a war against Rome. He bemoaned his fate and patiently endured what he was unable to prevent. But his young sons, strong of body but inexperienced in warfare, were not so ready to sit by and bear the trouble without a fight. So, Epiphanes and his brother, Callinicus, gathered their forces and met Petus. The difficult battle lasted until nightfall. The boys showed remarkable courage and lost none of their troops. Yet, at the battle's end, Antiochus would not stay at his camp, but took his wife and daughters and fled to Cicilia. Retreating as he did, Antiochus discouraged his own soldiers, who then revolted and went over to the Romans, thinking his kingdom finished. This defection forced Epiphanes and his men to leave the battlefield before they had no allies at all. Only ten horsemen were with him when Epiphanes crossed over the Euphrates safely to King Vologeses of Parthia. There, Epiphanes was treated respectfully, as if he still retained his authority.

3. After Antiochus had arrived in Tarsus of Cicilia, Petus ordered a centurion to go to him and send the ousted king to Rome in chains. But Vespasian would not allow the king to be brought to him like that, thinking it better to preserve the long friendship that had existed between him and Antiochus, rather than show anger at this pretense of war. Accordingly, he ordered that the king be loosed from his bonds while he was still on the road, and go instead to Lacedaemon to live. Vespasian also gave him a fine income so that Antiochus could not just live in plenty, but like the king that he was. When Epiphanes, who had feared for his father's life, heard of this he was greatly relieved. He also hoped that Caesar would also be reconciled to the family by way of Vologeses's intercession. For although they were very favorably endowed, he could not bear to live outside the Roman empire. So Caesar obliged Epiphanes, who went to Rome. His father quickly joined him from Lacedaemon. The Romans showed Antiochus every respect and there he remained.

4. We formerly mentioned the nation of Alans. They were Scythians and lived on Lake Meotis. The Alans had plans to plunder Media and beyond. So, they asked the king of Hyrcania to allow them passage through the

iron gates in his land that Alexander the Great had built. The king gave them permission, so a great multitude went through the gates and secretly attacked the Medes, plundering their highly populated and prosperous country. Not one Mede gave resistance. The Median king Pacorus had fled for his life into a stronghold and had given all his possessions to the Alans. He had only with difficulty saved his wife and concubines. Captured by the Alans, Pacorus had to ransom them for 100 talents. The Alans, therefore, easily plundered Media without opposition, proceeding as far as Armenia and laying waste to everything in their path.

Tirades was king of Armenia. He met the Alans and fought them, but was almost captured in the battle. One of the Alan soldiers cast a net over him at a distance, and would have soon drawn the king into imprisonment. But Tirades cut the rope with his sword and escaped. So the Alans, being even more angered by this sight, devastated Armenia and took with them a great multitude of its men and an enormous quantity of the plunder from the two kingdoms. They then retreated back to their own country.

Chapter 8

Concerning Masada and the Sicarii who Held it; and how Silva Planned the Siege of that Citadel. Eleazar's Speeches to the Besieged

1. Flavius Silva succeeded Bassus as procurator in Judea when Bassus died. Silva realized that warfare had ended in all of the country except for one stronghold still in rebellion. He, therefore, assembled his detached armies and set out against the stronghold called Masada. A strongman named Eleazar, commander of the Sicarii, occupied Masada. Eleazar was a descendant of Judas, the man who had persuaded many Jews not to submit to the taxation imposed by Cyrenius. It was at that time that the Sicarii had come to oppress those who were willing to submit to the Romans, treating them as enemies in every respect. They looted their treasures, drove their cattle away, and set fire to their homes. The Sicarii considered their submissive countrymen as foreigners for so cowardly betraying the freedom that was worthy of their utmost sacrifice. Those submitting to Roman taxation showed that they preferred slavery to Rome before death. Of course, this prevarication was no more than a pretense—a cloak, under which their barbarity hid, covering their greed. Their avarice was made

obvious by later activities. For the people who joined with the Sicarii to fight against Rome only suffered worse atrocities at Sicarii hands. When members of the Sicarii were later convicted of their lies and cover-up, they battered those who had incriminated them with added atrocities.

For that was indeed a time of manifold wickedness. No evil deeds were left undone. No one could invent any new form of iniquity, so deeply were they all infected, striving to act against one another to go to greater length, either privately or publicly, in their sin against God and injustice toward their neighbors. Men in power oppressed those under authority, and the people under authority earnestly worked to destroy the powerful. One faction wanted to tyrannize others while others practiced violence and plundered those wealthier than themselves. The Sicarii had first practiced these abominations and were the first to practice brutality against their partners. They left no insulting words unsaid, and no treacherous words untested, in order to destroy those whom they oppressed.

Yet John demonstrated that the Sicarii paled in comparison to his atrocities. He not only killed those who gave him good counsel, but even treated them worst of all—as his most bitter enemies among all of Jerusalem's citizens. John filled the entire country with 10,000 instances of his wickedness, such as a man who was already hardened in his sin toward God would naturally do. He ate unlawful food and rejected the purification rites his country had established. So it wasn't surprising that he, who was so insanely at enmity with God, would also shun any rules of gentleness and affection toward others.

Again, what mischief did Simon ben Gioras not undertake? Or, what kind of abuses did he not commit against the free men who had set him up as a despot? What friendships or brotherly affections were there which did not make him more belligerent in his daily murders? For they thought of misdeeds against strangers as beneath their courage, thinking that valor's real proof would be in the despicable mistreatment of close friends.

The Idumaens also contended with these men as to who could be guilty of the greatest madness! For these vile wretches cut the throats of high priests, to the end that no part of any devout honor to God might be preserved. Then they proceeded utterly to destroy any remnant of a political government, introducing instead a complete lawlessness.

Under this influence, the people called the Zealots emerged. They indeed lived up to their name, imitating every wicked deed, zealously avoiding the pursuit of any iniquity their memories recalled had been previously committed. Although the Zealots had taken on the name as meaning "zeal for good things," it only served as a cruel irony. For they thought their greatest good was to treat others with wild and brutal injustice. Accordingly, they met with the punishments that God deservedly brought upon them. They received every kind of misery under which man is capable of suffering, until the final minutes of their lives, when appointments with death came in various ways. Still, one might say that they suffered less than they deserved, because it would be impossible for them to be punished according to what they had earned. But this is not the proper place to mourn for those who suffered under the barbarity of these men. I will therefore return what remains of my present narration.

2. So, the Roman general, Silva, now came and led his army against Eleazar and the Sicarii, who held the fortress at Masada. Silva took control of the entire adjoining country placing garrisons at proper intervals. He built a wall completely around Masada so that none of those so enclosed might easily escape. He also placed his troops on guard at several points. Silva then pitched camp near the best place to make siege, where the rock on which the fortress was built came closest to a neighboring mountain. Masada wasn't a place easily supplied with provisions. Silva had to bring food for his army from a great distance, which was difficult for those Jews chosen for that purpose. Water also had to be brought in, as no spring was anywhere near the camp. When Silva had all these necessities in place, he began his siege. The siege promised to demand much skill and effort, principally due to the strength of the fortress, which I shall now describe.

3. Masada sat on a gigantic rock, surrounded by vastly deep valleys—so deep the eye could not reach their extremities. The cliffs were steep, so abrupt that no animal could walk upon them, except in two places. In those places the steepness decreases, allowing ascent, yet not without effort. Of the two ways leading to the top, one begins from Lake Asphaltites toward the rising sun, and the other rises from the west. The latter is a less difficult path. This western passageway is called "The Serpent," as it resembles that creature, being narrow and continually winding. Its course stops at various escarpments and frequently turns back upon itself. It then proceeds forward little by little, but with great difficulty. The trail is so narrow that one must place his foot one in front of the other. If a foot

slips, immediate destruction follows. On each side, a vast chasm looms, sufficiently terrifying to destroy anyone's courage. When a man has gone up this way for almost four miles, rest comes only at the top. The end comes not at a peak but a wide plain.

Upon this butte, the high priest, Jonathan, first built a fortress called Masada. Following that beginning, King Herod rebuilt the place to a great degree. He also constructed a mile-long wall around the entire top. It was eighteen feet high and twelve feet wide, and built using white stone. The wall contained thirty-eight towers, each of them rising seventy-five feet. Out of the towers one passed into smaller buildings on the interior of the entire wall. The king had reserved the hilltop for agriculture. Its soil was rich and better than any valley, so that those who occupied the fortress might never run out of food, in case they could not import it from elsewhere. Additionally, Herod built a palace on the western slope. It was within and beneath the walls of the fortress, but rose higher on the north side. The palace walls were strong and very high, and had ninety-foot high towers at each of its four corners. The furnishings of these buildings and their cloisters and baths were varied in design and very costly. Supported by single stone pillars on every side, the buildings had walls and floors paved with stones of several colors. Herod also cut out of the rocks many large reservoirs or cisterns for catching and holding water. Each habitation had one. They were above and around the palace and near the wall. By this means, Herod filled the need for water, as if Masada had natural springs.

A sunken road extended from the palace, leading to the very top of the mountain. Those outside the wall could not see the road. Indeed, enemies could not use any of the paths to the plain. The way from the east—as we've already said—could not be traversed because of its steepness. At the terminus of the western path, and at its narrowest point, Herod built a large tower at about 1,500 feet from the top of the hill. No one could bypass the tower, nor could it be taken. The citadel was so fortified that even those who approached on foot could not get near. Both nature and men's hands joined to frustrate any enemy attack.

4. The fortress's furnishings were wonderful as to both their splendor and their antiquity. Here, corn was stored in large quantities so as to guarantee food for a long time. There also existed wine and oil in abundance, with all kinds of pulses (dried beans and peas) and dates, all heaped together.

Eleazar found all of these supplies when his Sicarii deceitfully took possession of the fortress. The fruits were fresh and fully ripe and in no way inferior to those newly picked, even though they had been stored for a little less than one hundred years, since the time Herod brought them in and until the Romans took the place over. When the Romans took possession of these leftover fruits, they found them uncorrupted. It would be no mistake to say that the air here caused their endurance. The fortress was very high and the air free from pollution. The Romans also found a large store of weapons that had been accumulated by Herod. The weapons could have armed a force of 10,000 men. There was cast-iron, brass, and tin, which proved that Herod had taken great pains to ensure that everything was ready for any situation.

The report goes that Herod had prepared the fortress for himself as a refuge against two types of danger. First, he feared that the Jews might try to depose him, restoring their former kings to power. Second—and the supposed greater danger than the first—was the threat imposed by the Egyptian Queen Cleopatra. She did not conceal her intentions, but often spoke to Mark Anthony regarding her desire to cut Herod off, all the while begging Antony to bestow the kingdom of Judea on her. It is certainly a wonder that Anthony never complied with her demands, seeing as how he had been so enslaved by his passion for her. No one would have been surprised if he had granted her request. So, the fear of these dangers caused Herod to rebuild Masada and thereby leave it for the ultimate Roman stroke in the Jewish War.

5. So, as we've already said, the Roman general Silva had now built a wall around the entire mountain. By doing so, he had provided an enclosure to prevent any Jews from running away. Then he undertook his siege, though he only found one place that was capable of supporting his siege banks. Behind the tower that secured the western path leading to the palace and into the top of the hill, was a very wide and prominent rock about four hundred fifty feet below Masada's highest point. It was called the "White Promontory." Accordingly, Silva got up on that rock and ordered the army to bring earth. They then eagerly began their work, and with many men working together, a solid bank was raised three hundred feet high. Yet it was still not high enough to use the engines that were to be set upon it. So, another elevated bank was raised, using great stones compacted together. This elevated bank added another seventy-five feet

and was equidistant in breadth.

The machines were now ready. They were like those first developed for sieges by Vespasian and then modified by Titus. Another tower was raised at ninety feet in height and was completely covered in iron plates. From this high platform, the Roman engines threw darts and stones into the fortress. Soon, those who defended the walls were forced to retire, since they could not even show their heads above the wall. At the same time, Silva ordered the great battering ram to be brought up and set against the wall. The ram continually hammered the wall and with some difficulty a portion of the wall finally fell.

However, the Sicarii had hastened to build another wall, encompassing the breakthrough. The new wall was softer and yielding and able to successfully receive the terrible blows that devastated the first wall. The new wall was framed using great beams of wood laid lengthwise end-to-end in the same way they were cut. Two rows of beams paralleled each other and were separated from each other by the breadth of the wall. The Sicarii filled the space between the beams with dirt and rocks. To retain the earthen fill, the Sicarii further crisscrossed the wall with more beams and bound them together with the beams laid lengthwise. This work was like a strong edifice, for when the ram was applied to it the blows were weakened by its yielding nature. As the materials were shaken by the ram's concussions, they grew closer together, and the compaction caused the wall to become firmer than before.

When Silva realized this, he thought it best to destroy the new wall by setting it on fire. So he ordered his men to throw a great number of burning torches upon it. Made principally of wood, the wall soon caught fire. Once aflame, fire spread through its cavity and erupted into a mighty blaze. When the fire began, a north wind began to blow, which proved horrific for the Romans. It brought the flames down upon them until they were almost in despair of ultimately being successful, fearing their war machines would be burned up. But soon the north wind shifted to the south and blew strongly in the other direction, carrying the flame back against the wall that now blazed through its entire thickness. So the Romans, having now received God's help, return joyfully to their camp, resolving to rejoin the attack the next day. That night they set their watches on high alert lest some of the Jews should try to escape.

6. But Eleazar had no thought of escaping, nor would he permit anyone

else from doing so. But when he saw the new wall consumed by fire and could think of no other way of escape, he began to lose courage. Eleazar then spoke to his men, setting before their eyes what the Romans would do to their wives and children once overpowered. He spoke of suicide. Judging that this was the best thing to do under the circumstances, he gathered the bravest of his companions around him and made this encouraging speech:

"Generous friends, since we resolved long ago never to be servants of Rome, nor to any other than God, who alone is the true and just Lord of all mankind, the time is now come to put our former resolution into practice. Let us not now bring a rebuke upon ourselves by contradicting our earlier vows. When we said that we would not undergo slavery, it was in a time without danger. But now, along with slavery, we must also choose such other punishments that are intolerable—that the Romans will bring us under their power while we are still alive. We were first to revolt against them, and now we are the last to fight against them. But I don't believe that God has granted this to us as a favor. It is still within our power to die bravely and in freedom. This has not been the case with others who were conquered unexpectedly.

"It is clear that we will be overrun within the next day. But we still have time to die in a suitable fashion, together with our dearest companions. Our enemies cannot stop us, even though they will want to take us alive. We are now incapable of fighting them or defeating them. It would have been better to have formed an opinion of God's purposes much sooner, at the outset, when we were adamant of defending our liberty. In those early days, we treated each other poorly, and were treated even worse by our enemies. We now understand that the same God who had once taken our Jewish nation into his favor has condemned us to destruction. For if God had either continued to favor us or had been less displeased with us, he would not have overseen the destruction of so many men, nor delivered up his holy city to be burned and demolished by our enemies. We certainly had feeble hopes of surviving alone as free men, as if we were innocent of sin against God, or hadn't been partners to the sins of others. We also taught men to fight to preserve their liberty. But now we can see that God has convinced us that our hopes were in vain, by bringing this distress upon us to our state of desperation, beyond what we could have expected. For this unconquerable fortress has not proven to be our deliverance, and even though we still have a great abundance of food and a great quantity of arms and more necessities than we can use, God has now openly deprived us of all hope of escape.

"The fire that was driven down against our enemies did not turn back upon the wall we built of its own accord. This was the effect of God's anger against us for our many sins, of which we have been guilty in the most arrogant and excessive way against our own countrymen. But let's not receive punishment from the Romans but from God himself, executed by his own hands. For God's punishment will be easier than that of man. Let our wives die before they're abused, and our children before they have tasted slavery. And after we have killed them, let us then bestow upon one another as our funereal monument that same glorious benefit, preserving ourselves as freemen. But first let us destroy our money and this fortress by fire. For I am confident that not being able to capture us alive will bring the Romans real grief, and that they shall also have failed to take our money. Let us spare nothing but our provisions. They will be a testimony following our deaths that we did not fail for lack of supplies, but that according to our original resolution we have preferred death over slavery."

7. This was Eleazar's speech to his men. But all did not agree with him. While some were very zealous to put his advice into practice and, in a way, were filled with pleasure at thinking death a good thing, others had a more sympathetic pity upon their wives and families. As these men became emotional over their own certain death, they looked regretfully at one another. The tears in their eyes openly declared their disagreement with Eleazar's opinion. When Eleazar saw the fear in these men, and the other souls dejected at his proposal, he was afraid that these cowardly men would discourage others by their cries and tears from what he had said. So, Eleazar kept up his expectations, stirring himself up and recalling appropriate arguments to increase their courage. He began to speak rapidly and loudly to them regarding the soul's immortality. He moaned loudly and fixed his eyes on those who wept, saying:

"Truly, I was greatly mistaken when I thought I was helping brave men struggle hard to maintain their liberty, who were resolved to either live honorably or else to die. But I now find you people are no better than any others, either in virtue or courage. You are afraid of death, even though death will deliver you from even greater agonies. You ought to make no delay in this matter, nor wait for someone else to give you better advice. For, our ancient country's laws and our forefathers' actions—and even God himself—have since we were old enough to be rational taught us that catastrophes belong to the living and not to the dead. For to die is perfect liberty for our souls and sends them to a place of purity, where they will be insensible to all sorrow. While souls are tied down to mortal bodies, they partake of its misery, and to be perfectly honest, they are dead.

"For the union of the divine to the mortal is distasteful. It is true that the soul's power is great even when imprisoned in a mortal body, for it makes the body its sensible mechanism, moving it in invisible ways. The soul causes the body to accomplish more than it would otherwise be capable. However, freed from the weight that ties it to earth, the soul obtains its own proper place, and then partakes of those unhindered blessed powers and abilities. The soul is invisible to men's eyes, as is God Himself—for certainly we can't see the soul residing in the body. It is invisible, and it remains invisible when released. Yet it is the cause of the bodily change, for anything touched by the soul lives and flourishes. Then if removed, the body withers and dies. Such is its immortality.

"Let me speak of sleep as an evident demonstration of the truth of which I speak. Souls, when not distracted by the body, have the sweetest rest. The soul depends on itself and converses with God in its alliance with him. In sleep, the soul goes everywhere, foretelling what is to come in the future. So, why are we afraid of death while we are pleased with the rest we have when asleep? How absurd it is to pursue liberty while we're alive, and yet to deny it to ourselves when it will be eternal! We, therefore, who have been raised up in our own discipline, ought in our readiness to die become an example to others.

"Do we need foreigners to support us in this matter? Think about the Indians who profess their philosophy. These good men unwillingly submit their lifetimes, looking upon it as a necessary slavery, and hurry to see their souls loosed from their bodies. Even when no misfortune presses them upon them nor drives them to death, they have such a yearning for immortality that they tell others in advance that they are about to depart. Nobody stops them. Everyone thinks they're happy and gives them letters to carry to their dead friends. They're so firmly convinced that souls converse with one another in the afterlife. So, when these Indians have heard all the commands given to them, they deliver up their bodies for burning. And, in order to have their souls separate from their bodies in the greatest purity, they die in the midst of hymns commending them. For their dearest friends conduct them to their deaths more enthusiastically than the rest of mankind, and send them off on a long journey. At the same time, they mourn for themselves, but look upon the departed as happy persons, soon to partake of their immortal order of being. Are not we, therefore, ashamed to have lower ideas than the Indians? By our own cowardice, we have insulted our country's traditions that the rest of mankind covet and imitate.

"But, suppose we had been brought up under another persuasion, and taught that life is man's greatest good and death is a disaster. Well, then our current circumstances ought to induce us to bear the disaster of death courageously, since it is God's will that we must die. For it now becomes clear that God has decreed that the entire Jewish nation be deprived of the life, which we knew we would not use well. Don't you ascribe this occasion of our present circumstance to your own making? Do you think the Romans caused this war that has become so destructive to us all? These things have not come to pass because of Rome's power, but by the intervention of a much more powerful cause, making them merely appear to be our conquerors.

"Let me ask you, what Roman weapons killed the Jews at Caesarea? On the contrary, when the Jews had no thought of rebellion but were keeping the Sabbath festivals, they did not so much as lift a finger against Caesarea's citizens. Yet the Caesarean mob ran upon the Jews in scores, cutting their throats and the throats of their wives and children, without any regard to the Romans, who never believed we were their enemies until we revolted against them.

"But someone might say that the people of Caesarea always quarreled with the Jews living among them, and that when the opportunity arose they merely satisfied their hatred. What then shall we say about those of Scythopolis, who went to war against us because of the Greeks? But then they refused to join us in our war against the Romans. Can't you see how little our goodwill and fidelity toward them profited us? They and their entire families were slain in the most inhumane way. This was payback for the assistance they had afforded others. For the very same destruction they stopped from falling on others fell upon themselves, as if they were their own enemies.

"It would take too long for me to speak now about every devastation we have faced. You have to realize there was not one city in Syria that didn't kill its Jewish inhabitants. They were more bitter enemies against us than the Romans! Even those in Damascus, when they were unable to come up with a believable lie against us, filled their city with the most barbarous slaughter of our people, cutting the throats of 18,000 Jews, together with their wives and children. And as to the multitude that were tortured and killed in Egypt, we have heard they numbered more than 60,000. Of course, they were in a foreign country, and so had no natural defenses against their enemies. But they were all slaughtered in the same way.

"As for all those who have waged war against the Romans in our own country, did we not do so with high hopes of victory? For we had arms and walls and fortresses prepared to make Roman siege very difficult. We had the bravery not to fear danger in the cause of liberty, which encouraged all of us to revolt against Rome. But these advantages sufficed for only a short time, and simply raised our hopes while they became the origin of our misfortunes. For everything we had has now been taken from us, and all have fallen under our enemies. It's as if our earlier advantages made Rome's victory even more glorious and were not meant to preserve us.

"And as far as those already dead in the war, it is reasonable that we should consider them blessed, for they died defending, not betraying, their liberty. But as far as those who are now under Roman subjection, who can but pity their situation? And who would not hurry to die before being subjected to their same agonies? The Romans put some of them on the rack and were tortured with fire and beatings, and so died. Some have been half-devoured by wild beasts and yet saved alive to be eaten the second time, in order to bring laughter and sport to our enemies. Those who remain alive are still looked upon as being the most miserable, who, though desirous of death, could not welcome it.

"And where is that great city, the metropolis of the Jewish nation, so fortified by many walls around it with so many fortresses and towers to defend it, which could hardly contain the armaments of war for its tens of thousands of men to fight for it? Where is the city that was believed to be God's home? It is now demolished to its very foundations and has nothing but a monument of it preserved—I mean the camp of those who destroyed it, which still sits upon its ruins. Some unfortunate old men now lie among the temple's ashes, and the enemy, for our bitter shame and reproach, keeps a few women alive. Who is he who bears these tragedies to mind, yet can bear the sight of the sun, so he might live in peace? Who is there so much his country's enemy, or so unmanly, or so desirous of living, that will not repent that he is still alive?

"I cannot wish that we had all died before seeing our holy city demolished by our enemy's hands, or the foundations of our holy temple dug up in such a profane way. But since we had a generous hope that only deluded us, as if it provided us with the power to avenge ourselves on our enemies. Since our hope is now utmost arrogance and has left us in our distress, let us make haste to die bravely. Let us pity our children, our wives, and ourselves while it is within our power to have pity upon them. For we were born to die, as are those we have fathered. Nor is it in the power of the happiest of our race to avoid death. But for abuses, and slavery, and the sight of our wives and children being led away in the most sordid way, these evils are neither natural nor necessary for men. For those who prefer these miseries over death, when it is in their power to do otherwise, are destined to undergo them because of their own cowardice. We rebelled against Rome with pretensions to courage, and when at last they invited us to save ourselves, we would not comply. Who will, therefore, believe the Romans will not receive us alive with great rage? You young men, who by the strength of your bodies are able to sustain many torments, will be miserable. Miserable also will you older men be, who cannot bear up under the same agonies! A man whose hands are bound will be obliged to hear his son's voice pleading for his father's help.

"But now our hands are at liberty and hold swords. Let them now serve us in our glorious strategy. Let us die before we become slaves under our enemies. Let us go out of this world together with our wives and children in the state of freedom. This is what our laws command us to do. This is what our wives and children want from our hands. God himself has brought this obligation upon us, while the Romans want the contrary, and are afraid that any of us should die before they capture us. Let us, therefore, make haste, and instead of affording them such pleasure, leave them an example causing them to be astonished at our deaths, and in admiration of our courage to die."

Chapter 9

How the People in the Masada Fortress were Convinced by Eleazar's Words. Only Two Women and Five Children Survive after All others are Killed by One Another

1. While Eleazar's exhortation was continuing, his listeners cut him off short. They proceeded to do the work he called for as if full of a ravenous need. They moved with a demoniacal fury, one endeavoring to go before another as if their eagerness demonstrated courage and good conduct, or as if to avoid being left to be the last to strike. They moved with great zeal to slay their wives and children, and even themselves. Indeed, when going about the work, their courage did not fail them, as one might imagine it would have. But they held fast to the same resolve without wavering, encouraged by Eleazar's speech. Yet every man still retained his natural love for his family, because their reasons for doing what they did seemed very just, even when taking the lives of those dearest to them. The husbands tenderly embraced their wives and took their children into their arms, giving them long, loving kisses with eyes full of tears. Yet, at the same time they finished the task, they had resolved to finish it as if their loved ones had been executed by a stranger's hand. They had no other comfort but the necessity to perform these executions, to avoid the prospect of suffering agony at their enemies' hands.

In the end, not one man failed to carry out his part in the horrible executions. Every one of them dispatched his most beloved and dearest relations. Miserable men they were indeed! Their distress forced them to kill their wives and children with their own hands, as the lightest of the evils that faced them. So, unable to bear their grief any longer, and thinking it injurious to those slain to live even the shortest time after them, each man laid all his possessions on the ground and set fire to the pile. They then chose ten men by lot to slay all the rest. Every one of those to die first laid down beside his wife and children and threw their arms around them, and then offered their necks to the sword stroke of those chosen by lot to that sad task.

When the ten had fearlessly executed the rest, they made the same rule of casting lots for themselves. He whose lot it was to kill the other nine would then turn his sword upon himself. Accordingly, each man had enough courage to not lag behind in either doing or suffering. So, at last the nine offered their necks to the executioner. He who was last of all, lest perchance one or more among them who were not yet dead should need a final thrust, perceived that they were all dead and set fire to the palace. Then, with the force of his hands, ran himself through and fell down dead next to his family. So, all these people died for this purpose: that not one person be left alive to be subject to the Romans.

Yet an old woman and another woman who was Eleazar's relative—and who was superior to most women in wisdom and education—had concealed themselves in underground caverns. They had taken water with them to drink and had hidden there while the rest were intent on killing each other. Those dead numbered one hundred sixty, including women and children. This terrible slaughter occurred on the 15th of the month Xanthicus [Nisan] (May, 73 AD).

2. The Romans, meanwhile, expected to fight in the morning. Accordingly, they donned their armor and laid plank bridges on their ladders from the banks, and made an assault upon the fortress. But no enemy was to be seen, only an awful silence on every side. Perfect quiet reigned at Masada, except for a fire that raged.

At length, the soldiers tried to raise someone with shouts that resounded like a battering ram's blow. The two women heard the noise and came out of their underground hiding place. They informed the Romans what had taken place. The second woman clearly described how it was all accomplished. Even though the Romans listened to their description of such a desperate undertaking, they did not believe the women. They also tried to douse the fire quickly, cutting a way through it. Then the Romans came into the palace and were met with the sight of all the dead. They took no pleasure in it, even though it had been done to their enemies. They could do nothing but wonder at the Jewish courage and resolve, and the unmovable contempt of death that so many of them had shown as they carried out such an enterprise.

Chapter 10

Many of the Sicarii Flee to Alexandria only to Find more Danger there; on which Account the Temple Built by Onias, the High Priest, was Destroyed

1. When he had taken Masada, General Silva left a garrison in the fortress while he went to Caesarea. No enemies now remained in the country. The Romans had conquered the Jews in the long war. Yet repercussions and precarious disorders were found in places far away from Judea. There were still pogroms against the Jews in Alexandria that killed many. Many Sicarii were able to escape into Egypt from their agitations in Judea. They

were not content to have survived the Judean war but began to create new disturbances elsewhere. The Sicarii persuaded many Egyptian Jews to assert their liberty and think of the Romans as equals, no better than themselves. They would look only upon God as their sole Lord and Master. But when some of the more prominent Jews in Alexandria opposed the Sicarii, the Sicarii assassinated some and continued to put pressure on the others to revolt against Rome.

When the senior Jewish elders realized the Sicarii madness rising, they reasoned that it was no longer safe to overlook them. So they assembled all the Jews together. Accusing the Sicarii, the elders verified that they had been the architects of the evil that had now come upon Egypt. They said, "These men who have run away from Judea have no certain hope of ultimate escape. Because their evil will soon come to the light, and the Romans will destroy them, they came here and tried to make us partners in calamities that belong to them alone. We have not been accessories to any of their crimes." Accordingly, the elders exhorted the assembled Jews to be careful, lest they, too, be destroyed along with the Sicarii, and to make restitution to the Romans by turning these men over to them. The Jews, understanding the great danger they faced, agreed with the elders' proposals and attacked the Sicarii violently, immediately capturing six hundred of them.

But it wasn't long until all those who had fled into Egypt at Thebes were captured. The Jews were amazed at the courage—or rather madness—and purposeful resolution of the Sicarii. For, when they suffered every torture and aggravation against their bodies, they could not get a single one to confess that Caesar was their Lord. To a man, the Sicarii would not yield to the agony. It was as if their bodies were insensitive to the pain inflicted by torture and fire, and instead inspired a rejoicing soul underneath. But what was most astonishing to those who saw this was the children's courage. Not one of the Sicarii children would succumb to the torture and confess Caesar as Lord. The strength of their souls continued to prevail over their bodies' weaknesses.

2. Lupus was governor of Alexandria at the time and sent Caesar word of this disturbance. Knowing of the Jewish restlessness and tendency to cause uprisings, and fearing a general mutiny as others joined them, Vespasian ordered Lupus to demolish the Jewish temple in the Egyptian region known as Onias. The region of Onias had been colonized in the

following way: Onias ben Simon was one of the high priests who had fled from Antiochus, the Syrian king, when he fought against the Jews. Onias arrived in Alexandria, where Ptolemy, because of his own hatred of Antiochus, received him warmly. Onias assured Ptolemy that if the king sympathized with his proposal, he would bring all the Jews to assist him against Antiochus. When the king agreed to do it as far as he could, Onias asked him to allow him to build a Jewish temple somewhere in Egypt, and there to worship God according to Jewish customs. For the Jews would be most prepared to fight against Antiochus, who had destroyed the Jews' temple in Jerusalem, and they would come to Ptolemy's aid more eagerly, if they were granted liberty of conscience. If the proposal were accepted, many would join him.

3. So, Ptolemy accepted Onias's proposal and gave him land about twenty-three miles from Memphis, in the Nomos of Heliopolis. There, Onias built a fortress and a temple, but unlike the temple in Jerusalem. It resembled a tower constructed of large stones, and rose to a height of ninety feet. He constructed its altar after its counterpart in Judea, which in like manner he adorned with furnishings, except for the candlestick. He did not make a candlestick, but had only a single lamp hammered out of a piece of gold. A wall of fired brick with gates of stone surrounded the entire temple. The king also gave Onias a large land area for revenue, so that the priest might be substantially provided for and that much abundance would be available for God's worship.

But Onias had ulterior motives for the temple. He was determined to become a rival party for those in Jerusalem, for he couldn't forget the insults he had received when banished from the city. And so, he thought that by building this temple he might draw away large numbers of Jews to himself. Also, a certain ancient prophet named Isaiah had made a prediction six hundred years before that a Jew in Egypt would build this temple. (Josephus is probably referring to Isaiah 19:19 and following.) So, this is the history of the Onias Temple.

4. So when Lupus, Alexandria's governor, received Caesar's letter he went to the temple and took some of the gifts that had been dedicated to it, and shut the temple down. Lupus died shortly thereafter and was succeeded by Paulinns. This man took all the gifts remaining there and threatened the priests severely if they did not give him everything. Paulinns did not

allow anyone to worship there or even to go near the sacred place. The building had stood three hundred forty years before the time it was closed.

Chapter 11

Concerning Jonathan, one of the Sicarii who Stirred up Trouble in Cyrene and Falsely Accused the Innocent

1. Meanwhile, the Sicarii madness had spread like a contagious disease, reaching as far as the city of Cyrene. A vile man named Jonathan, a weaver by trade, went to Cyrene and convinced a number of the poorer people to listen to him. He then led them into the desert, promising to show them many signs and wonders. Jonathan concealed his treachery from the other Jews of Cyrene while misleading them. But one of the more prominent Jews informed Catullus, governor of the Libyan Pentapolis, of Jonathan's excursion into the desert and of the preparations he had made for it. So, Catullus sent cavalry and infantry units out after him and easily overran the unarmed pilgrims. Many died but a few more were taken alive and brought to Catullus. Jonathan got away at this time, but following a long and diligent search throughout the country, he was captured. When he was brought to Catullus, not only did he escape punishment, but had devised a plan whereby Catullus might create his own mischief. He falsely accused the wealthiest Jews of the city, saying that they had been the cause of what he had done.

2. Catullus eagerly believed Jonathan's lie. He then proceeded to exaggerate the importance of the matter, so that it might appear that he had taken part in winning the Jewish war. But even worse yet, Catullus not only believed the concocted story, but also taught the Sicarii to falsely accuse other men. He, therefore, ordered Jonathan to accuse one named Alexander, with whom Catullus had previously quarreled and against whom he had open hatred. He also got Jonathan to name Alexander's wife, Bernice, as a co-conspirator. Catullus then had the couple executed. Following them, he executed all the wealthy Jews of the city, numbering some 3,000. Catullus thought he could get away with these murders because he confiscated all their property and put it in with Caesar's revenues.

3. Concerned that Jews living in other places might convict him of his crimes, Catullus made further false accusations. He persuaded Jonathan

and others captured with him to make accusations of sedition against other prominent Jews living in Alexandria and Rome. One of those accused was Josephus, the author of these books. However, this plot contrived by Catullus was not destined to work in the way he had hoped. For, although he came to Rome and brought Jonathan and his companions with him in chains—believing that their appearance would prove the substance of his false accusations, Vespasian suspected foul play and set up an inquiry to get at the truth. When he found that the accusations were all fabrications, he cleared the Jews of the charges against them. Titus was particularly concerned with the situation, and brought a deserving punishment upon Jonathan. He was first tortured and then burned alive.

4. But the emperors were gentle with Catullus. They did not punish him severely at the time, but it wasn't long before he became very ill and died miserably. Not only was his body afflicted, but his mind also suffered even more. He was mentally disturbed, and continually cried out that he saw the ghosts of those he had slain standing before him. Seeing these visions, he was not able to contain himself but leapt out of his bed as if torture and fire had been brought upon him. His illness grew progressively worse, to the end that his intestines became so corrupted that they fell out of his body, and he died. Thus, Catullus became one of the great examples of Divine Providence, demonstrating that God punishes wicked men.

5. And here we shall put an end to our history. We promised to deliver it accurately to those who wanted to understand the way in which the war was fought between the Romans and the Jews. What I have written, both in scope and style, must be judged by its readers. But, as for its agreement with the facts, I shall boldly say that truth has been my goal throughout its entire composition.

MAPS & CHARTS

1. Map of Israel in New Testament Times:

428 The Jewish Wars

2. Map of the Cities of Decapolis:

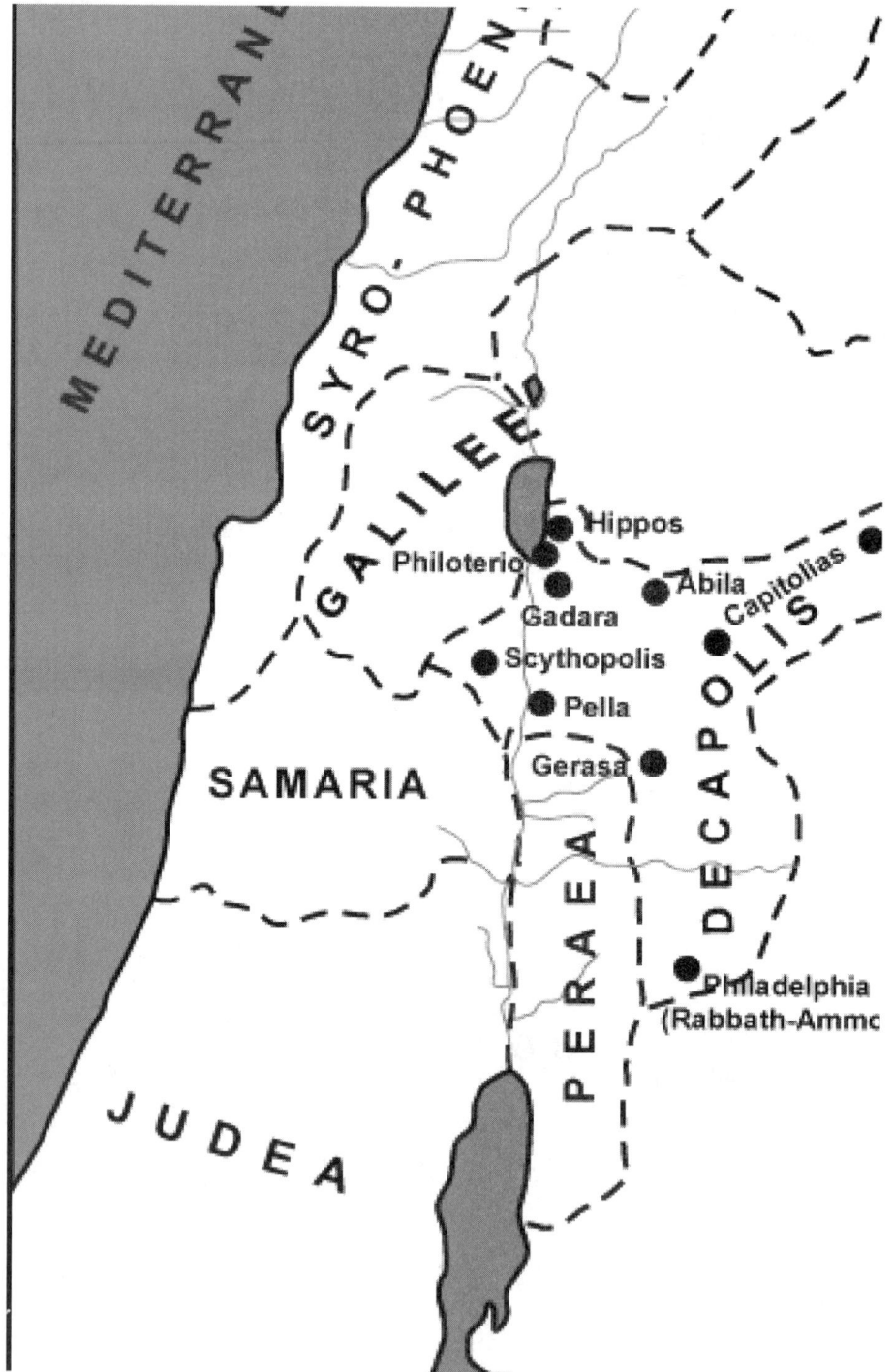

3. First Century Jerusalem.

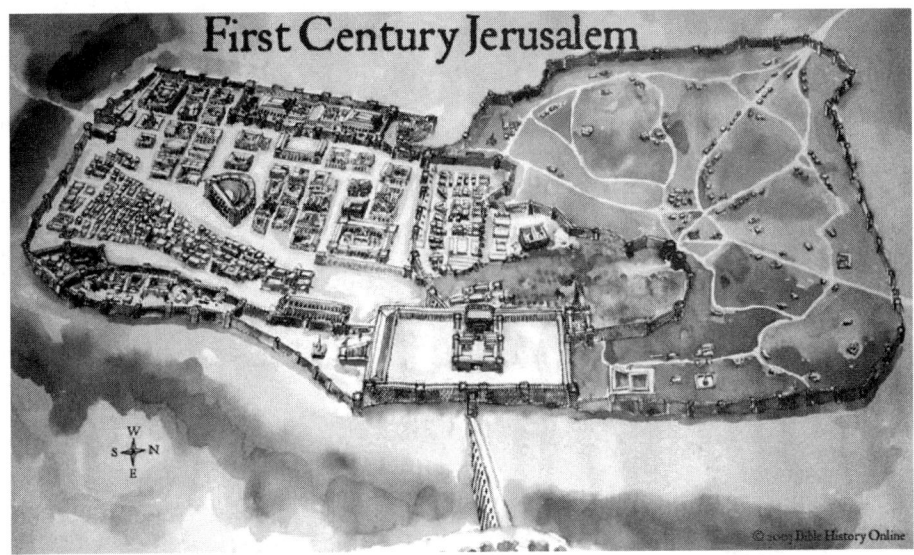

4. Herod's Temple: (All maps and illustrations are courtesy of www.bible-history.com)

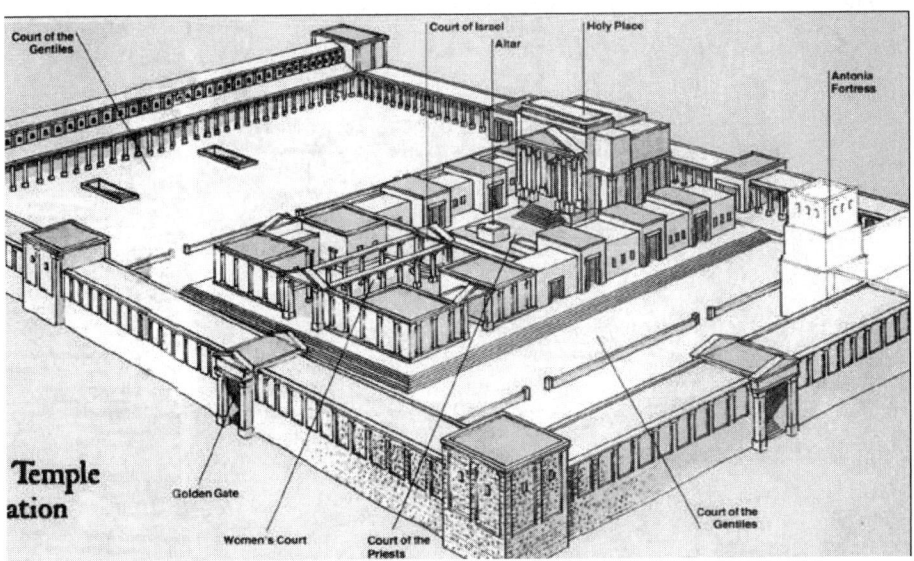

5. The Hasmonaean Family Tree

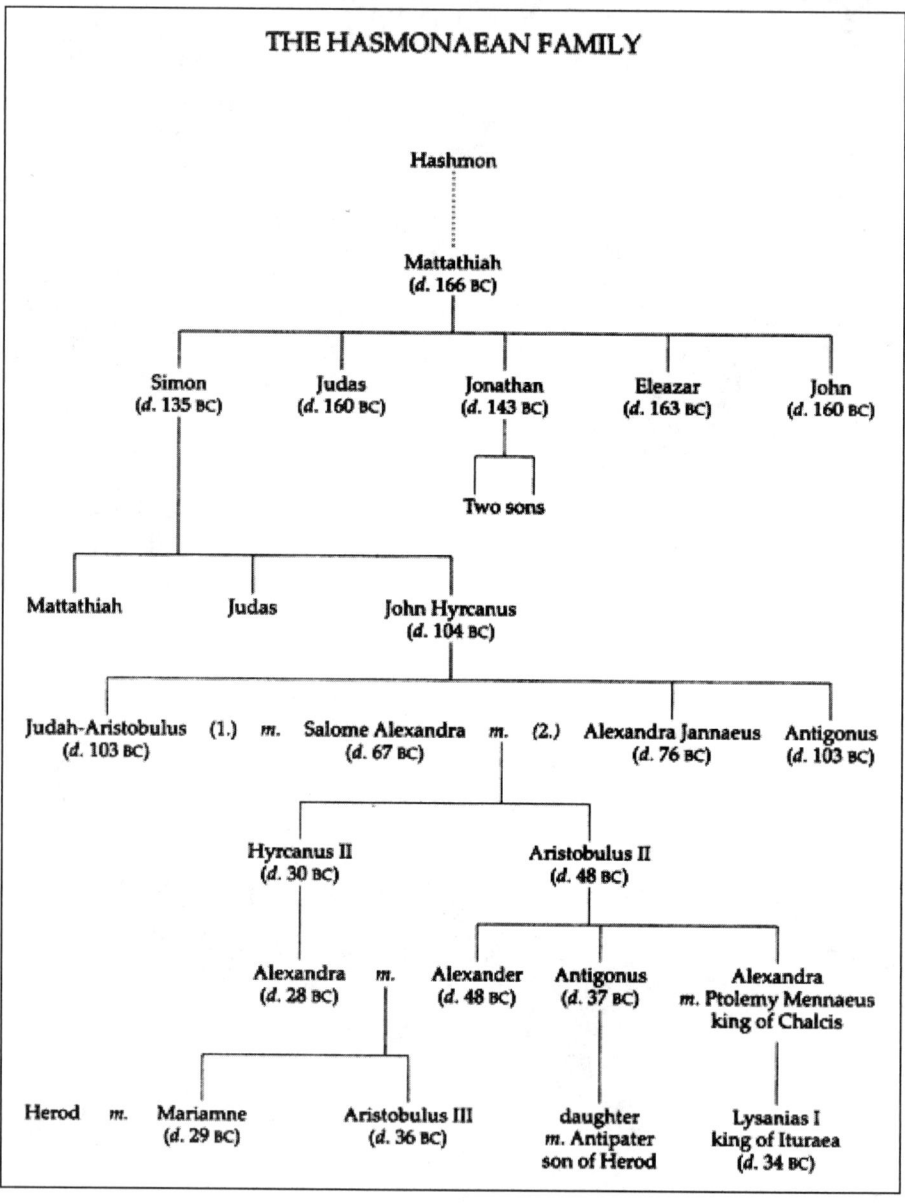

6. The Herodian Family Tree:

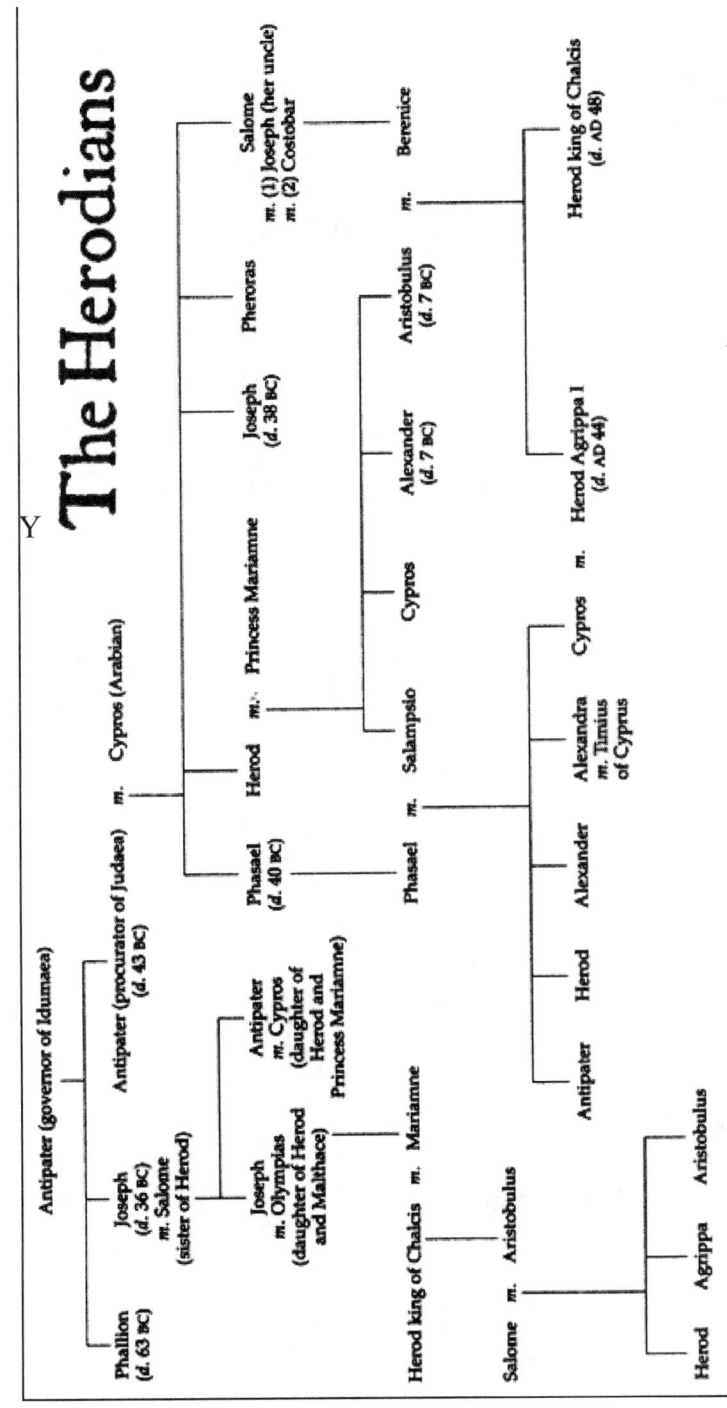

432 The Jewish Wars

7. Descendents of Herod and Miriamne:

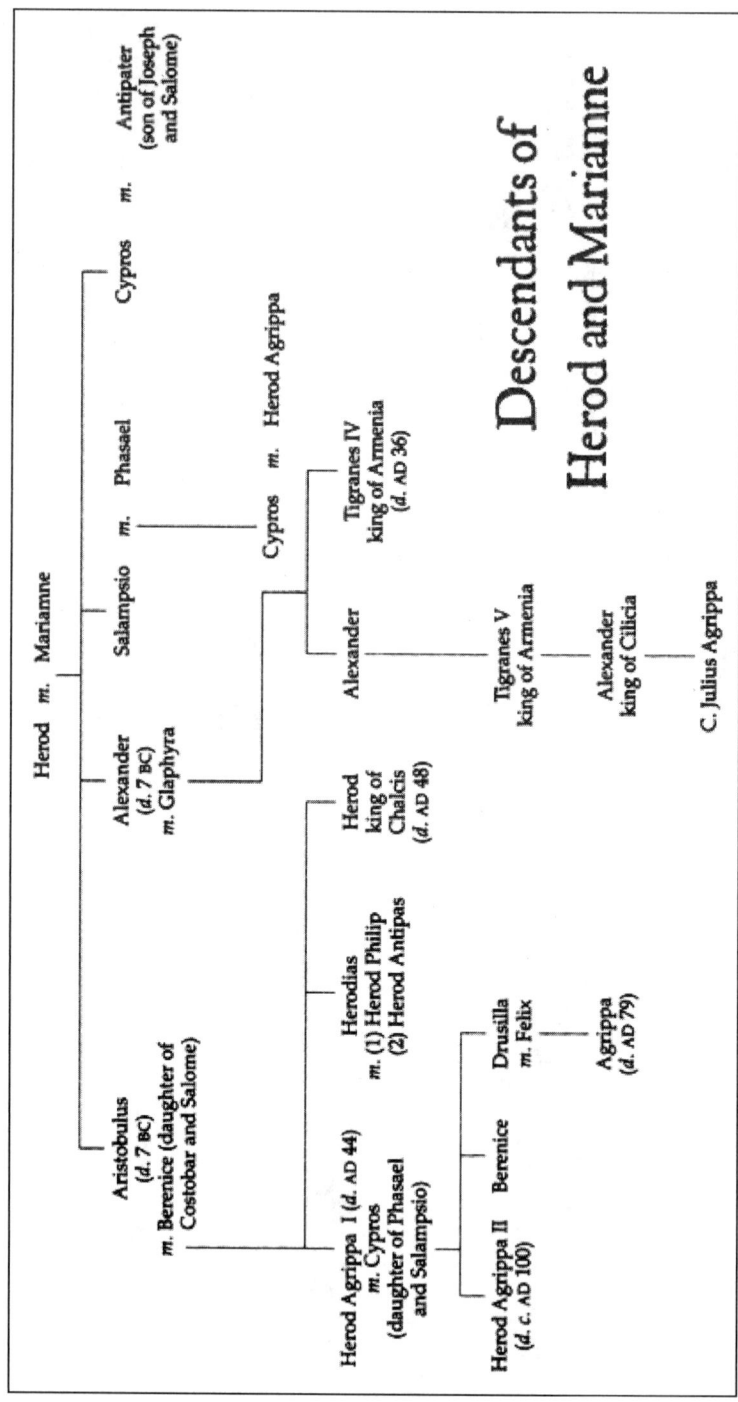

8. Descendents of Herod and His Other Wives:

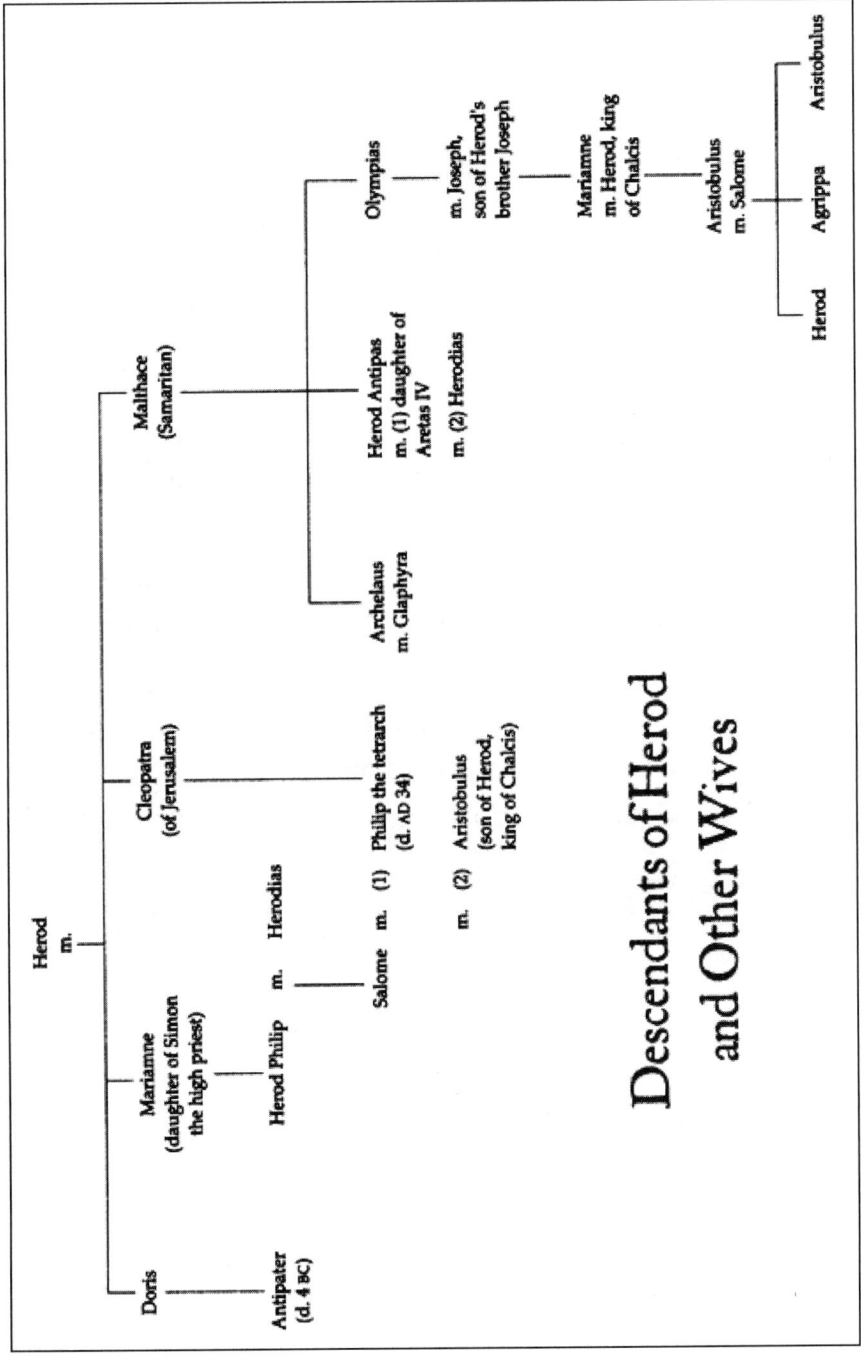

Would you like to review
The Jewish Wars – A Paraphrase?

It's easy!

Simply go to www.amazon.com and enter
"The Jewish Wars - Bob Beasley" in the search box.

When you are redirected to the correct page,
scroll down until you see "Customer Reviews."

Click on the button "Write a Customer Review."
Then leave your review, 1 to 5 stars,
plus a headline and a short (or long) review of the book.

We will greatly appreciate your feedback,
and so will other readers!

The Life of Josephus - A Paraphrase:
Now Available as an eBook